MW00837176

High-Speed Digital Design

HIGH-SPEED DIGITAL DESIGN

A Handbook of Black Magic

HOWARD W. JOHNSON, PH.D.
Signal Consulting, Inc.

MARTIN GRAHAM, PH.D.
University of California at Berkeley

For book and bookstore information

http://wwwphptr.com

Prentice Hall PTR, ***Upper Saddle River, New Jersey 07458***

Library of Congress Cataloging-in-Publication Data

Johnson, Howard W.
High-speed digital design : a handbook of black magic / Howard W. Johnson, Martin Graham.
p. cm.
Includes bibliographical references and index.
ISBN 0–13–395724–1
1. Digital electronics. 2. Logic Design. I. Graham, Martin, (date). II. Title:
TK7868.D5J635 1993
621.381--dc20 93–27
CIP

Editorial/production supervision: bookworks
Manufacturing buyer: Mary McCartney
Cover design: Bruce Kenselaar

Prentice-Hall, Inc.
A Simon & Schuster Company
Upper Saddle River, New Jersey 07458

The publisher offers discounts on this book when ordered in bulk quantities. For more information, contact:

Corporate Sales Department
Prentice Hall PTR
1 Lake Street
Upper Saddle River, New Jersey 07458

Phone: (201) 236-7152
Fax: (201) 236-7141
Printed in the United States of America

20 19 18 17 16 15

ISBN 0-13-395724-1

Prentice-Hall International (UK) Limited, *London*
Prentice-Hall of Australia Pty. Limited, *Sydney*
Prentice-Hall Canada Inc., *Toronto*
Prentice-Hall Hispanoamericana, S.A., *Mexico*
Prentice-Hall of India Private Limited, *New Delhi*
Prentice-Hall of Japan, Inc., *Tokyo*
Simon & Schuster Asia Pte. Ltd., *Singapore*
Editora Prentice-Hall do Brasil, Ltda., *Rio de Janeiro*

Contents

Preface

This is a book for digital designers. It highlights and explains analog circuit principles relevant to high-speed digital design. Teaching by example, the authors cover ringing, crosstalk, and radiated noise problems which commonly beset high-speed digital machines.

None of this material is new. On the contrary, it has been handed down by word of mouth and passed along through application notes for many years. This book simply collects together that wisdom. Because much of this material is not covered in standard college curricula, many practicing engineers view high-speed effects as somewhat mysterious, ominous, or daunting. For them, this subject matter has earned the name "black magic." The authors would like to dispel the popular myth that anything unusual or unexplained happens at high speeds. It's simply a matter of knowing which principles apply, and how.

Digital designers working at low speeds do not need this material. In low-speed designs, signals remain clean and well behaved, conforming nicely to the binary model.

At high speeds, where fast signal rise times exaggerate the influence of analog effects, engineers experience a completely different view of logic signals. To them, logic signals often appear hairy, jagged, and distorted. For their products to function, high-speed designers must know and use analog principles. This book explains what those principles are and how to apply them.

Readers without the benefit of formal training in analog circuit theory can use and apply the formulas and examples in this book. Readers who have completed a first year class in introductory linear circuit theory may comprehend this material at a deeper level.

Chapters 1–3 introduce analog circuit terminology, the high-speed properties of logic gates, and standard high-speed measurement techniques, respectively. These three chapters form the core of the book and should be included in any serious study of high-speed logic design.

The remaining chapters, 4–12, each treat specialized topics in high-speed logic design and may be studied in any order.

Appendix A collects highlights from each section, listing the most important ideas and concepts presented. It can be used as a checklist for system design or as an index to the text when facing a difficult problem.

Appendix B details the mathematical assumptions behind various forms of rise time measurement. This section helps relate results given in this book to other sources and standards of nomenclature.

Appendix C lists standard formulas for computing the resistance, capacitance, and inductance of physical structures. These formulas have been implemented in MathCad and are available from the authors in magnetic form.

ACKNOWLEDGMENTS

Many people have contributed to this book, and we would like to thank them all. To our teachers, employers, fellow workers, clients, customers, and students, we thank you for motivating us to learn, for showing us problems we could not solve, and for occasionally humbling us when we acted like we knew too much.

The authors would like to thank individually the following people for the generous contributions they have made to the writing of this book: For meticulously reviewing the text and for offering many, many good suggestions we thank Dan Nitzan, Jim Pomerene, Joel Cyprus, Ernie Kim, Tim Ryan, and Charlie Adams.

For her efficient and cheerful assistance in preparing the figures, we thank our assistant, Pamela Moore.

Dr. Johnson would like to thank the former officers and management of ROLM corporation, particularly Ken Oshman, Bob Maxfield, and Gibson Anderson for giving him a big head start in the electronics industry.

For having a profound effect on his approach to problem-solving, and on his teaching career, Martin Graham wishes to acknowledge his mentor of long ago, Professor William McLean.

Of course, we owe a big debt of gratitude to Tektronix for loaning us a Tek 11403 digitizing oscilloscope. Their scope produced all the fine waveform displays you see in the book. Each waveform was captured, stored in memory, and then plotted directly to hard copy. Thank you, Leo Chamberlain and Jim McGoffin.

Last, and certainly not least, to our wives, Elisabeth and Selma, for their devotion and untiring support, we express our heartfelt appreciation and thanks.

A NOTE TO THE READER

To those of you who will undoubtedly report the discovery of technical errors in the manuscript, thank you for your attention and for taking the time to write to us about it. The authors will personally send a certificate of appreciation to the first person to document each substantive technical error in the book. Please send your comments to:

Howard W. Johnson, Ph.D.
Signal Consulting, Inc.
16541 Redmond Way, Suite 264
Redmond, WA 98052

High-Speed
Digital Design

1

Fundamentals

High-speed digital design, in contrast to digital design at low speeds, emphasizes the behavior of passive circuit elements. These passive elements may include the wires, circuit boards, and integrated-circuit packages that make up a digital product. At low speeds, passive circuit elements are just part of a product's packaging. At higher speeds they directly affect electrical performance.

High-speed digital design studies how passive circuit elements affect signal propagation (ringing and reflections), interaction between signals (crosstalk), and interactions with the natural world (electromagnetic interference).

Let's begin our study of high-speed digital design by reviewing some relationships among frequency, time, and distance.

1.1 FREQUENCY AND TIME

At low frequencies, an ordinary wire will effectively short together two circuits. This is not the case at high frequencies. At high frequencies, only a wide, flat object can short two circuits. The same wire which is so effective at low frequencies has too much inductance to function as a short at high frequencies. We might use it as a high-frequency inductor but not as a high-frequency short circuit.

Is this a common occurrence? Do circuit elements that work in one frequency range normally not work in a different frequency range? Are electrical parameters really that frequency-sensitive?

Yes. Drawn on a log-frequency scale, few electrical parameters remain constant across more than 10 or certainly 20 frequency decades. For every electrical parameter, we must consider the frequency range over which that parameter is valid.

Exploring further this idea of wide frequency ranges, let's first consider very low frequencies, corresponding to extremely long intervals of time. Then we will see what happens at very high frequencies.

A sine wave of 10^{-12} Hz completes a cycle only once every 30,000 years. At 10^{-12} Hz, a sine wave of transistor-transistor logic (TTL) proportions varies less than 1μV in a day. That is a very low frequency indeed, but not quite zero.

Any experiment involving semiconductors at a frequency of 10^{-12} Hz will (eventually) reveal that they do not function. It takes so long to run an experiment at 10^{-12} Hz that the circuits turn to dust. Viewed on a very long time scale, integrated circuits are nothing but tiny lumps of oxidized silicon.

Given this unexpected behavior at 10^{-12} Hz, what do you suppose will happen at the opposite extreme, perhaps 10^{+12} Hz?

As we move radically up in frequency, to very short intervals of time, other electrical parameter changes occur. For example, the electric resistance of a short ground wire measuring 0.01 Ω at 1 kHz increases, due to the skin effect, to 1.0 Ω at 1 gHz. Not only that, it acquires 50 Ω of inductive reactance!

Big changes in performance always occur in electric circuit elements when pushed to the upper end of their operating frequency range.

How high a range of frequencies matters for high-speed digital design? Figure 1.1 answers this question with a graph illustrating the relationship between a random digital pulse train and the important part of its frequency spectrum.

The digital signal illustrated in Figure 1.1 is the output of a D flip-flop clocked at rate F_{clock}. The data value during each clock interval randomly toggles between 1 and 0. In this example the 10–90% rise and fall time, called T_r, comprises 1% of the clock period.

The spectral power density of this signal, plotted in Figure 1.1, displays nulls at multiples of the clock rate, and an overall –20-dB/decade slope from F_{clock} up to the frequency marked F_{knee} (the *knee frequency*). Beyond F_{knee}, the spectrum rolls off much faster than 20 dB/decade. At the knee frequency, the spectral amplitude is down by half (–6.8 dB) below the natural 20-dB/decade rolloff.[1] The knee frequency for any digital signal is related to the rise (and fall) time of its digital edges, but not to its clock rate:

$$F_{\text{knee}} = \frac{0.5}{T_r} \qquad [1.1]$$

where F_{knee} = frequency below which most energy in digital pulses concentrates

T_r = pulse rise time[2]

Shorter rise times push F_{knee} higher. Longer rise times pull F_{knee} lower.

The important time domain characteristics of any digital signal are determined primarily by the signal's spectrum below F_{knee}. From this principle we may deduce two important qualitative properties of digital circuits:

(1) Any circuit which has a flat frequency response up to and including F_{knee} will pass a digital signal practically undistorted.

[1]Applies to gaussian pulse shapes described in Appendix B.

[2]For now, think of this as the 10–90% rise time. See Appendix B for an exhaustive discussion of the many ways rise time can be defined.

(2) The behavior above F_{knee} of a digital circuit will have little effect on how it processes digital signals.

Note that F_{knee} is defined only by the signal rise time and bears no direct relation to other frequency domain parameters. This simple definition makes F_{knee} easy to use and easy to remember.

When you apply F_{knee}, keep in mind that it is an imprecise measure of spectral content. Used as a guidepost, F_{knee} can help classify frequency-sensitive effects as totally insignificant, merely worrisome, or completely devastating. For most digital problems, that is exactly what we wish to know.

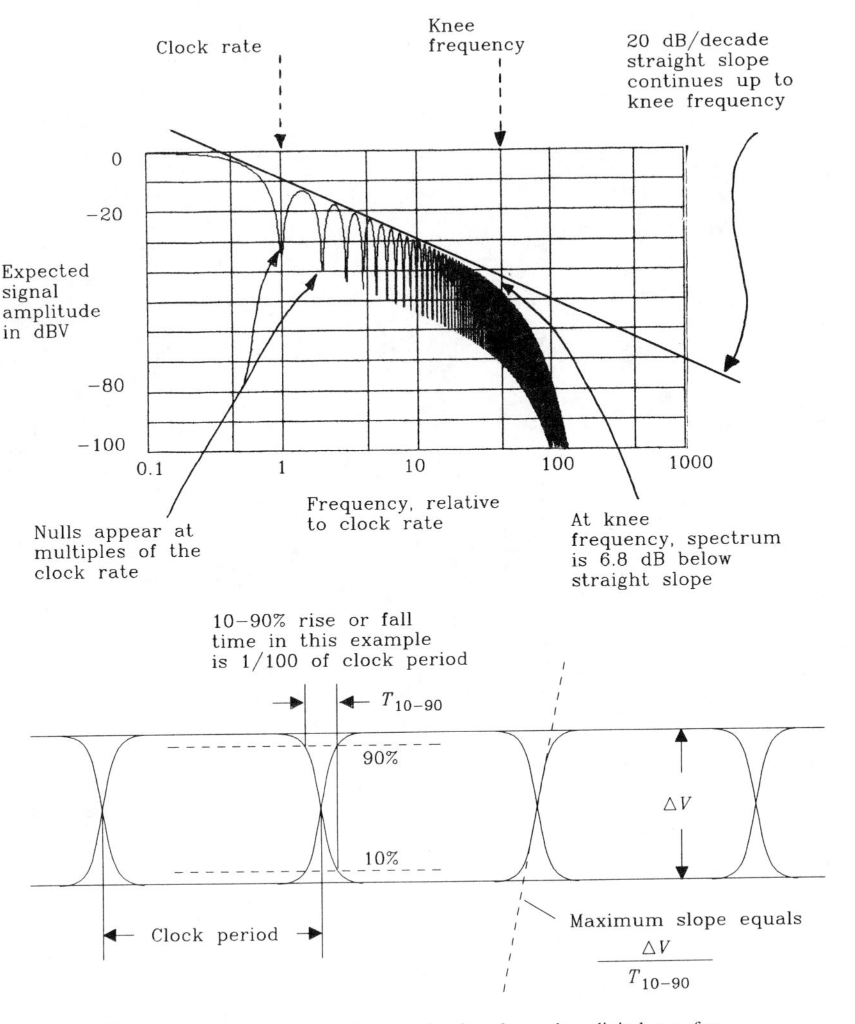

Figure 1.1 Expected spectral power density of a random digital waveform.

Of course, F_{knee} has limitations. F_{knee} cannot make precise predictions about system behavior. It doesn't even define precisely how to measure rise time! It is no substitute for full-blown Fourier analysis. It can't predict electromagnetic emissions, whose properties depend on the detailed spectral behavior at frequencies well above F_{knee}.

At the same time, for digital signals, F_{knee} quickly relates time to frequency in a practical and useful manner. We will use F_{knee} throughout this book as the practical upper bound on the spectral content of digital signals. Appendix B contains additional information, for those interested, about various measures of rise time and frequency.

Referring back to deduction (1) above, if a system has a nonflat frequency response below F_{knee}, how will it distort a digital signal? Here is an example.

We know the response of a circuit at high frequencies affects its processing of short-time events (like a rise time). The response of a circuit at low frequencies affects its

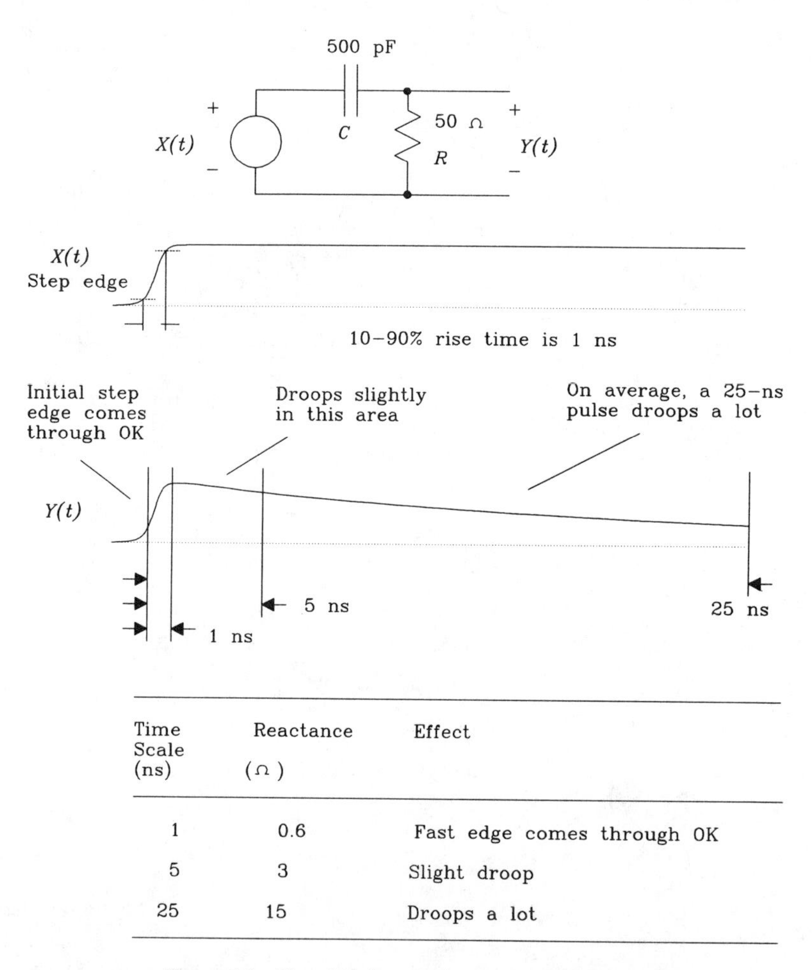

Time Scale (ns)	Reactance (Ω)	Effect
1	0.6	Fast edge comes through OK
5	3	Slight droop
25	15	Droops a lot

Figure 1.2 Time domain analysis of a simple *RC* filter.

processing of longer-term events (such as a long, steady pulse). Figure 1.2 shows a circuit having different characteristics at high and low frequencies. This circuit passes high-frequency events (the rising edge) but does not pass low-frequency events (the long, steady part).

Let's start our analysis of Figure 1.2 at one particular frequency: F_{knee}. At the frequency F_{knee}, capacitor C has a *reactance* (i.e., impedance magnitude) of $1/C2\pi F_{knee}$.

We can calculate the reactance using this formula or substitute rise time for F_{knee}:

$$X_C = \frac{1}{2\pi F_{knee} C} = \frac{T_r}{\pi C} = 0.6\Omega \qquad [1.2]$$

where T_r = rise time of step input, s
F_{knee} = highest frequency in step input, Hz
C = capacitance, F

Equation 1.2 shows how to estimate the reactance of a capacitor using either the knee frequency or rise time.

A 0.6-Ω reactance acts as a virtual short in the circuit in Figure 1.2. The full amplitude of the leading edge, corresponding to a frequency of F_{knee}, will come blasting straight through the capacitor.

Over a time interval of 25 ns, corresponding roughly to a frequency of 20 MHz, the capacitive reactance increases to 15 Ω, causing the coupled signal to droop noticeably.

POINTS TO REMEMBER:

The response of a circuit at high frequencies affects its processing of short-time events.

The response of a circuit at low frequencies affects its processing of long-term events.

Most energy in digital pulses concentrates below the knee frequency:

$$F_{knee} = \frac{0.5}{T_r}$$

The behavior of a circuit at the knee frequency determines its processing of a step edge.

The behavior of a circuit at frequencies above F_{knee} hardly affects digital performance.

1.2 TIME AND DISTANCE

Electrical signals in conducting wires, or conducting circuit traces, propagate at a speed dependent on the surrounding medium. Propagation delay is measured in units of picoseconds per inch. Propagation delay is the inverse of propagation velocity (also called propagation speed), which is measured in inches per picosecond.

TABLE 1.1 PROPAGATION DELAY OF ELECTROMAGNETIC FIELDS IN VARIOUS MEDIA

Medium	Delay (ps/in.)	Dielectric constant
Air (radio waves)	85	1.0
Coax cable (75% velocity)	113	1.8
Coax cable (66% velocity)	129	2.3
FR4 PCB, outer trace	140–180	2.8–4.5
FR4 PCB, inner trace	180	4.5
Alumina PCB, inner trace	240–270	8–10

The propagation delay of conducting wires increases in proportion to the square root of the dielectric constant of the surrounding medium. Manufacturers of coaxial cable often use dielectric insulators made of foam or ribbed material to reduce the effective dielectric constant inside the cable, thus lowering the propagation delay and simultaneously lowering dielectric losses. The difference between the two coax cables listed in Table 1.1 is their dielectric insulation.

The delay per inch for printed circuit board traces depends on both the dielectric constant of the printed circuit board material and the trace geometry. Popular FR-4 printed circuit board material has a dielectric constant at low frequencies of about 4.7 ± 20%, which deteriorates at high frequencies to 4.5. For propagation delay calculations, use the high-frequency value of 4.5.

Trace geometry determines whether the electric field stays in the board or goes into the air. When the electric field stays in the board, the effective dielectric constant is bigger and signals propagate more slowly. The electric field surrounding a circuit trace encapsulated between two ground planes stays completely inside the board, yielding an effective dielectric constant, for typical FR-4 printed circuit board material, of 4.5. Traces laid on the outside surface of the printed circuit board (outer traces) share their electric field between the air on one side and the FR-4 material on the other, yielding an effective dielectric about halfway between 1 and 4.5. Outer-layer PCB traces are always faster than inner traces.

Alumina is a ceramic material used for constructing very dense circuit boards (up to 50 layers). It has the advantage of a low coefficient of thermal expansion and machines easily in very thin layers, but it is very expensive to manufacture. Microwave engineers like the slow propagation velocity (large delay) of alumina circuits because it shrinks the size of their resonant structures.

POINTS TO REMEMBER:

Propagation delay is proportional to the square root of the dielectric constant.

The propagation delay of signals traveling in air is 85 ps/in.

Outer-layer PCB traces are always faster than inner traces.

1.3 LUMPED VERSUS DISTRIBUTED SYSTEMS

The response of any system of conductors to an incoming signal depends greatly on whether the system is smaller than the effective length of the fastest electrical feature in the signal, or vice versa.

The effective length of an electrical feature, like a rising edge, depends on the time duration of the feature and its propagation delay. For example, let's analyze the rising edge of a 10KH ECL signal. These gates have a rise time of approximately 1.0 ns. This rising edge, when propagating along an inner trace of an FR-4 printed circuit board, has a length of 5.6 in.:

$$l = \frac{T_r}{D} \qquad [1.3]$$

where l = length of rising edge, in.
T_r = rise time, ps
D = delay, ps/in.

Figure 1.3 depicts a series of snapshots of the electric potential along a straight trace 10 in. long. A rising edge of 1-ns duration impinges on the left end of the trace. Evidently, as the pulse propagates along the trace, the potential is not uniform at all points. The reaction of this system to the incoming pulse is distributed along the trace, which we label a *distributed system*. The snapshot taken at 4 ns shows the physical length of this rising edge is 5.6 in.

Systems physically small enough for all points to react together with a uniform potential are called *lumped systems*. Figure 1.3 illustrates the response of a 1-in. trace, carrying the same 1-ns rising edge, which behaves as a lumped system. The voltage on every part of this line is (almost) uniform at all times.

The classification of a system as distributed or lumped depends on the rise time of the signals flowing through it. The distinguishing characteristic is the ratio of system size to rise-time size. For printed circuit board traces, point-to-point wiring, and bus structures if the wiring is shorter than one-sixth of the effective length of the rising edges, the circuit behaves mostly in a lumped fashion.[3]

[3]Some authors use $l/\sqrt{2\pi}$; others use $l/4$. The idea is that small structures are lumped circuits, while big ones are distributed.

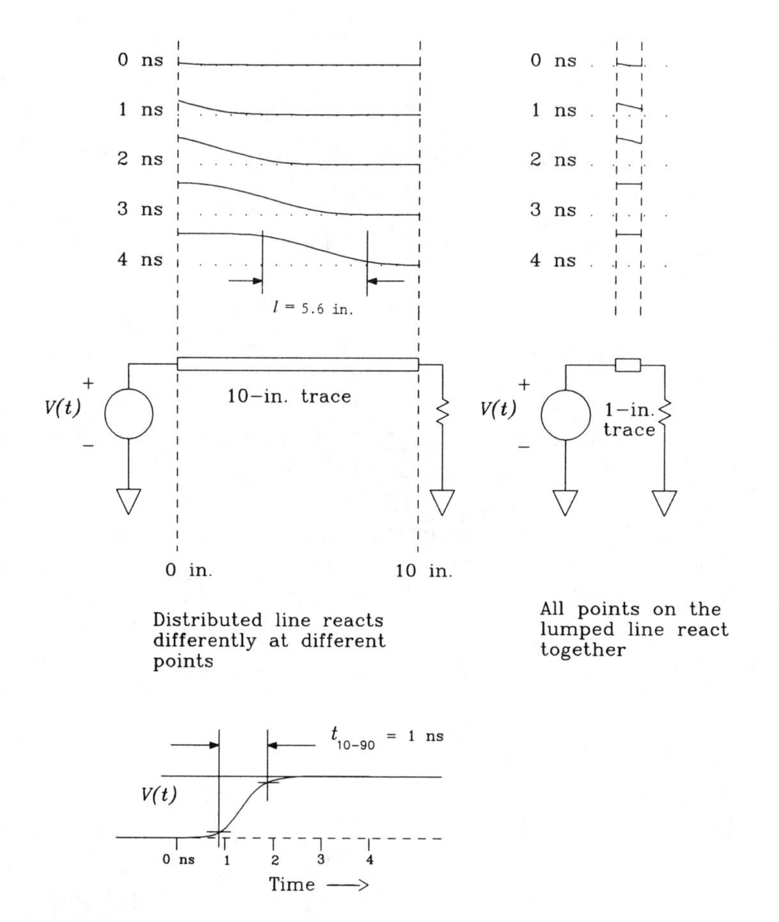

Figure 1.3 Snapshots in time of the electric potential on distributed and lumped transmission lines.

POINTS TO REMEMBER:

$$\text{Length of rising edge } l = \frac{\text{Rise time (ps)}}{\text{Delay (ps/in.)}} \qquad [1.4]$$

Circuits smaller than $l/6$ are lumped circuits.

1.4 A NOTE ABOUT 3-dB AND RMS FREQUENCIES

When translating specifications from the analog world to the digital world, one often must convert from frequency response to rise time.

For example, oscilloscope manufacturers quote a maximum operating bandwidth for each vertical amplifier and a corresponding maximum bandwidth for each probe. Depending on the manufacturer, they may quote a 3-dB bandwidth or an RMS (noise equivalent) bandwidth. In either case, the conversions between bandwidth and rise time will depend on the exact shape of the scope's frequency response curve.

Fortunately, we do not often need to compute an exact rise time. For the purposes of this book, we may devise an approximate relation that is easy to apply while ignoring complicated details of the exact frequency response shape. Appendix B provides justification for this approach, comparing exact calculations for several different pulse types.

The conversions listed below assume we are translating from frequency response to a 10–90% rise time. As explained in Appendix B, for the accuracy we require in diagnosing and fixing digital problems it hardly matters whether we define rise time using the 10–90% rise time, inverse of the slope at the center of the pulse, or standard deviation method.

$$F_{3\text{dB}} \approx \frac{K}{T_r} \qquad [1.5]$$

$$T_r \approx \frac{K}{F_{3\text{dB}}} \qquad [1.6]$$

where $F_{3\text{dB}}$ = frequency at which impulse response rolls off by 3 dB
T_r = pulse rise time (10–90%)
K = constant of proportionality depending on exact pulse shape; $K = 0.338$ for gaussian pulses; $K = 0.350$ for single-pole exponential decay

If we change our pulse type from gaussian to a single-pole exponential decay, the constant in Equation 1.6 changes from 0.338 to 0.350. For most digital designs, such a subtle distinction hardly matters.

Where a manufacturer quotes the RMS bandwidth, or equivalent noise bandwidth,[4] of a subsystem, the following relation computes the 10–90% rise time of that subsystem. Here the constant K changes from 0.36 to 0.55 depending on the pulse type, a slightly more significant change than in Equation 1.6.

$$T_r \approx \frac{K}{F_{\text{RMS}}} \qquad [1.7]$$

where F_{RMS} = RMS bandwidth
T_r = rise time (10–90%)
K = constant of proportionality depending on exact pulse shape; $K = 0.361$ for gaussian pulses; $K = 0.549$ for single-pole exponential decay

[4]The noise bandwidth of a frequency response *H(f)*, or RMS bandwidth, is the cutoff frequency at which a box-shaped frequency response would pass the same amount of white noise energy as *H(f)*.

When you look at a scope response to a very fast rising edge (much faster than the scope response), you can usually tell if it has a single-pole or a gaussian-type response. If the leading edge of the response has a sharp corner, suddenly taking off at a steep angle and blending into a long, sweeping tail, it is probably a single-pole response. If the pulse edge sweeps up gently, with symmetric leading and trailing edges, it is probably nearly gaussian. In between, use $K = 0.45$.

1.5 FOUR KINDS OF REACTANCE

Four circuit concepts separate the study of high-frequency digital circuits from that of low-speed digital circuits: *capacitance, inductance, mutual capacitance, and mutual inductance*. These four concepts provide a rich language for describing and understanding the behavior of digital circuit elements at high speeds.

There are many ways to study capacitance and inductance. A microwave engineer studies them using Maxwell's equations. A designer of control systems uses Laplace transforms. An advocate of Spice simulations uses linear difference equations. Digital engineers use the step response.

The step response measurement shows us just what we need to see: what happens when a pulse hits a circuit element. From the step response, if we wish, we can derive a curve of impedance versus frequency for a circuit element. In that sense the step response measurement is (at least) as powerful as any frequency-domain measure of impedance.

Our investigation of capacitance and inductance will focus on the step response of circuit elements.

Figure 1.4 illustrates a classic step response measurement for a two-terminal device. In this figure, we use a step source having an output impedance of R_S ohms. The step source is shunted by the device under test, across which we measure the voltage response. In practical measurements, the step input is repeated over and over again while the results are synchronously displayed on an oscilloscope.

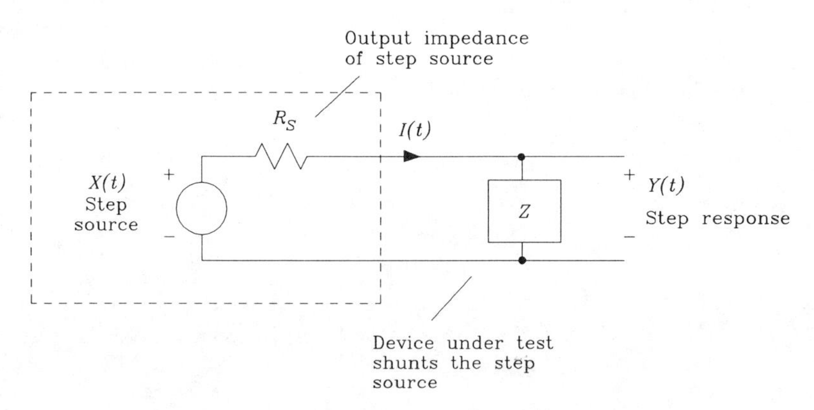

Figure 1.4 Step response test for a two-terminal device.

With practice, anyone can learn to instantly characterize the device under test by observing the step response and using these three rules of thumb:

(1) Resistors display a flat step response. At time zero, the output rises to a fixed value and holds steady.

(2) Capacitors display a rising step response. At time zero, the step response starts at zero but then later rises to a full-valued output.

(3) Inductors display a sinking step response. At time zero, the output rises instantly to full value and then later decays back toward zero.

To first order, we may characterize any circuit element according to whether, as a function of time, its step response stays constant, rises, or falls. We label these elements resistive, capacitive, and inductive, respectively.

Reactive effects (both capacitance and inductance) further subdivide into ordinary and mutual categories. The ordinary varieties of capacitance and inductance describe the behavior of individual circuit elements (two-terminal devices). The mutual capacitive and inductive concepts describe how one circuit element affects another. In digital electronics, mutual capacitance or inductance usually creates unwanted crosstalk, which we strive to minimize. Plain capacitance or inductance can be a help or a hindrance, depending on the circuit application.

We will define and use a special version of step response for characterizing mutual capacitive and mutual inductive circuit elements.

Our brief study of reactive concepts considers only lumped circuit elements, in this order:

- Ordinary capacitance
- Ordinary inductance
- Mutual capacitance
- Mutual inductance.

1.6 ORDINARY CAPACITANCE

Capacitance arises wherever there are two conducting bodies charged to different electric potentials. Two bodies at different electric potentials always have an electric field between them. The energy stored in their electric field is supplied by the driving circuit. Because the driving circuit is a limited source of power, the voltage between any two bodies takes a finite amount of time to build up to a steady-state value. The reluctance of voltage to build up quickly in response to injected power, or to decay quickly, is called *capacitance*. Structures enclosing a large amount of electric flux at low voltages, like two parallel plates, have lots of capacitance.

Figure 1.5 shows idealized current and voltage waveforms for a capacitor driven by a 30-Ω source.[5] The step response of a capacitor grows as a function of time. When a

[5]The 30-Ω source is an approximation to the drive capability of a standard TTL output.

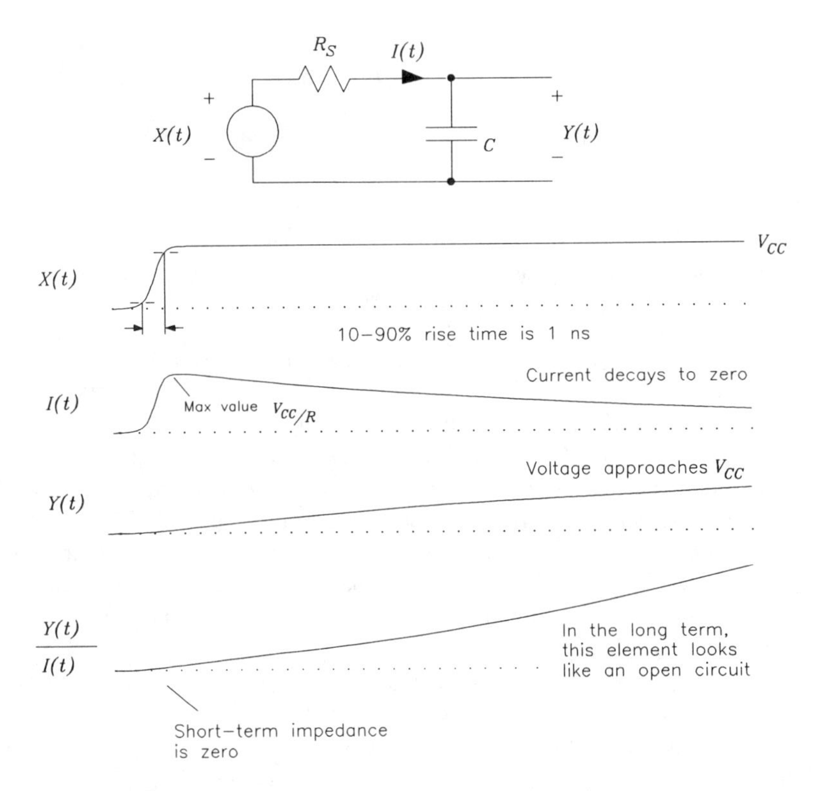

Figure 1.5 Step response of a perfect capacitor.

voltage step is first applied, lots of power flows into the capacitor to build up its electric field. The initial current into the capacitor is quite high and the ratio $Y(t)/I(t)$ very low. At short time scales, the capacitor looks like a short circuit.

Over time, the ratio $Y(t)/I(t)$ gets bigger. Eventually, the current falls to near zero and the capacitor starts to look like an open circuit. Finally, after the electric field surrounding the capacitor is fully formed, only a small leakage current flows as a result of the imperfect insulating qualities of the dielectric medium between the capacitor plates. The ratio $Y(t)/I(t)$ then is extremely high.

The step response of some circuit elements displays a capacitive character when viewed at one time scale, and an inductive behavior at a different time scale, or vice versa.[6] For example, the mounting leads on capacitors commonly have enough inductance to cause the overall component to look inductive at very high frequencies. The step response of such a capacitor shows a tiny pulse of perhaps a few hundred picoseconds at time zero (corresponding to the inductance), then a drop to zero, and then a normal upward-moving capacitive ramp.

If the rise time of the step source is too slow, the output trace won't show the inductive spike. Because the spike is so short, it is also easy to miss if the scope time base

[6]Equivalently, the element behaves differently at different frequencies. Our focus here is on time domain response.

is set for too slow a sweep. It is interesting to contemplate the idea that by adjusting the rise time and setting the time base sweep, we can cause the step response measurement to emphasize the character of a circuit element in a particular frequency range. Roughly speaking, if the step rise time is T_r, the step response near time zero is related to the impedance magnitude of the circuit element near frequency F_A:

$$F_A \approx \frac{0.5}{T_r} \quad [1.8]$$

where T_r = rise time of step source
F_A = approximate analysis frequency

By visually averaging the step response over a time interval, we can estimate the impedance magnitude at lower frequencies. Use Equation 1.8 to compute the approximate analysis frequency that corresponds to an averaging interval of T_r.

The final value of the step response indicates the impedance magnitude at DC.

With a step rise time of T_r, we cannot infer much about the behavior of the component at frequencies greater than F_A. Make sure your step source is fast enough to reveal what you want to see.

Figure 1.6 shows a measurement setup ideal for characterizing capacitors of a few picofarads over a time interval of nanoseconds. This arrangement is ideal for characterizing the capacitance of circuit traces, gate inputs, bypass capacitors, and other common digital circuit elements. This method drives the capacitor under test with a known resis-

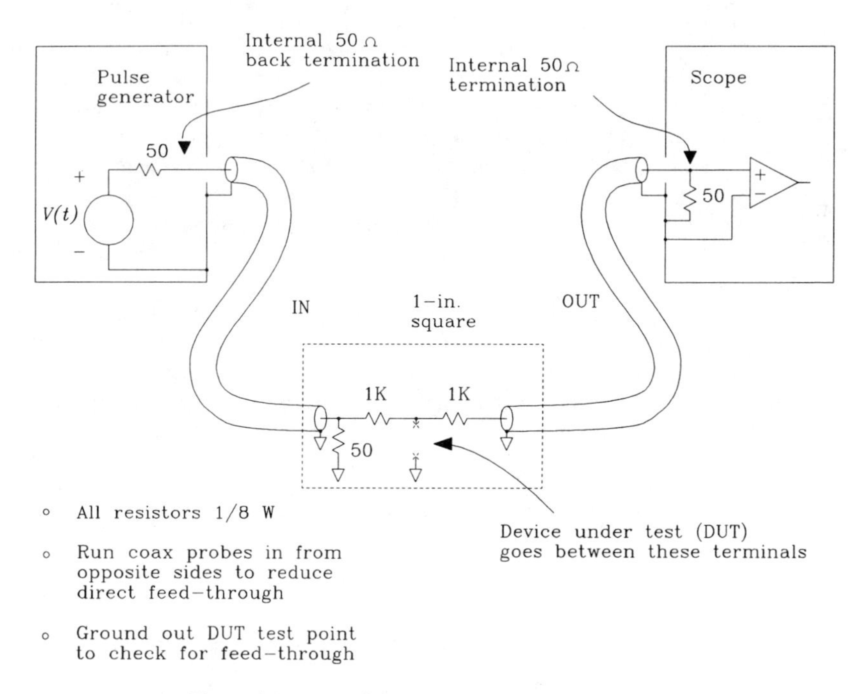

Figure 1.6 A 500-Ω lab setup for measuring capacitance.

tance. By measuring the rise time of the resulting waveform, we can infer a value for the capacitance. In comparison to techniques used at audio frequencies, this setup is very complicated. The complication follows from the difficulty of containing and directing electromagnetic field energy at high frequencies. The coaxial cables are used to direct the test signals and measured results into and out of a solid ground plane area no more than 1 in.^{2} where the actual measurement takes place. Limiting the measurement footprint to 1 in.^{2} ensures that the circuit will behave in a lumped fashion.

Example 1.1: Measurement of a Small Capacitance to Ground

The device under test (DUT) in this example (Figure 1.6) is a parallel plate capacitor, 0.5 in. × 0.75 in., printed in $1\frac{1}{2}$-oz copper on an epoxy FR-4 printed circuit board, nominally 0.008 in. above a solid ground plane. This structure forms a capacitor with extremely low parasitic series inductance.

The measurement setup consists of two RG-174 coaxial cables, IN and OUT. The IN cable, terminated with 50 Ω to ground, includes a 1K drive resistor connected from the terminated output to the device under test. The 1K resistor isolates the signal source from the device under test, providing to the source a constant terminating impedance regardless of the impedance of the DUT. Isolation ensures consistent source rise time and amplitude performance regardless of the DUT load impedance.

The source pulse generator provides signals of similar amplitude and rise time as that expected in the actual circuit. When measuring passive components, the DC offset of the pulse generator does not much matter. On the other hand, when measuring gate inputs, always adjust the pulse source to span the input switching range and apply power to the gate under test. That biases the gate into its operating range for the test. Gates requiring lots of input drive current may need source resistors smaller than 1KΩ.

If your signal generator has a 50-Ω back-termination feature, engage it to reduce reflections on the IN cable. This feature inserts 50 Ω in series with the signal generator output, reducing reflections back and forth along the source cable between the unavoidable slight mismatch at the test jig and the output impedance of the signal source. Using the back termination, unwanted reflections from your source signal now are attenuated first when they bounce off the test jig and a second time as they bounce off the back termination resistor in the signal source on their return path to your measuring apparatus. The back termination reduces the available source drive amplitude by half but improves the system step response.

The OUT cable connects separately through 1K Ω to the circuit under test, running to an oscilloscope input internally terminated with 50 Ω. The 1K resistor acts as a 21:1 probe. The advantages of this sensing arrangement are detailed later in this book in the section on oscilloscope probing. Both IN and OUT cables are 3 ft long.[7]

The open-circuit response of this probe, when driven by a 2.6-V step input from the source with the DUT disconnected, appears in Figure 1.7. The top trace is recorded at 5 ns/division, and the bottom trace is an exploded view of the same signal at 500 ps/division.

The Tektronix 11403 scope used to record this waveform automatically computes a 10–90% rise time of 818 ps. The nominal step amplitude is 63 mV (the scope measures a

[7]Longer cables can be an advantage, in that reflections occur so late that they do not show on the scope. The disadvantage to longer cables is that they introduce more signal dispersion. At some length the cable response begins to deteriorate the observable rise time. In Figure 1.7 reflection effects from the 3-ft cables show up about 8 ns into the picture.

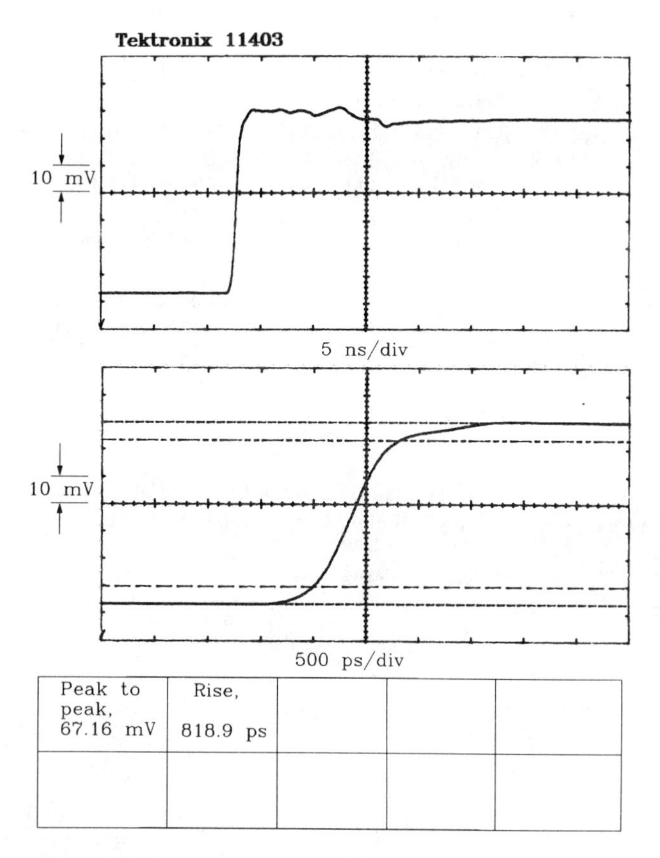

Figure 1.7 Open-circuit response of a 500-Ω capacitance test setup.

peak of 67 mV). Note that the measured step amplitude is $\frac{1}{21}$ the amplitude at the DUT of 1.3 V, which is in turn half the source drive voltage.

The Thevinen equivalent circuit for this test arrangement, as shown in Figure 1.8, includes the aggregate system rise time lumped into the source. It matters not whether the signal source or the scope contributes more to the slowness of the observed rise time. Any reasonable combination of source and scope having a similar open-circuit rise time will behave similarly under the influence of the DUT. It matters only that we know the aggregate rise time of the source scope combination. When measuring passive components, it similarly matters only that we know the observed step amplitude, details of the actual voltage at the DUT and the probe attenuation ratio being unimportant.

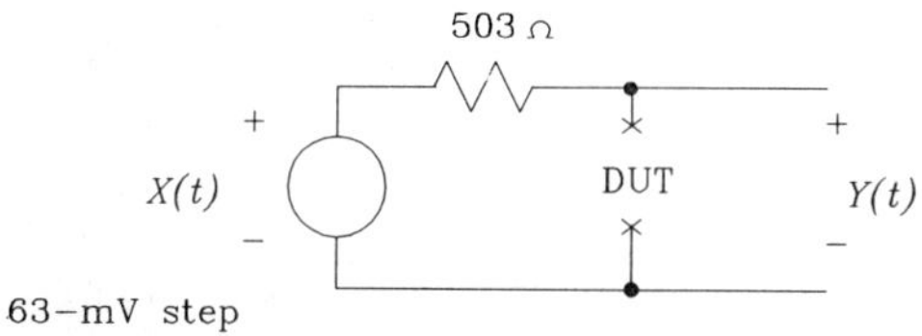

Figure 1.8 Thevenin equivalent of a 500-Ω capacitance test setup.

The source impedance is 503 Ω, as measured at the DUT terminals using an ohmmeter with the pulse source turned off but the back termination to 50 Ω still connected. This 503 Ω is the parallel combination of the 1K drive resistor and the 1K sense resistor.

With the DUT connected, the observed voltage displays a capacitive character, starting low and then rising (see Figure 1.9). A stored copy of the original drive waveform is superimposed on this picture for reference. Over the range of time scales observable with this probe, from 800 ps (the aggregate source-probe combination rise time) to 40 ns (the length of trace appearing in the scope photo) the DUT appears perfectly capacitive.

Cursors in Figure 1.9 mark the 63% point along the rise time where we can read off the *RC* time constant of 23.5 ns.[8] Knowing the drive resistance of 503 Ω, we can compute the capacitance of the DUT using the relation $C = \tau/R$:

$$C = \frac{23.5 \times 10^{-09}\,\text{s}}{503\ \Omega} = 46.7\ \text{pF} \qquad [1.9]$$

We can use the relationship between rise time and frequency to derive a rough idea of the reactance presented by a capacitor to the leading edge of a digital waveform. This

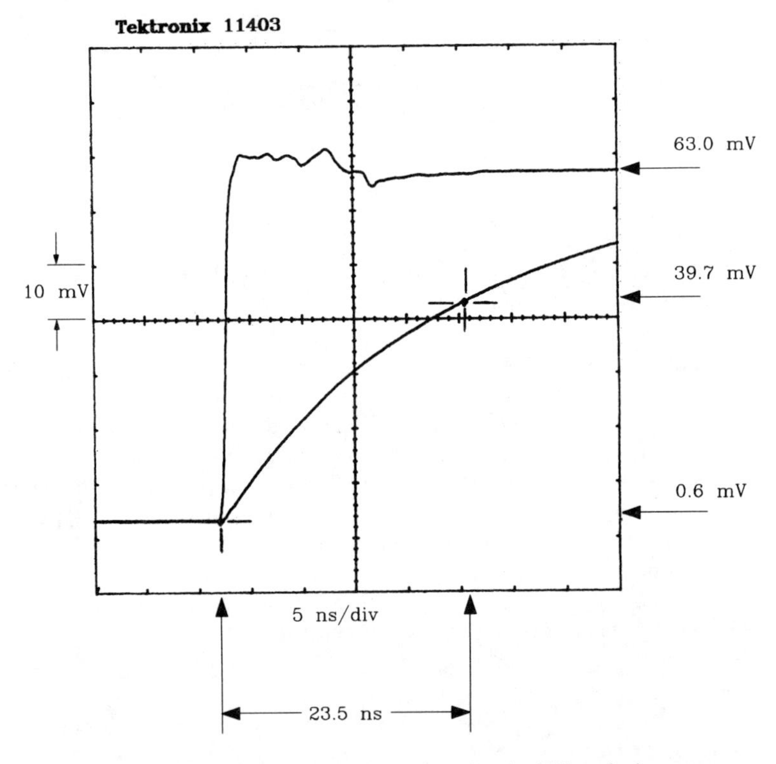

Figure 1.9 Finding a time constant using the 63% method.

[8]In one time constant, a rising edge progresses to 63% of its final value, and a falling edge deteriorates to 37% of its initial value.

approach is very useful when considering the distortion introduced in a digital waveform by a capacitive load.

$$X_C = \frac{T_r}{\pi C} \quad [1.10]$$

The capacitor in Example 1.1 has a reactance of 20.44 Ω to a rising edge of 3 ns. We therefore predict it will significantly distort (by slowing down) a 3-ns rising edge from a TTL driver having an output impedance of 30 Ω.

The current through a capacitor at any point in time is always related to the rise time of the voltage across it according to the general formula

$$I_{\text{capacitor}} = C\frac{dV_{\text{capacitor}}}{dt} \quad [1.11]$$

Equation 1.11 will help us later when calculating crosstalk due to capacitance between circuits.

POINT TO REMEMBER:

A capacitance test jig is easy to build using a pulse source and an oscilloscope.

1.7 ORDINARY INDUCTANCE

Inductance arises wherever there is electric current. Electric current creates a magnetic field, with the energy stored in the magnetic field supplied by some driving circuit. Because any driving circuit is a limited source of power, current always takes a finite amount of time to build up to a steady-state value. The reluctance of current to build up quickly, or to decay quickly, is called *inductance.*

Figure 1.10 shows idealized current and voltage waveforms for an inductor driven by a 30-Ω source. The step response of an inductor shrinks as a function of time. When a voltage step is first applied, almost no current flows, making the ratio $Y(t)/I(t)$ very high. At short time scales, the inductor looks like an open circuit.

Over time, the ratio $Y(t)/I(t)$ decays. Eventually, the voltage drops to near zero and the inductor starts to look like a short circuit. Later, after the magnetic field surrounding the inductor is fully formed, the current is limited only by the DC resistance of the inductor. The ratio $Y(t)/I(t)$ is then extremely low.

Figure 1.11 shows a measurement setup optimized for characterizing inductors of a few nanohenries. This arrangement is ideal for measuring the inductance of ground traces or short lengths of connecting wiring.

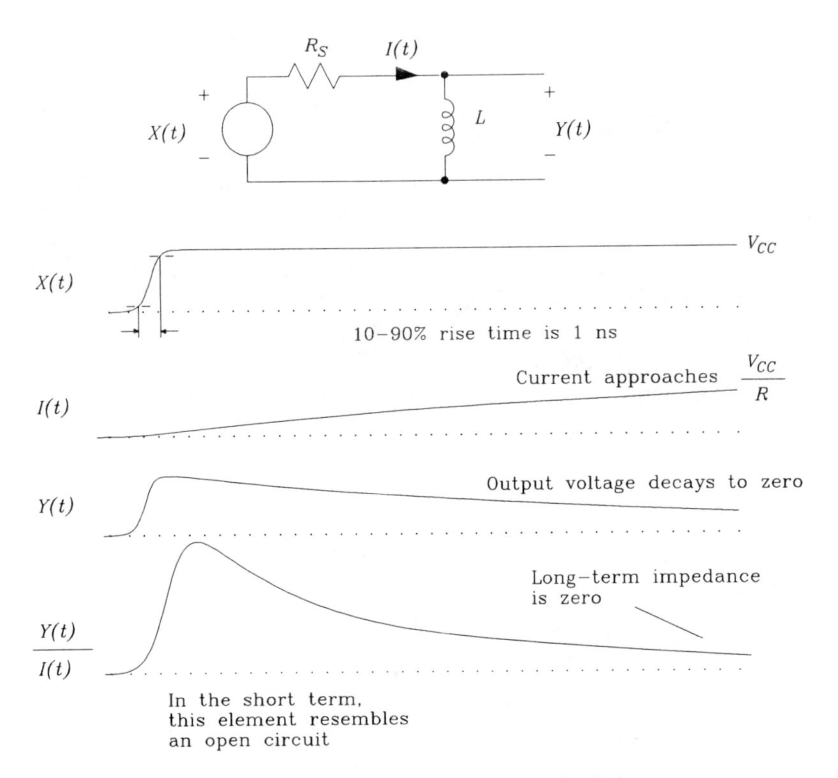

Figure 1.10 Instantaneous resistance of a perfect inductor.

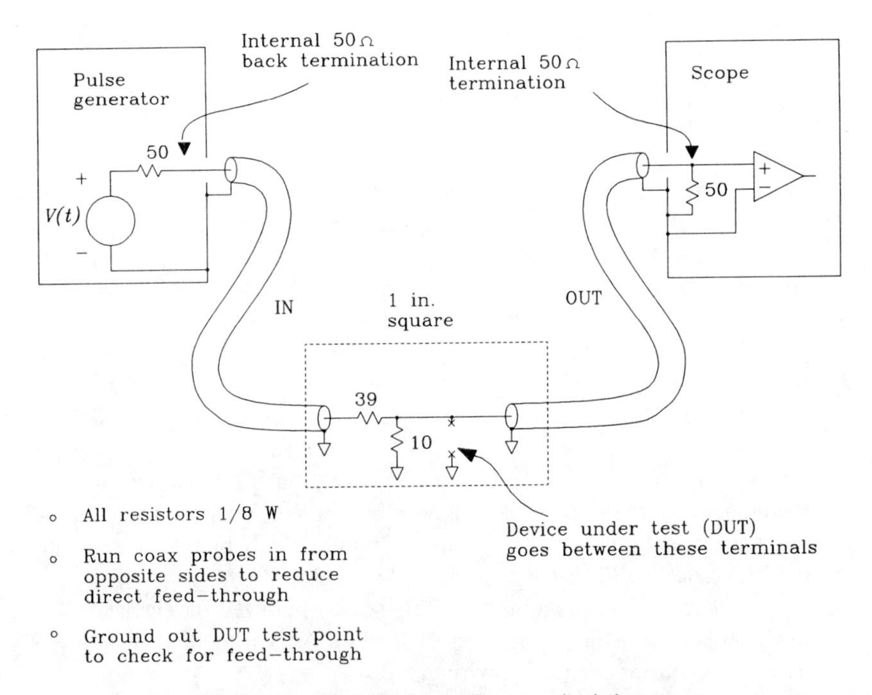

Figure 1.11 A 7.6-Ω lab setup for measuring inductance.

EXAMPLE 1.2: Measurement of a Small Inductance to Ground

The device under test (DUT) in this example (Figure 1.11) is a short 1-in. circuit trace printed in $1\frac{1}{2}$-oz copper on an epoxy FR-4 printed circuit board. The trace nominally rides 0.008 in. above a solid ground plane and is 0.010 in. wide. The far end of the trace is shorted to ground with a 0.035-in.- diameter via. This structure has a parasitic capacitance to ground of 2 pF when open-circuited, and half that when the far end is shorted.[9] The calculated inductance is about 9 nH.

We plan to characterize this circuit using an 800-ps rise time. First check that the parasitic capacitive reactance at that speed is much larger than the inductive reactance we wish to see.

$$X_C = \frac{T_r}{\pi C} = 254\ \Omega \qquad [1.12]$$

$$X_L = \frac{\pi L}{T_r} = 35\ \Omega \qquad [1.13]$$

The capacitive reactance, which appears in parallel with our measurement, is eight times bigger than the expected inductive reactance. The effect of this capacitance will be to increase our observed value of L/R by about 12%.

The measurement setup consists of two RG-174 coaxial cables, IN and OUT. The IN cable, terminated in a total of 49 Ω to ground, includes a tap of 10 Ω for driving the DUT. In this jig the source is not as well isolated from the DUT as in the capacitance test jig. The terminating impedance seen by the source varies from 39 to 49 Ω under various DUT load conditions.[10] Because we expect reflections off the mismatch at the DUT, do not forget to back-terminate your pulse generator.

The generator is adjusted for no DC offset. The inductor would short out any DC offset anyway.

The source impedance is 7.6 Ω as measured at the DUT with the source turned off but the back termination connected. This is a parallel combination of the 50 + 39-Ω source impedance, the 10-Ω tap resistor, and the 50-Ω probe impedance.

We have arranged for a low source impedance at the DUT to exaggerate the L/R decay time. Had we used a test jig with a 500-Ω Thevenin equivalent source resistance, the expected L/R time would be only 0.018 ns. With a 7.6-Ω source we expect a 1.2-ns L/R decay constant.

The OUT cable connects directly to the DUT in this experiment and runs to an oscilloscope input internally terminated with 50 Ω. Both IN and OUT cables are 3 ft long.

The open-circuit response of this 7.6-Ω test setup, when driven by a 2.4-V step input, appears in Figure 1.12. The scope automatically computes a 10–90% rise time of 788 ps. The step amplitude is 417 mV. The probe is a 1:1 arrangement, so the voltage at the DUT is actually 417 mV.

Figure 1.13 shows the Thevenin equivalent circuit of this 7.6-Ω test setup.

[9]For those familiar with short transmission line theory, according to the pi model *(C+L+C)* for a short transmission line shorting the far end just shorts out one of the two capacitors. The result is a tank circuit composed of the full inductance and half the open circuit capacitance.

[10]Because the short-term impedance of the inductor is high, the best choice for the terminating network is 39 and 10 Ω, giving a 49-Ω initial value. This terminates the initial rising step edge with 49 Ω. Were we measuring the low-inductance capacitor of Example 1.1, the best choice would be 50 and 10 ohms, because the short-term impedance of a capacitor is initially zero.

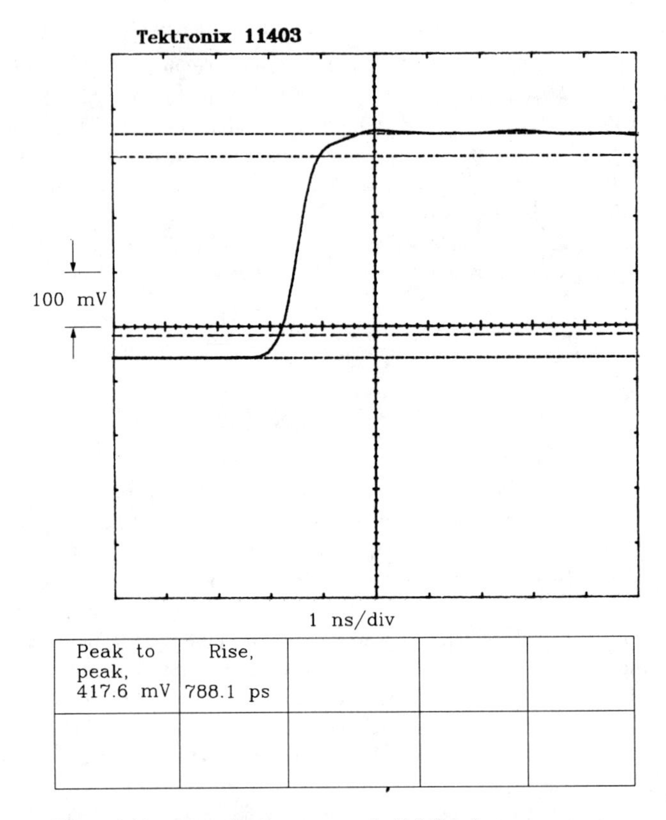

Figure 1.12 Open-circuit response of a 7.6-Ω inductance test setup.

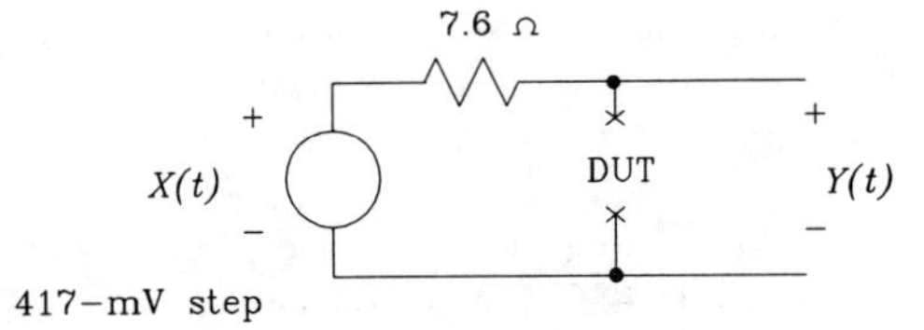

Figure 1.13 Thevenin equivalent of a 7.6-Ω inductance test setup.

With the DUT connected (Figure 1.14) the observed voltage manifests an inductive character, rising quickly along with the input signal and then decaying later toward zero. Over the range of time scales observable with this setup, from 800 ps (the aggregate source-scope combination rise time) to 7 ns (the length of trace appearing in the scope photo), the DUT is inductive. The exponential decay time is 1.36 ns as measured by the two cursors carefully positioned at voltages separated by a multiplicative factor of *e*.

From the measured decay constant, using the relation $L = R\tau$, we may compute the inductance of the DUT:

$$L = (1.4 \times 10^{-09})(7.6\ \Omega) = 10.6\ \text{nH} \qquad [1.14]$$

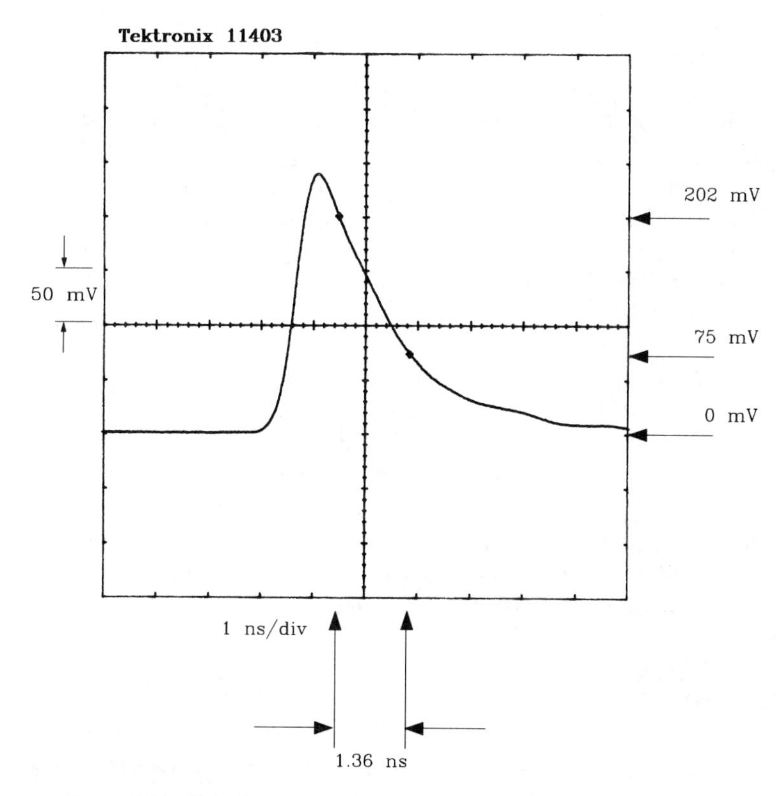

Figure 1.14 Decaying exponential response of a 7.6-Ω inductance test setup.

We can use the relationship between rise time and frequency to derive a rough idea of the reactance presented by an inductor to the leading edge of a digital waveform. This approach is very useful when considering the ground bounce introduced in a poor ground connection by a parasitic series inductance.

$$X_L = \frac{\pi L}{T_r} \qquad [1.15]$$

The inductor of Example 1.2, a trace 1 in. long, has a reactance of 9.4 Ω to a rising edge of 3 ns. If this trace is used to ground a 50-Ω terminator for 3-ns rising edges, the composite termination value will be off by 20%. If this trace is used to ground a bank of eight 50-Ω terminators, the parallel impedance of eight terminators ($\frac{50}{8} = 6\ \Omega$) is actually less than the trace impedance. If all eight terminated lines switch simultaneously, the terminating bank won't work.

The voltage across an inductor at any point in time is always related to the rise time of the current through it according to the general formula

$$V_{\text{inductor}} = L\frac{dI_{\text{inductor}}}{dt} \qquad [1.16]$$

We will use Equation 1.16 later when calculating crosstalk due to inductance between circuits.

On the subject of what does and doesn't work as a short circuit, consider two common ways of shorting a digital line to ground: a knife blade and a pair of needle-nosed pliers.

During the debugging process, one often needs to short a signal to ground in order to test a hypothesis about how the circuit functions (or doesn't function). If the shorting instrument has too high an inductance, narrow pulses will slip through unshorted. Clock lines and asynchronous interrupt lines are especially susceptible to this narrow-pulse problem.

The inductance of a knife blade when used to short together two circuit nodes located 0.300 in. apart is on the order of a couple of nanohenries. To a 1-ns rising edge, the knife blade presents an impedance magnitude of about 6 Ω (Equation 1.15).

The inductance of a pair of needle-nosed pliers used to short the same nodes (one plier tip on each node) is on the order of 10–20 nH. The trip up one leg of the pliers, through the joint, and back down the other side introduces an order of magnitude more inductance than a small knife blade. To the same 1-ns rising edge, the pliers present an impedance magnitude of at least 30 Ω. Thirty ohms is not low enough to ground out a short TTL pulse. Enough said.

1.8 A BETTER METHOD FOR ESTIMATING DECAY TIME

In the inductance test jig, the ratio of the expected characteristic decay time $T_{L/R}$ to the open-circuit rise time of the test setup T_{open} is not very great:

$$\frac{T_{L/R}}{T_{\text{open}}} = \frac{1.2 \text{ x } 10^{-09}}{0.8 \times 10^{-09}} = 1.5 \qquad [1.17]$$

This low ratio means that before the initial step rise completes, the test waveform has already begun to decay. The measured output waveform is not a clean exponential shape but is more complex. Careful observation of the peak amplitude in Figure 1.14 reveals that it reaches a maximum of only 250 mV, compared to the open-circuit asymptote of 417 mV. This evidence suggests that the exponential time constant measured in Example 1.2 may not accurately reflect the true inductance. Were we to go further out on the waveform to measure the decay constant, away from the initial step, the actual wave would decay more exponentially. Unfortunately, we can't go much further out because parasitic coupling, reflections, and other noise make the waveform lumpy as we move toward the right of the screen.

1.8.1 Measuring Total Area Under a Response Curve

We need a more reliable way to use the shape of the curve in Figure 1.14 to estimate inductance. We would like something that considers the whole waveform, not just the values at two cursor positions, and is immune to distortion generated by the limited rise time of the measuring equipment. Our next approach uses the measure of total area under the curve in Figure 1.14.

Figure 1.15 invokes an area measurement feature on the Tektronix 11403 scope to compute the area, in picovolt-seconds, under the curve from Figure 1.14. The total computed area is 495.7 pVs. In the absence of automatic assistance, a simple trapezoidal approximation based on seven points is an effective way to measure area by hand. Position points at the beginning, halfway up the rising edge, and at the top. Then sprinkie the rest along the decay curve.

We now show the simple mathematical relation between the area under the curve and the inductance L.[11]

First, the voltage across an inductor at any point in time is always related to the rise time of the current through it according to the general formula

$$V_{\text{inductor}} = L\frac{dI_{\text{inductor}}}{dt} \qquad [1.18]$$

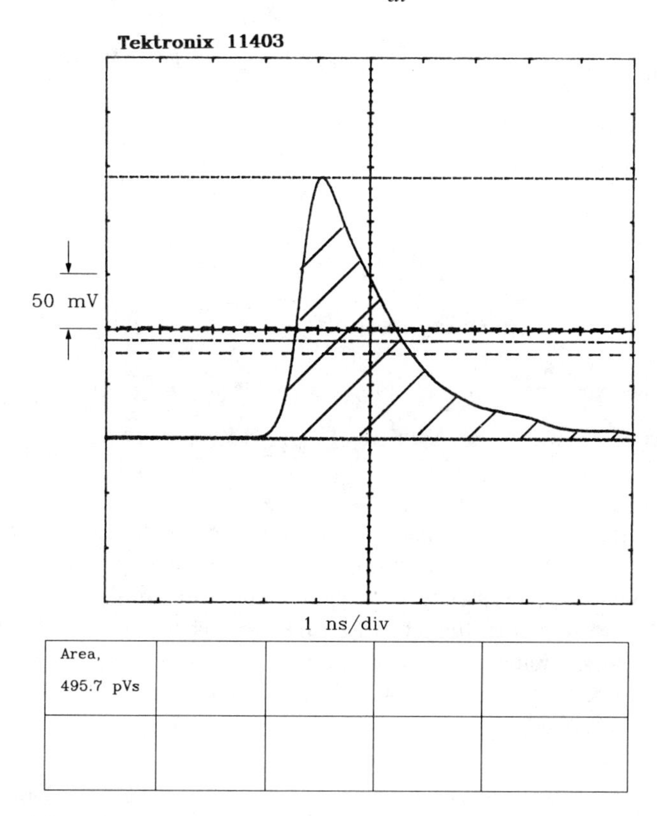

Figure 1.15 Area under the response of a 7.6-Ω inductance test setup.

[11]This explanation uses a little calculus but you do not need calculus to use the result.

Integrating both sides of Equation 1.18, we get

$$\int_0^{\infty} V_{\text{inductor}}(t)dt = L\int_0^{\infty} \frac{dI_{\text{inductor}}(t)}{dt}\,dt \qquad [1.19]$$

The right side of Equation 1.19, the integral of a derivative, is just equal to the difference between the final and starting values of $I(t)$:

$$\int_0^{\infty} V_{\text{inductor}}(t)dt = L[I(\infty) - I(0)] \qquad [1.20]$$

The value on the left side of Equation 1.20 is exactly the area under the probe response curve from Figure 1.14. This link completes the connection between area and inductance:

$$\text{area} = L[I(\infty) - I(0)] \qquad [1.21]$$

Or, restated using ΔI to represent the step change in current from initial to final value:

$$L = \left[\frac{\text{area}}{\Delta I}\right] \qquad [1.22]$$

Knowing the source impedance R_S, we can convert ΔI into $\Delta V/R_S$, putting Equation 1.22 into its final form:

$$L = \frac{(\text{area})(R_S)}{\Delta V} \qquad [1.23]$$

1.8.2 Application to Figure 1.15

For the experiment shown in Figure 1.15, we get

$$L = \frac{(\text{area})(R_{\text{Thevenin}})}{\Delta V_{\text{open circuit}}} = \frac{(495\ \text{pVs})(7.6\ \Omega)}{418\ \text{mV}} = 9.0\ \text{nH} \qquad [1.24]$$

Because the area method uses the whole waveform rather than a measurement culled from two cursor locations, it is more immune to noise and waveform distortion than the cursor measurement technique. Noise immunity accrues because noise tends to average to zero area. Immunity to waveform distortion due to the limited rise time of the observing equipment is a more interesting property. This property derives from the fact that whatever the step response of the measurement setup, the area under the curve doesn't change.[12]

[12]By evaluating the area under the response curve, we are in fact evaluating the frequency response at DC. Any perturbations in the frequency response of the scope or signal generator not affecting the DC response will also not affect the area under the response curve. One may resort to Fourier integrals or to the definition of signal convolution to prove these remarks.

POINTS TO REMEMBER:

The area under an exponential L/R decay provides an accurate measure of the decay time constant.

A slow pulse generator rise time, or a slow scope, does not change the area measured with our inductance test jig.

1.9 MUTUAL CAPACITANCE

Wherever there are two circuits, there is mutual capacitance. Voltages in one circuit create electric fields, and these electric fields affect the second circuit. Every two circuits interact electrically, with the coefficient of interaction decaying rapidly with increasing distance. The coefficient of electrical interaction between two circuits is called their *mutual capacitance*, units of which are farads, or amp-seconds/ volt. A mutual capacitive coupling between two circuits is simply a parasitic capacitor connected from circuit A to circuit B.

A mutual capacitance C_M injects a current I_M into circuit B proportional to the rate of change of voltage in circuit A according to this rule:

$$I_M = C_M \frac{dV_A}{dt} \qquad [1.25]$$

Equation 1.25 is a simple approximation to the actual coupled noise current. A complete formula would use the difference in voltages between circuits A and B and the loading effect of capacitor C_M on both circuits. The simple approximation (Equation 1.25) works under these three assumptions:

(1) The coupled current flowing in C_M is much smaller than the primary signal current in circuit A. Capacitance C_M therefore does not load circuit A.

(2) The coupled signal voltage in circuit B is smaller than the signal on A. We can ignore the small coupled voltage on B when calculating the noise current and assume the voltage difference between circuits A and B simply equals V_A.

(3) Assume the capacitor is a large impedance compared to the impedance to ground of circuit B. We will calculate the coupled noise voltage as I_M times the impedance of circuit B to ground. This procedure ignores interactions between the mutual capacitance and the secondary circuit.

When the coupled noise voltage is less than 10% of the signal step size, these approximations are accurate to about one decimal place. That's accurate enough to tell which effects are worth pursuing. When the coupled noise is greater than 10%, the approximation is worse, but then with 10% crosstalk a digital circuit probably won't work and improving the accuracy of the answers doesn't help.

1.9.1 Relation of Mutual Capacitance to Crosstalk

Given a known amount of mutual capacitance C_M, a fixed circuit rise time T_r, and a known impedance in the receiving circuit equal to R_B, we may estimate crosstalk as a fraction of the driving waveform V_A.

First derive the maximum change in voltage per unit time of waveform V_A, where ΔV is the driving waveform step height and T_r is the driving waveform rise time:

$$\frac{dV_A}{dt} = \frac{\Delta V}{T_r} \qquad [1.26]$$

Next, compute the mutual capacitive current which flows from circuit A to circuit B using Equation 1.27:

$$I_M = C_M \frac{\Delta V}{T_r} \qquad [1.27]$$

Multiply the interfering current I_M by R_B to find the interfering voltage, and divide by ΔV to express this result as a fractional interference level:

$$\text{Crosstalk} = \frac{R_B I_M}{\Delta V} = \frac{R_B C_M}{T_r} \qquad [1.28]$$

In situations involving multiple interfering sources (e.g., EMI filter layouts involving many parts crammed together inside a connector shell), estimate the mutual capacitances between each pair of elements separately and then sum the fractional crosstalk from each source to each receiving circuit. An interference level as low as 2%, summed over five nearby sources, gives 500 mV of interference in a TTL system. That's more than typical TTL noise margins and represents a serious problem.

EXAMPLE 1.3: Measurement of Mutual Capacitance

Figure 1.16 depicts a situation involving mutual capacitive coupling. Two 1/4-W carbon composition resistor bodies are mounted with 0.1 in. centerline separations on a 0.063-in.-thick epoxy PCB. The PCB has a solid ground plane on the solder side only and is blank on the circuit side. The resistors seat firmly against the circuit side of the PCB, 0.063 in. above the ground plane.[13] Inject the test signal at one end of resistor R_2 and measure the coupled current at the opposite end of resistor R_3. This arrangement separates the IN and OUT coaxial cable connections, reducing direct feed-through. The terminating resistor R_1 is a 1/8-W resistor mounted on the solder side of the circuit board. The back termination on the pulse generator is engaged as is the scope terminator.

[13] An internal ground plane on a multilayer card would be closer to the surface than 0.063 in., lowering the mutual capacitive coupling.

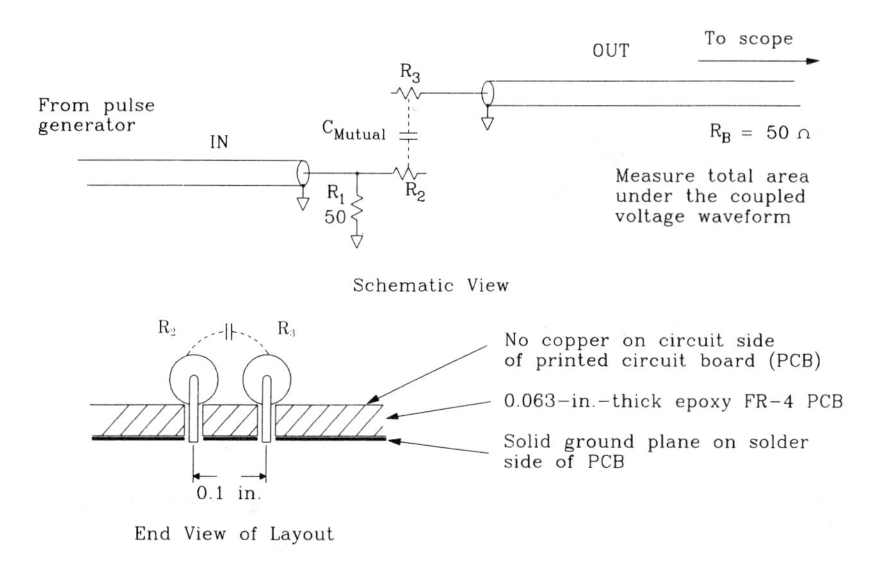

Figure 1.16 Mutual capacitive coupling.

Figure 1.17 depicts the measurement results from this setup. In the top section of the screen, recorded at 5 ns/ division, the drive waveform (at 1 V/division) and the interfering voltage (at 20 mV/division) both appear. The driving waveform has a rise time of about 800 ps. The bottom section of the screen shows only the interfering voltage on a scale of 500 ps/division.

We can estimate the mutual capacitance using an area formula similar to Equation 1.23. Here it is the integrated current, area/R_B, that equals the step change in voltage times the mutual capacitance.[14] The mutual capacitance is equal to

$$C_M = \frac{\text{area}}{R_B \Delta V} = \frac{56.48 \text{ pV} - \text{s}}{(50\ \Omega)\ (2.7\ \text{V/div})} 0.4 \text{ pF} \qquad [1.29]$$

Using Equation 1.28, we can check the expected peak interference level at a rise time of 800 ps:

$$\text{Crosstalk} = \frac{R_B C_M}{T_r} = \frac{(50)(0.4)\text{ pF}}{800 \text{ ps}} = 0.025 \qquad [1.30]$$

Compare the prediction of Equation 1.30, based only on the measured area, with the actual measured peak interference

$$\text{Crosstalk} = \frac{(3.8 \text{ div})(20 \text{ mV/div})}{(2.7 \text{ div})(1 \text{ V/div})} = 0.028 \qquad [1.31]$$

1.9.2. Mutual Capacitance Between Terminating Resistors

What happens if we ground the resistors in Example 1.3?

[14]Since integrated current is just total transferred charge, we recognize that the total charge induced in circuit B, due to a change in voltage ΔV in circuit A, is equal to $\Delta V C_M$.

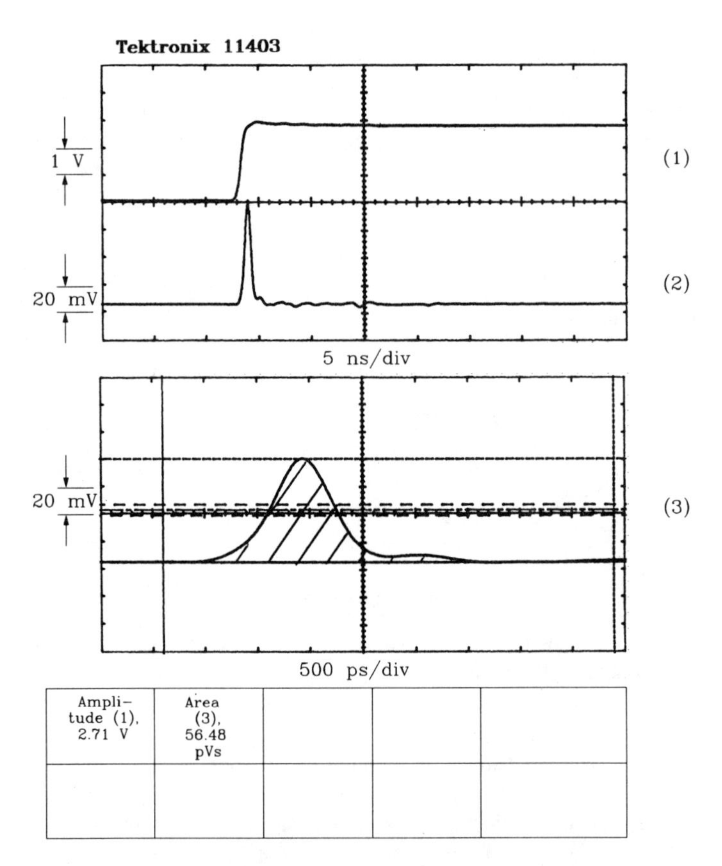

Figure 1.17 Mutual capacitance of two 1/4-W resistors.

If we ground one end of each resistor in Example 1.3, the capacitively coupled noise voltage divides approximately by 6. Thinking intuitively, if we represent the mutual capacitance by a parasitic capacitor connected between the centers of the two resistors, grounding one end of resistor R_A divides the driving voltage in half. The ground on one side of resistor R_B allows the coupled current to split, with two-thirds of the current flowing via one side of resistor R_B directly to ground, and one-third of the current flowing the other way. The other path leads through the opposite half of resistor R_B, through the coax cable to the scope, and on to ground. The product of one-half the drive voltage and one-third the receive sensitivity is one-sixth. When we use resistors in the configuration in Figure 1.16 as terminators for a signal having a rise time of 800 ps, the coupled crosstalk due to mutual capacitance will be about 0.025/6 = 0.004.

We will see in the next section that mutual inductive coupling is more of a problem than mutual capacitance in digital designs.

1.10 MUTUAL INDUCTANCE

Wherever there are two loops of current, there is mutual inductance. The current in one loop creates a magnetic field, and that magnetic field affects the second loop. Every two loops interact, with the coefficient of interaction decaying rapidly with increasing distance. The coefficient of interaction between two loops is called their *mutual inductance*, units of which are henries, or volt-seconds/amp. A mutual inductive coupling between two circuits acts the same as a tiny transformer connected between circuit A and circuit B as shown in Figure 1.18. Anywhere we see two nearby loops of current, the two currents interact like the primary and secondary of a transformer, and we get mutual inductance.

A mutual inductance L_M injects a noise voltage Y into circuit B proportional to the rate of change of current in circuit A according to this rule:

$$Y = L_M \frac{dI_A}{dt} \qquad [1.32]$$

Quick changes in current in loop A induce large voltages in loop B, hence the importance of mutual inductive coupling to high-speed design.

Equation 1.32 is a simple approximation to the actual coupled noise voltage. A complete formula would use the difference in currents between primary and secondary and the loading effect of the primary and secondary windings on both circuits. The assumptions surrounding the use of Equation 1.32 are similar to those for Equation 1.25, namely:

(1) The induced voltage across L_M is much smaller than the primary signal voltage. Attaching L_M therefore does not load circuit A. Noise voltages coupled by mutual inductance in digital products are always smaller than the source signal.

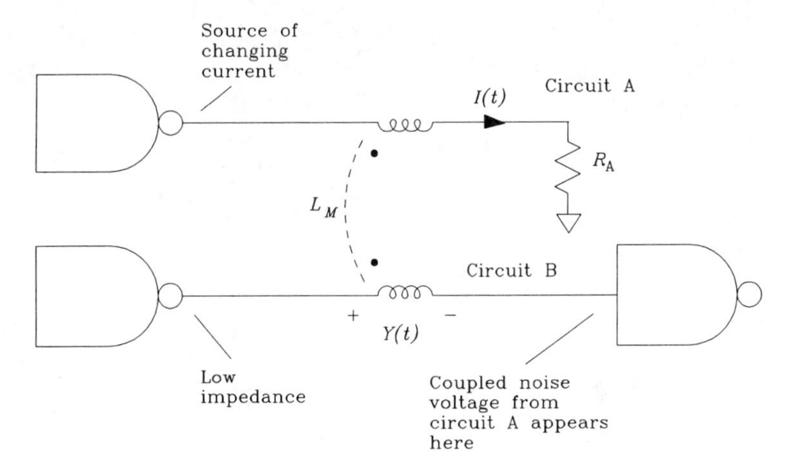

Figure 1.18 Mutual inductance lumped-circuit model.

(2) The coupled signal current in circuit B is smaller than the current in A. We can ignore the small coupled current in B and assume the difference between primary and secondary currents in the coupling transformer just equals I_A.

(3) Assume the secondary impedance is small compared to the impedance to ground of circuit B. Just add the coupled noise voltage to the voltage otherwise present in circuit B. This procedure ignores interactions between the mutual inductance and the secondary circuit.

In digital circuits mutual inductance, like mutual capacitance, usually induces unwanted crosstalk between circuits.

Figure 1.19 illustrates the exact process by which mutual inductive coupling operates:

(1) Any current in loop A produces a pattern of magnetic field energy. At stronger currents, more magnetic energy is stored per unit volume in the space surrounding loop A.

(2) Over the area subtended by loop B, we can compute the total strength of the magnetic field from A. The total magnetic field strength over the area of loop B, called the *magnetic flux* in B, is a function of the distance between loops A and B, their physical proportions, and their relative orientations and is directly proportional to the current in A. More current in A produces more flux in loop B.

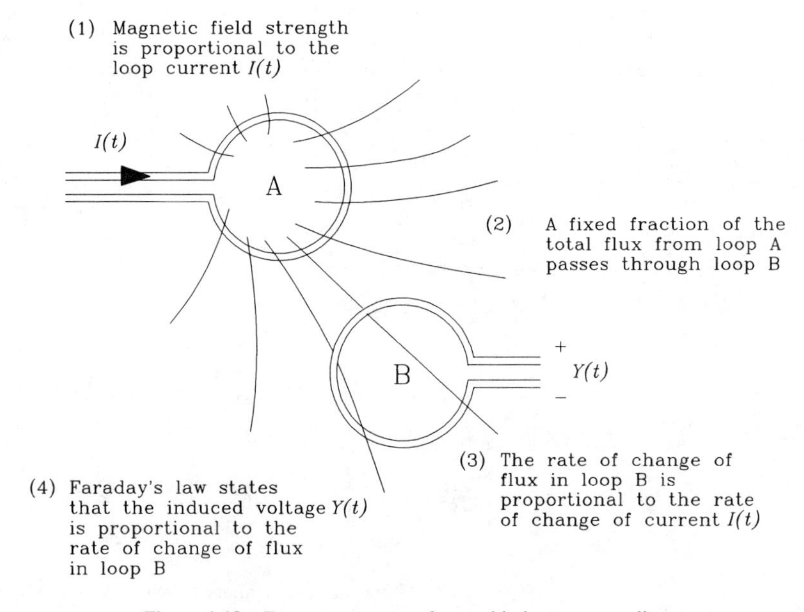

Figure 1.19 Four-step process of mutual inductance coupling.

(3) Changes in the loop current of A produce corresponding proportional changes in the magnetic flux passing through loop B.

(4) Faraday's law tells us that the voltage induced in loop B is proportional to the rate of change of magnetic flux passing through loop B.

Linking together ideas (1) through (4), we see that the voltage induced in loop B is proportional to the rate of change of current in loop A. The constant of proportionality is called the mutual inductance between circuits A and B.

Because a magnetic field is a vector quantity, flipping over loop B has the effect of reversing the polarity of flux coupling. The coupled noise voltage reverses polarity. Flipping over loop A has the same effect. Orienting loop B parallel to the local lines of magnetic field strength results in zero net flux passing through B and zero noise coupling. Mutual inductive coupling, unlike mutual capacitive coupling, is capable of inducing crosstalk with a polarity opposite that of the driving signal. It is also highly sensitive to loop orientation.

1.10.1 Relation of Mutual Inductance to Crosstalk

Given a known amount of mutual inductance L_M, a fixed circuit rise time T_r, and a known source impedance in the driving circuit A equal to R_A, we may estimate crosstalk as a fraction of the driving waveform V_A.

First derive the change in voltage per unit time of waveform V_A, where ΔV is the driving waveform step height and T_r is the driving waveform rise time:

$$\frac{dV_A}{dt} = \frac{\Delta V}{T_r} \qquad [1.33]$$

Next, assume that loop A is resistively damped by resistance R_A, and so the current and voltage are in proportion to each other. This situation commonly arises when we drive a voltage $V(t)$ through a transmission line terminated resistively. In most cases we can relate changes in current to changes in voltage using some well-defined resistance R_A:

$$\frac{dI_A}{dt} = \frac{\Delta V}{R_A T_r} \qquad [1.34]$$

Next, compute the mutual inductive interference Y, which appears in circuit B due to changes in current in circuit A, by substituting Equation 1.34 into Equation 1.32:

$$Y = L_M \frac{\Delta V}{R_A T_r} \qquad [1.35]$$

Divide by ΔV to express this result as a fractional interference level:

$$\text{Crosstalk} = \frac{L_M}{R_A T_r} \qquad [1.36]$$

In situations involving multiple interfering sources (e.g., a number of lines sharing a common ground return path), estimate the mutual inductances between each pair of elements separately and then sum together the fractional crosstalk from each source to each receiving circuit. An interference level as low as 2%, summed over five nearby sources, will contribute 500 mV of interference in a TTL system. That's more than typical TTL noise margins and represents a serious problem.

EXAMPLE 1.4: Measurement of Mutual Inductance

Figure 1.20 depicts a simple measurement of mutual inductance.

Two carbon-composition resistor bodies are mounted as in Example 1.3 on 0.1-in. centerlines. The right-hand ends of both resistors are grounded, and the measuring cables, IN and OUT, connect to the left end of each resistor, respectively. Resistor R_A serves as a termination for the signal source. The signal source has a rise time of 800 ps.

The cables IN and OUT connect at right angles to the resistor bodies. This right-angle connection separates the cables as much as possible, reducing the direct feed-through. The back termination on the pulse generator is engaged.

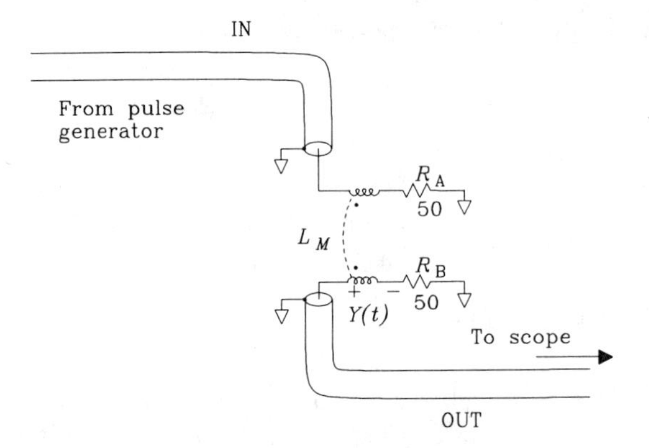

Figure 1.20 Measurement setup for mutual inductance.

Figure 1.21 shows the magnetic field pattern emanating from resistor R_A. Some of the magnetic field lines encircle resistor R_B, and some don't. The proportion of the total flux from R_A that encircles resistor R_B is a constant determined solely by the physical dimensions and positions of the two resistor bodies.

Magnetic field lines that encircle resistor R_B are said to penetrate the loop formed by resistor R_B. When we refer to the loop formed by resistor R_B, imagine a current loop starting at the grounded end of R_B. From there, travel through R_B into the coaxial probe, down the coax to the internal terminating resistor R_T in the scope, through that resistor back to the scope chassis, back down the coax shield to the local ground plane, and across that local ground plane back to R_B. Any change in the total magnetic flux passing through this loop induces a voltage around the loop.

With equal-value resistors at positions R_B and R_T, the induced voltage divides equally between them, so we expect only half the total induced voltage at the scope. If resistor R_B were a 0-Ω resistor of the same physical dimensions, we would see the full induced potential at the scope.

From the measured result in Figure 1.22 we can estimate the mutual inductance using Equation 1.23 (all the same reasoning applies) and remembering to multiply by 2 to account for the halving of the received signal:

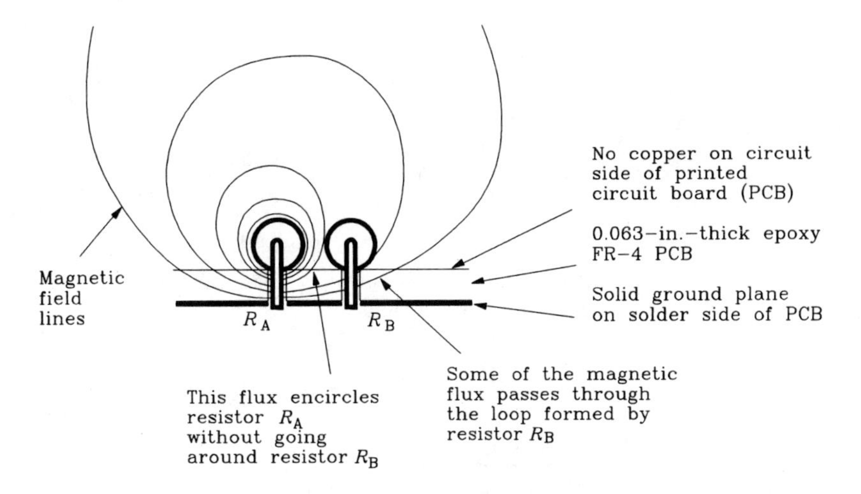

Figure 1.21 Example of mutual inductive coupling.

$$L_M = \frac{(\text{area})(2R_A)}{\Delta V} \approx 3.0 \text{ nH} \qquad [1.37]$$

where area = 80 pVs (from Fig. 1.22)
ΔV = 2.7 V (from Fig. 1.22)
R_A = 50 Ω (from Fig. 1.20)

A more accurate method of determining inductance subtracts from the measured area that portion which we believe is due to mutual capacitive coupling and then uses the corrected area to calculate inductance. The mutual capacitive interference area, from Example 1.3, is 56/6 pVs, correcting the measured result in Figure 1.17 by a factor of 6 to account for the grounding of each resistor.

The corrected area is

$$\text{area'} = 80 - \frac{56}{6} = 71 \text{ pVs} \qquad [1.38]$$

The corrected mutual inductance is then

$$L_M = \frac{(\text{area'})(2R_A)}{\Delta V} = 2.6 \text{ nH} \qquad [1.39]$$

Let's work backward now from the deduced area to see if we can predict the peak value of interference expected in Figure 1.22. Use Equation 1.38, remembering to divide by 2 to account for the splitting of the induced signal between R_B and R_T,

L_M = 2.6 nH (from Equation 1.39)
T_r = 800 ps
R_A = 50 Ω

$$\text{Crosstalk}_{\text{induc}} = \frac{L_M}{2R_A T_r} = 0.032 \qquad [1.40]$$

Add to this the capacitive crosstalk (corrected by a factor of six) from Example 1.3:

$$\text{Crosstalk}_{\text{cap}} = \frac{0.025}{6} = 0.004 \qquad [1.41]$$

$$\text{Crosstalk}_{\text{total}} = \text{Crosstalk}_{\text{induc}} + \text{Crosstalk}_{\text{cap}} = 0.036 \qquad [1.42]$$

Compare the prediction of Equation 1.42, based on the measured area, with the actual measured peak interference from Figure 1.22:

$$\text{Crosstalk} = \frac{(4.6\ \text{div})(20\ \text{mV/div})}{(2.7\ \text{div})(1\ \text{V/div})} = 0.034 \qquad [1.43]$$

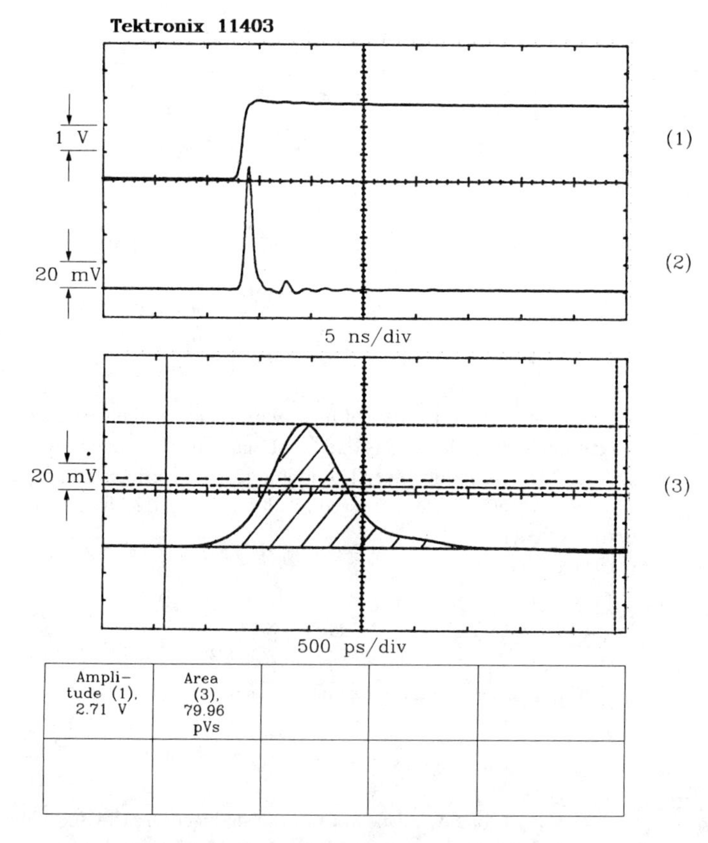

Figure 1.22 Mutual inductance of two 1/4-W resistors.

1.10.2 Reversing a Magnetically Coupled Loop

Let's test one of the theories about mutual inductive coupling, namely, that if we reverse one of the loops, the coupling reverses sign.

First, go to the measurement setup in Figure 1.20 and rewire the OUT cable to the opposite end of R_B. Then connect the left side of R_B to ground. We have effectively reversed the leads on the transformer-like inductive coupling between R_A and R_B.

Figure 1.23 shows the result: we get a negative pulse, having a total area of 59 pVs. This reversed pulse equals one-half of the inductive coupling *minus* one-sixth of the capacitive coupling from Example 1.3. The inductive and capacitive couplings now have reversed signs, and so their effects subtract instead of adding. We must correct the area measurement from Figure 1.23 by adding back in the capacitive influence before using it in Equation 1.23:

$$\text{area}' = 59 + \frac{56}{6} = 68 \text{ pVs} \qquad [1.44]$$

The corrected mutual inductance is

$$L_M = \frac{(\text{area}')(2R_A)}{\Delta V} = 2.5 \text{ nH} \qquad [1.45]$$

The figure 2.5 nH compares favorably to the 2.6 nH computed in Equation 1.39.

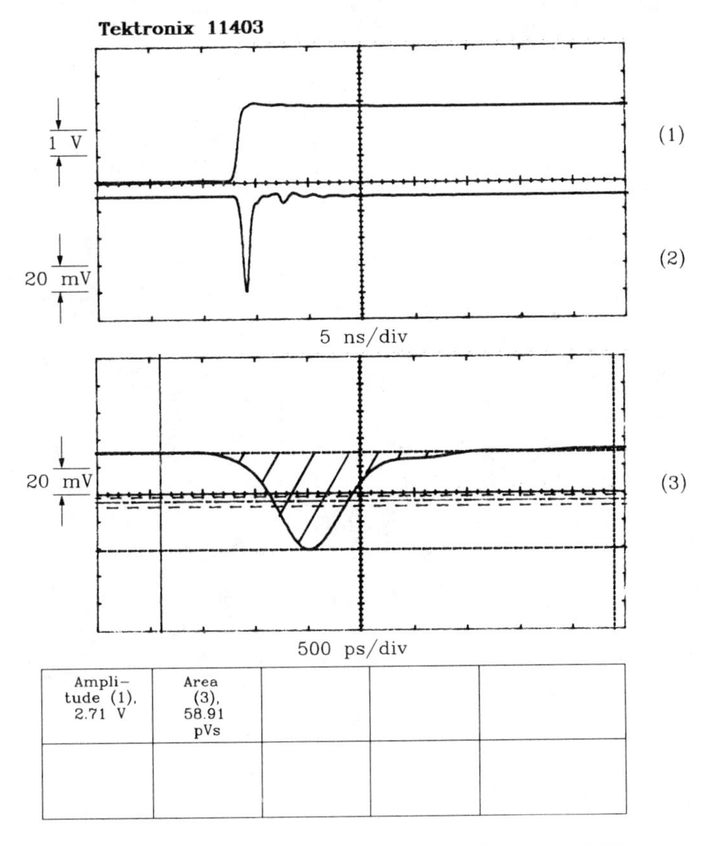

Amplitude (1), 2.71 V	Area (3), 58.91 pVs			

Figure 1.23 Reversing the inductive pickup from Figure 1.22.

1.10.3 Ratio of Inductive to Capacitive Coupling

With both resistors grounded, the relative capacitive coupling amounts to 0.004, while the inductive crosstalk is 0.032. This ratio is typical for circuits operating at a 50-Ω impedance level. Higher-impedance circuits, involving larger *dV/dt* and relatively smaller *dI/dt* experience a relatively greater amount of capacitive coupling.

The problem with inductive coupling is exaggerated in situations involving low-output impedance gates which directly drive transmission structures. In these cases the total inductively coupled signal energy ends up at the far termination rather than being split in two as in Example 1.4.

POINT TO REMEMBER:

Among high-speed digital circuits, mutual inductance is often a worse problem than mutual capacitance.

2

High-Speed Properties of Logic Gates

Power, speed, and packaging are paramount considerations in the design of any digital machine. Every designer wants power low, speed high, and packaging cheap. Unfortunately, no logic family satisfies users on all fronts. We are forced to choose from a variety of logic families, each tailored in some way to a particular application. Will this need for variety ever go away? Will we ever have a perfect logic family suited for all needs?

Historically, the answer has been no. Even when a new technology sweeps the field, superior in every way to its competition, users pressing for every advantage in their designs will still demand variety. All logic families exhibit tradeoffs among power, speed, and packaging, and all logic manufacturers do their best to exploit these tradeoffs.

Let's take a look at one very old digital technology called the wire spring relay to learn how these basic tradeoffs interact. The wire spring relay was the last (and best) generation of relays used for logic machines before electron tubes took over.

2.1 HISTORICAL DEVELOPMENT OF A VERY OLD DIGITAL TECHNOLOGY

The wire spring relay, developed in the late 1940s for use in automatic telephone exchanges by Western Electric, represented a big technological advance over earlier relays. The contact elements of the wire spring relay were supported on the ends of long thin wires which themselves formed the spring element of the relay. The small size, low mass, and simple construction of the wire spring relay made it a very fast-acting, cheap alternative to traditional relay designs which incorporated separate assemblies for the spring and the contact armature. Wire spring technology quickly swept the competition, and crossbar telephone exchanges were manufactured by Western Electric using wire spring relays as late as 1965.

Wire spring relay technology encompassed more than just relay construction, it affected system packaging as well. These new relays were encapsulated in a rectangular package having all their electrical connections in a standard array at one end. The relays plugged into arrays of sockets on standard relay panels and stood shoulder to shoulder, saving space. Wire wrap pins protruded from the back side of each relay panel for interconnections.

With a standard relay package, manufacturers could use the same relay rack panel for many different applications, according to the pattern of the wire wrap connections on the back side. This compared to the earlier practice of designing individual relay-mounting locations for each device, often incorporating peculiar spring latches, actuators, or other mechanical contrivances which tightly coupled relay construction to the overall purpose and function of a digital machine. Wire spring relay designs separated the electrical and mechanical elements of a system. This packaging approach lowered overall design and manufacturing costs.

Standard packaging was cheap but sacrificed a lot of flexibility. The standard package held no more than one relay having 12 poles of double-throw contacts (12PDT). Users requiring larger monolithic relay operations had to split them up among multiple packages, dissipating more power with each additional package. Splitting up operations was inefficient.

For cost reasons, Western Electric engineers chose not to incorporate heat sink fins on each package. For reliability, they used simple convection cooling in all their equipment. These factors limited the total power dissipation allowed inside each relay package. This power limitation, coupled with the limited space inside the standard package, ultimately meant that Western Electric could place no more than two drive coils inside each package. The most dense wire spring relay configuration was a dual 5PDT relay (two independent relays each having five poles of double-throw contacts).

The relays operated from a standard 48-V power supply and were available with either 750- or 2400-Ω coils. Why two coils? The 750-Ω coil consumed much more drive current and therefore switched much faster than the 2400-Ω coil. On the other hand, the 2400-Ω coil took less power and generated less heat than its 750-Ω cousin. Because of their heat advantage, 2400-Ω coils could be packed closer together than 750-Ω coils. Maximum operating speed and maximum logic density were both indirectly determined by power dissipation.

Do these issues sound familiar? Are logic systems today still constrained by packaging, power dissipation, and speed tradeoffs?

Of course. We face today many of the same issues confronted by our predecessors. Power, speed, and packaging are all still heavily interrelated. The modern tradeoffs for high-speed design sound like this:

(1) Standard packaging of logic devices saves money in manufacturing but reduces flexibility. The initial investment required to support a new package type is staggering, and so most system designers stick with whatever packages their device vendors offer.

(2) Standard packaging limits both the number of gates and the number of pins per package. Both factors force designers to partition large systems among many device packages. Because signal connections between packages respond more

slowly and require more power than connections internal to one package, partitioning slows overall system performance and boosts power dissipation.

(3) The maximum allowable power dissipation per package is limited by a combination of package construction and the cooling system employed. The cooling properties of the package are independent of the semiconductor die placed inside the package. Packages with good cooling properties always cost extra.

(4) As individual logic elements shrink in size, the number of gates per package increases. Higher-density packaging has the benefit of cutting assembly cost and product size dramatically but often dissipates more total power per package. The maximum allowable power dissipation per package eventually constrains the number of gates per package.

(5) Speed and power in a given technology are somewhat interchangeable, with higher-speed devices usually dissipating more power. At the highest speeds, maximum power dissipation per package again becomes a limitation.

The next section presents the specific relations between power and speed for modern logic families.

POINT TO REMEMBER:

Today, just as in the days of relay logic, power and packaging have a big impact on system performance.

2.2 POWER

Actual power dissipation in a logic device is only indirectly related to the typical supply current I_{CC} on its data sheet. Typical power dissipation as rated by the manufacturer often ignores additional power dissipation which occurs at high speeds, plus power dissipated by driving heavy output loads. These effects can often cause the actual power supply current to sometimes far exceed the typical I_{CC} rating.

We will study the power consumption of high-speed logic circuits according to the four power categories illustrated in Figure 2.1. These categories are

- Input power
- Internal dissipation
- Drive circuit dissipation
- Output power.

Each of the four power categories further subdivide into active and quiescent dissipation.

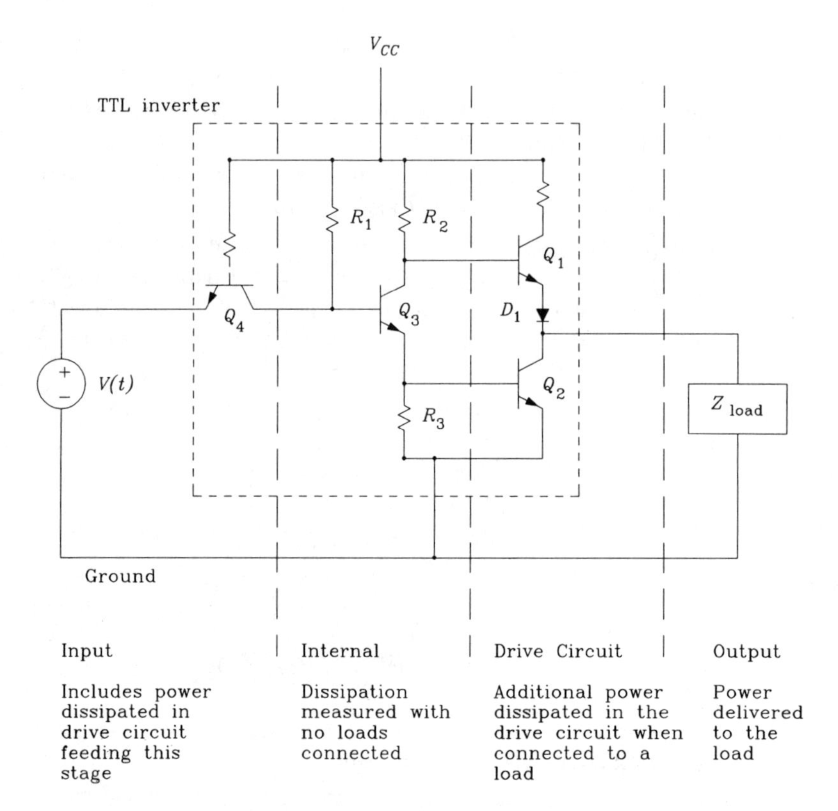

Figure 2.1 Categories of power dissipation in and around a logic device.

2.2.1 Quiescent versus Active Dissipation

Quiescent power dissipation is that power used to hold a circuit in one logic state or the other. This power is calculated by observing the current I and voltage drop V across each resistive element in the circuit, computing the power VI for each element, and summing the results. It is quiescent power, under no load, that we see most often quoted on data sheets.

In the examples that follow we simply average the power during high and low states when figuring quiescent dissipation. If your circuit spends more time in one state than the other, consider using a weighted-average, or worst-case, figure.

2.2.2 Active Power When Driving a Capacitive Load

Every time a logic circuit switches, it dissipates extra energy beyond its normal quiescent power dissipation. When cycled at a constant rate, the *active power dissipation* equals

$$\text{Power} = (\text{cycle frequency})(\text{excess energy used per cycle}) \qquad [2.1]$$

The two most common causes of active power dissipation are load capacitance and overlapping bias currents.

Figure 2.2 illustrates the situation when driving a load capacitance. At time t_1 switch A closes, charging the capacitor up to V_{CC}. As the capacitor charges, current surges through the finite charging resistance of the drive circuit. This current surge dissipates energy. At time t_2 switch B closes, discharging the capacitor through the finite discharging resistance of the drive circuit. This current surge also dissipates energy. If the experiment is repeated, it happens that the energy dissipated charging the capacitor exactly equals the energy dissipated discharging the capacitor, and the sum of the two equals

$$\text{Energy per cycle} = C{V_{CC}}^2 \qquad [2.2]$$

where C = capacitance, F
V_{CC} = charging voltage, V

If repeated at a cycle rate of F hertz,[1] the power *expended in the driving circuit* when charging up and down the capacitor equals

$$\text{Power} = FC{V_{CC}}^2 \qquad [2.3]$$

No net power is dissipated in the capacitor itself; all the energy is expended heating up the driving circuit.

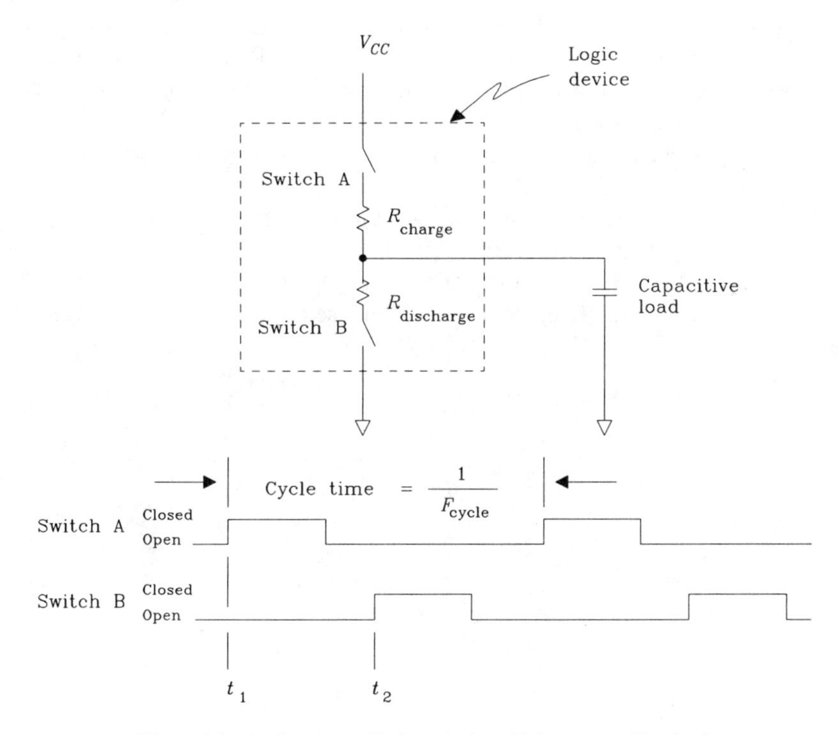

Figure 2.2 Active power dissipated when driving a capacitive load.

[1]In a clocked system with alternating transitions between 1 and 0, F would be equal to one-half the clock rate. In a system with random transitions, F equals one-fourth the clock rate.

The simple model in Equation 2.3 accounts for active power dissipation in the drive circuitry of either CMOS or TTL circuits.

2.2.3 Active Power Due to Overlapping Bias Currents

In Figure 2.1, the output drive circuit of a TTL inverter switches between HI and LO states by alternately placing either Q_1 or Q_2 in the conducting state, but not both. This arrangement of two active circuits, one that pulls the output voltage HI and another that pulls the output voltage LO, is called a totem-pole output stage. TTL and CMOS circuits commonly have totem-pole output stages.

Diode D_1 in Figure 2.1 ensures that when transistors Q_3 and Q_2 fully saturate, clamping the output at a LO state, transistor Q_1 will completely cut off. This feature prevents heavy currents that might otherwise flow if Q_2 and Q_1 simultaneously conducted. Every logic family having a totem-pole output contains some circuit feature meant to prevent both HI and LO output drivers from conducting simultaneously.

Experimentation with the TTL driver circuit depicted in Figure 2.1 reveals that as the device switches from one state to another, transistors Q_1 and Q_2 may simultaneously conduct for a brief instant. Any overlap in conduction generates a current surge from V_{CC} to ground which dissipates power, in the form of heat, inside transistors Q_1 and Q_2.

Before the introduction of Schottky TTL logic, when a TTL device switched from a LO to HI state, transistor Q_2 tended to stay in saturation, draining off its stored base charge through resistor R_3, long after transistor Q_1 began to conduct. The stored base charge effect created a fixed period of overlap. New Schottky circuits do not saturate transistor Q_2 and therefore exhibit less overlap current.

CMOS circuits, depicted in Figure 2.3, may exhibit overlapping conduction between field effect transistors Q_1 and Q_2, depending on the critical gate-to-source voltage V_{GS} of the two transistors. The exact value of the V_{GS} parameter is highly dependent on the manufacturing process, so it is unwise to generalize knowledge gained from mea-

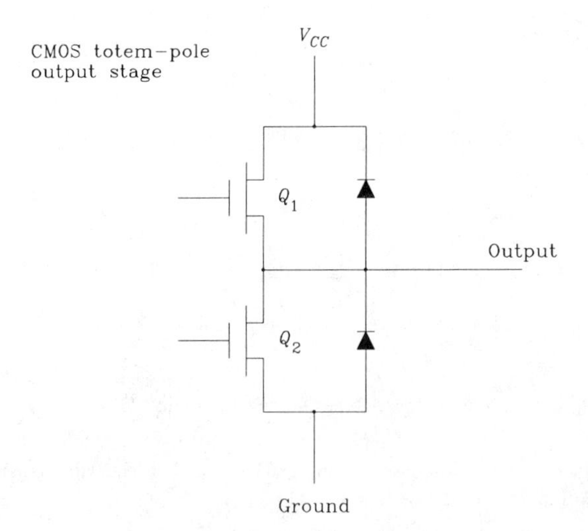

Figure 2.3 CMOS totem-pole output.

surement of a few CMOS devices. Figure 2.4 illustrates the typical DC power supply current drawn by a 74HC00 gate as a function of input voltage.[2] With CMOS devices that do exhibit overlap, slowing down the input switching time tends to lengthen the overlap period, as the internal circuitry responds more slowly and spends more time near the voltage where Q_1 and Q_2 both conduct.

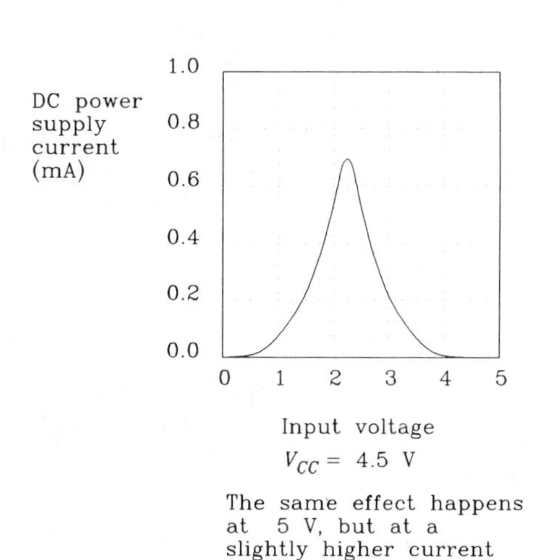

Figure 2.4 DC consumption of Signetics 74HC00-type circuit versus input voltage. (Permission granted by Philips Semiconductors-Signetics.)

With a fast input transition, the size and shape of the overlap current pulse is consistent on every cycle and dissipates a consistent amount of energy per cycle. The extra power dissipation due to overlapping bias currents is therefore proportional to the switching cycle rate. Unlike power dissipation due to load capacitance, the power due to overlapping drive currents does not grow with the square of power supply voltage.

As shown in Figure 2.4, the overlap current for a 74HC00 circuit (1 mA) is not very large compared to the maximum drive current that this type of gate can produce (10–20 mA).

For TTL circuits, the overlap effect is more pronounced. If you take a TTL inverter and connect its input to its output, it will bias itself into the overlap region and dissipate a lot of power. You can feel the circuit warm up. TTL circuits are therefore not good candidates for use as linear, small signal processing elements (like oscillators) because they draw excess power in the linear state. Emitter-coupled logic (ECL) circuits, on the other hand, draw no additional current at the crossover region and make excellent linear processing elements.

2.2.4 Input Power

Input power comes into a chip from other devices. It is required to bias and activate the input circuits.

[2] *Signetics High-Speed CMOS Manual,* Signetics Company, Sunnyvale, Calif., 1988.

Table 2.1 compares the quiescent and active input characteristics of four different logic families: Signetics 74HCT CMOS, Texas Instruments 74AS TTL, Motorola 10KH ECL, and GigaBit Logic 10G GaAs.

In each case, the quiescent input power is determined by multiplying the required input current by the power supply voltage. *This sums the actual power dissipated inside the receiving logic device with power dissipated in the driving device.*

For active input power calculations, we plug the input capacitance, the typical input voltage swing, and the operating frequency into Equation 2.3. This computes the total power dissipated in any circuit which drives this input.

These input power figures are relatively low. They gain significance only for nets having unusually large fan-out, or for systems which must operate on extremely low power.

TABLE 2.1 INPUT CHARACTERISTICS

	74HCT00	74AS00	10H101	10G001
I_{in}HI (mA)	0	+0.020	+0.425	+0.400
I_{in}LO (mA)	0	−0.500	+0.0005	−0.100
$P_{quiescent}$ (mW)	0	1.3	1.1	1.3
C_{in} (pF)	3.5	3	3	1.5
ΔV_{in} (V)	5.0	3.7	1.0	1.5
P_{active} (mW)				
$F = 1$ MHz	0.09	0.04	0.003	0.003
$F = 10$ MHz	0.9	0.4	0.03	0.03
$F = 100$ MHz			0.3	0.3
$F = 1000$ MHz				3.0

2.2.5 Internal Dissipation

Internal power is used to bias and switch nodes internal to a logic device. Internal power includes both quiescent and active power dissipation.

Quiescent internal power is defined with no loads connected and with the inputs in a random state.[3] Average over all possible input states to find the quiescent power dissipation.

The internal active power dissipation constant K_{active} is measured by cycling the inputs at some predetermined frequency F. Leave the output pins disconnected. Measure the total power P_{total} at cycle rate F hertz and calculate the active power dissipation constant:

$$K_{active} = \frac{P_{total} - P_{quiescent}}{F} \qquad [2.4]$$

The active power dissipation constant tells us how many additional watts are dissipated for every 1-Hz increase in the cycle rate. The power dissipation constant K_{active} may now be used to estimate the total power dissipation at any other frequency F':

[3]Measure the device power dissipation by connecting a current meter in series with the device power input pins, and using the relationship power = (current)(voltage).

$$P'_{\text{total}} = P_{\text{quiescent}} + F' K_{\text{active}} \qquad [2.5]$$

Equation 2.5 accounts for the extra energy dissipated per cycle inside a logic device but does not include extra energy dissipated in the driver section caused by a connected load (remember that we did this experiment with no loads connected).

CMOS devices exhibit a clear linear relationship between internal power dissipation and cycle frequency over a very wide frequency range. That relationship is easy to see because the internal quiescent power drain of CMOS circuits is incredibly low. TTL devices exhibit the same phenomenon, but their large quiescent power drain masks the effect until the cycle frequency approaches the maximum operating frequency of the device. Figure 2.5 shows the internal power per gate versus operating frequency for several different TTL logic variants. Above 10 MHz, the active power consumed becomes larger than the quiescent power, and the total power curve looks proportional to frequency. Below 1 MHz, the active power is less than the quiescent power and the total power curve looks flat with respect to frequency.

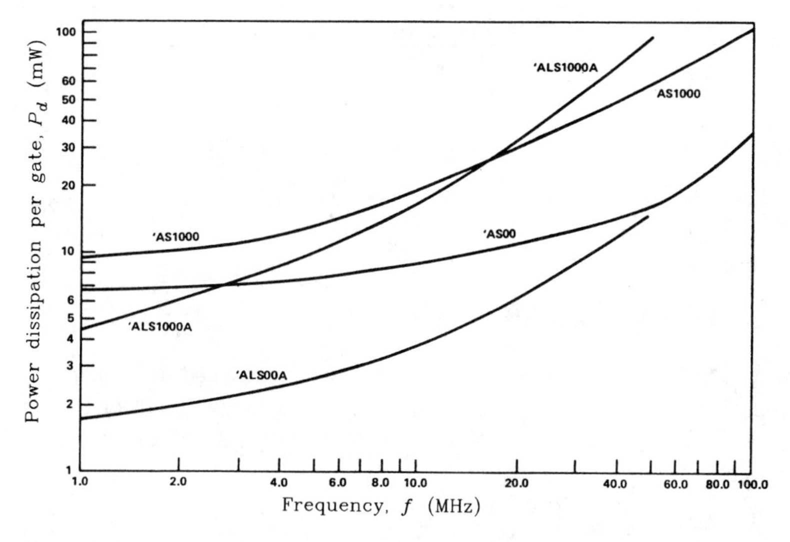

Figure 2.5 Internal power dissipation per gate versus frequency. (Reprinted by permission of Texas Instruments.)

ECL and GaAs families, switching over smaller voltage ranges than their TTL and CMOS counterparts, exhibit little increase in power with frequency. Remember that the voltage swing V is squared in Equation 2.3, so an ECL voltage swing of 1.0 V driving a capacitor of size C dissipates much less drive circuit energy than a TTL 5.0-V swing across the same capacitor. Equations 2.6–2.8 show just how much less.

$$P_{\text{active ECL}} = FC(\Delta V_{\text{ECL}})^2 = FC(1.0)^2 \qquad [2.6]$$

$$P_{\text{active TTL}} = FC(\Delta V_{\text{TTL}})^2 = FC(5.0)^2 \qquad [2.7]$$

where F = cycle rate, Hz
C = capacitance, F
ΔV_{ECL} = ECL switching voltage, V
ΔV_{TTL} = TTL switching voltage, V

The ratio of active power dissipated in ECL to active power dissipated in TTL driving identical load capacitors is 0.04.

$$\frac{P_{\text{active ECL}}}{P_{\text{active TTL}}} = \frac{FC(1.0)^2}{FC(5.0)^2} = \frac{(1.0)^2}{(5.0)^2} = 0.04 \qquad [2.8]$$

The active dissipation of ECL and GaAs devices, compared to their quiescent power, is relatively much smaller than in TTL or CMOS circuits.

Some CMOS devices work over a wide range of operating voltages. The data sheets for these CMOS devices[4] rate their internal power dissipation in terms of an equivalent capacitance C_{PD}. Using this model, the internal power dissipation of a CMOS gate, operated at a cycle rate of F hertz and a power supply voltage of V, would be

$$\text{CMOS internal dissipation} = C_{PD}V^2F \qquad [2.9]$$

where C_{PD} = equivalent power dissipation capacitance, F
V = switching voltage, V
F = switching frequency, H

This model lumps together both the effects of internal capacitance and the effect of overlapping bias current even though bias current effects are not strictly proportional to the square of voltage.

2.2.6 Drive Circuit Dissipation

Most of the power used by a logic device is dissipated in its output drive circuitry. The amount of power dissipated in output drive circuitry depends on the output circuit configuration, the logic levels, the output load, and the speed of operation. The four popular output configurations we will consider here are

- Totem pole
- Emitter follower
- Open collector
- Current source.

Because the characteristics of each output will be important later in understanding transmission lines, we delve into a lot of detail here.

2.2.6.1 Quiescent Power Dissipated in a Totem-Pole Output Circuit

Once a totem-pole output circuit has fully switched, the quiescent power dissipation is equal to the source (or sink) current times the residual voltage across the conduct-

[4]Such as the 74HC family.

ing arm. We will compute the power in both LO and HI states and average the two values.

The idealized TTL driver illustrated in Figure 2.6 shows the power dissipation in both LO and HI states. The voltage drop V_{LO}, for standard TTL, is fixed by the saturation of Q_2 at about 0.3 V. Schottky TTL logic has a slightly higher low-level output voltage under load of about 0.4 V. In the HI state, the voltage drop ($V_{CC}-V_{HI}$) is fixed by the V_{BE} of Q_1 and the forward-biased diode D_1 at about 1.4 V. Note that Q_1 does not go into saturation because its base never rises more positive than its collector. The average total quiescent power dissipated in the drive circuit of a Schottky TTL device is approximately:

$$P_{quies} = \frac{0.4 I_{sink} + 1.0 I_{source}}{2} \qquad [2.10]$$

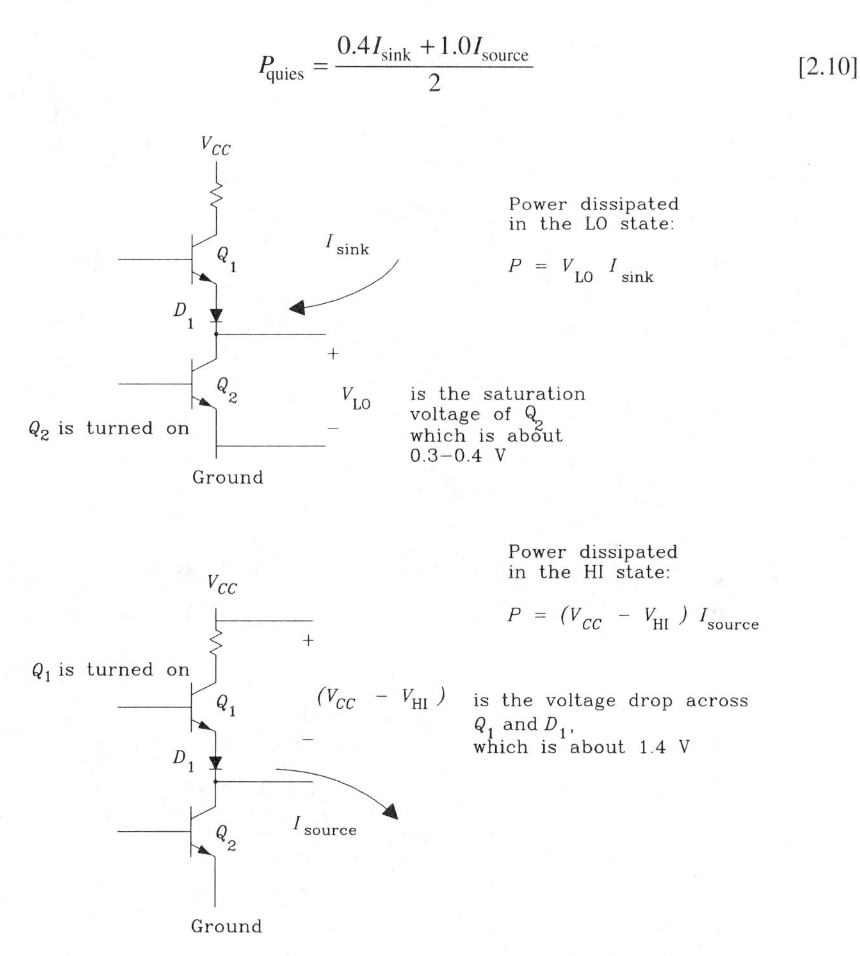

Figure 2.6 Quiescent power dissipated in a TTL totem-pole output circuit.

CMOS drivers more closely resemble the circuit in Figure 2.7. From a CMOS data sheet you can usually get an idea of the values for R_A and R_B by examining the output voltage versus output current specifications, as explained in Example 2.1.

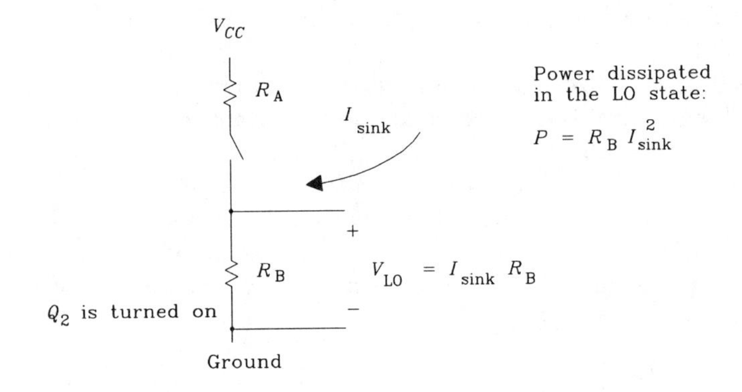

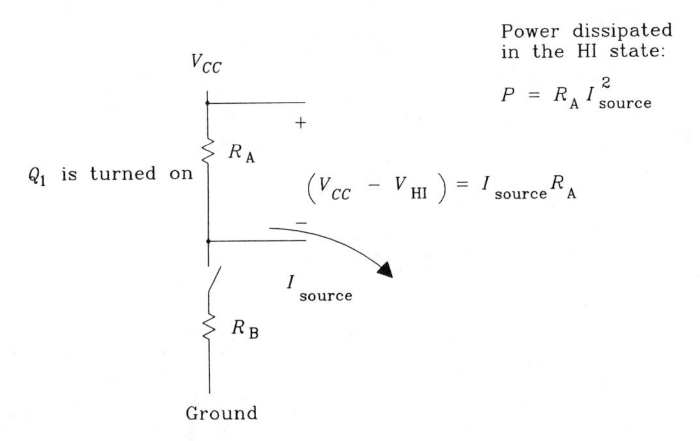

Figure 2.7 Quiescent power dissipated in a CMOS totem-pole output circuit.

EXAMPLE 2.1: Determining Output Resistance of CMOS Driver

The Signetics HCT family standard output driver has the following ratings at a power supply voltage of 4.5 V:[5]

V_{OL} (I_0 = 4.0 mA)	
Typical at 25°C	0.15
Max. –40 to +85°C	0.33
V_{OH} (I_0 = –4.0 mA)	
Typical at 25°C	4.32
Min. –40 to +85°C	3.84

The voltage drop in the low state ranges from 0.15 to 0.33 V at a current of 4 mA. The low-state resistance therefore ranges from

$$R_{\text{low state typ}} = 0.15 / 0.004 = 37\ \Omega \quad [2.11]$$

$$R_{\text{low state max}} = 0.33 / 0.004 = 83\ \Omega \quad [2.12]$$

[5]Don't ask why they make this specification at 4.5 V instead of 5 V; nobody knows for sure, but we can expect the output resistance to be a little smaller at 5.0 V.

The voltage drop in the high state between the 4.5-V power supply and the output ranges from 0.18 to 0.66 V, at a current of 4 mA. The high-state resistance therefore ranges from

$$R_{\text{high state typ}} = 0.18 / 0.004 = 45\ \Omega \tag{2.13}$$

$$R_{\text{high state max}} = 0.66 / 0.004 = 165\ \Omega \tag{2.14}$$

At different power supply voltages, the output resistance of a CMOS driver changes a lot. This effect shows in the specifications for HC (not HCT) logic, which runs at power supply voltages anywhere between 2 and 6 V.[6] Output resistance in the HC family gets smaller at larger operating voltages. Consequently, HC logic runs faster at higher voltages.

The total quiescent power dissipated in the drive circuit of a CMOS device is approximately

$$P_{\text{quies}} = \frac{R_{\text{B}} I_{\text{sink}}^{2} + R_{\text{A}} I_{\text{source}}^{2}}{2} \tag{2.15}$$

Note that the output current appears as a squared term.

2.2.6.2 Active Power Dissipated in a Totem-Pole Output Circuit

Designers are often tempted to load up a totem-pole output circuit to its maximum DC fan-out capability based solely on the DC input current requirements of the attached inputs. This is particularly tempting when designing CMOS bus structures, as the theoretical fan-out capacity is unlimited. Heavily loaded bus structures suffer two disadvantages: The rise time will be slow, and the driver power dissipation will be high.

Example 2.2 shows realistic rise-time and power dissipation calculations for a heavily loaded CMOS bus.

EXAMPLE 2.2: Performance of CMOS Bus

We are constructing a large bus for the shared-memory subsystem of a parallel computer, as shown in Figure 2.8. The bus connects 20 small CPUs, any of which may access an 8-bit-wide random-access memory. The entire system fits on one large circuit card.

The bus is implemented using 50-Ω controlled-impedance traces that are 10 in. long. Figure 2.8 shows that the bus propagation length is much shorter than the rise time of a 74HCT640 gate, and so no terminators are used at either end of the bus.

We expect, based on the DC fan-out parameters, that each bus driver should easily be able to drive 20 other circuits. Given the 9-ns maximum propagation delay of each transceiver, we plan to operate the bus on a 30-ns cycle (33 MHz).

To check out this design, compute the load capacitance on each trace and compare that to the drive resistance of the three-state outputs. Then figure the *RC* rise time of the bus. Last, calculate the power dissipation inside each driver.

Load capacitance Each driver, when switched to its OFF state, still presents a capacitive load. This I/O load capacitance for each driver is specified by the manufacturer at 10 pF. We

[6] HC logic may be powered anywhere from 2 to 6 V. Its switching threshold lies halfway between V_{CC} and ground. HCT logic, designed for compatibility with TTL, has its switching threshold offset toward ground (like TTL) and is designed for operation only near 5 V.

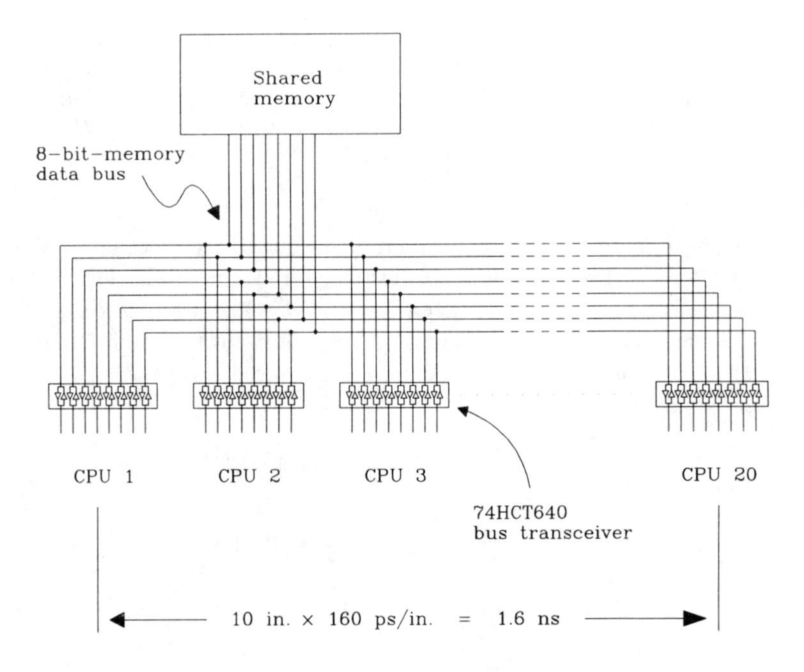

Figure 2.8 Shared-memory bus.

have 20 loads, so that's 200 pF of load. Add to that value the 2-pF/in. load of the backplane traces, and we get

$$C_{\text{load}} = (10\ \text{pF/driver})(20\ \text{drivers}) + (2\ \text{pF/in.})(10\ \text{in.}) \\ = 220\ \text{pF} \tag{2.16}$$

Output resistance of 74HCT640 The *Signetics High-Speed CMOS Data Manual* lists the following specifications (the high-side drive transistor is the worst):

$$V_{CC} = 4.5\ \text{V}$$
$$V_{OH} = 3.84\ \text{V}$$
$$I_{\text{out}} = 6.0\ \text{mA}$$

The high-side drive resistance of an HCT bus driver output is

$$\frac{V_{CC} - V_{OH}}{I_{\text{out}}} = 110\ \Omega \tag{2.17}$$

***RC* rise time** As the output switches from low to high, the charging time constant is roughly equal to the driver output resistance times the output load capacitance.[7]

[7]This approximation ignores the fact that output resistance is a dynamic function of output voltage. Modeling the resistance is a difficult task not generally worth the rewards. Some data books claim their output drivers act like perfect current sources, which is also not true. Whichever method you follow, corroborate your calculations with a physical measurement.

$$T_{RC} = (110\ \Omega)(220\ \text{pF}) = 24\ \text{ns} \quad [2.18]$$

The number T_{RC} is the time required for the output to go from a low state to a level that is 63% high. It takes a little more than twice as long to get to 90% high. The 10–90% rise time of a simple *RC* circuit is 2.2 times the *RC* product:

$$T_{10-90} = 2.2T_{RC} = 53\ \text{ns} \quad [2.19]$$

What a surprise! We thought we were using drivers with a 9-ns maximum propagation delay, but the actual delay turns out to be 53 ns! If we run this bus at 33 MHz, the data bits won't have time to fully rise or fall before the next bit interval. Let's slow the bus down to 16 MHz to give more time between data cells.

Power dissipation in each driver

$V_{CC} = 5.5$ V(worst case supply voltage)
$C = 220$ pF (load capacitance)
$F_{\text{clock}} = 16$ MHz (we lowered the clock frequency)
$F_{\text{data}} = 8$ MHz (worst cycle rate is one-half of clock)

Applying Equation 2.3, calculate the dissipation per driver:

$$P_{\text{driver}} = (8.0 \times 10^{6})(220 \times 10^{-12})(5.5)^2 = 0.053\ \text{W} \quad [2.20]$$

Multiplying by eight drivers per package, we get a total package dissipation of

$$P_{\text{total}} = 8(0.053) = 0.424\ \text{W} \quad [2.21]$$

Section 2.4.3 will show that this is a lot of dissipation for a 20-pin plastic package. This bus design is impractical both because the rise time is too slow and because the driver power dissipation is too high. We must derate this bus below 16 MHz.

2.2.6.3 Quiescent Power Dissipated in Emitter Follower Output Circuit

Figure 2.9 illustrates an ECL or GaAs emitter follower output circuit. This circuit sources current in both HI and LO states.

The logic HI and LO output voltages are similar for both the 10KH and 10G families, although subtle differences exist in the temperature-tracking characteristics of various ECL and GaAs emitter-coupled logic families. These families are normally powered by a –5.2-V power supply. The output HI voltage (most positive) is nominally –0.9 V, while the output LO voltage (most negative) is –1.7 V.

Emitter-coupled logic circuits require a pull-down resistor, usually terminated either to –5.2 V or to an intermediate supply voltage of –2.0 V. We present here calculations for both cases.

When pulled down to V_T volts by a Thevenin equivalent resistance *R*:

$$P_{\text{quies}} = \frac{1}{2}\frac{(V_{CC} - V_{\text{HI}})(V_{\text{HI}} - V_T) + (V_{CC} - V_{\text{LO}})(V_{\text{LO}} - V_T)}{R} \quad [2.22]$$

When ECL logic powered by –5.2 V is pulled down by resistor *R* to –5.2 V, Equation 2.22 may be reduced by substituting these numeric values:

$V_{CC} = 0$ (positive supply voltage)
$V_{HI} = -0.9$ (nominal logic HI level)
$V_{LO} = -1.7$ (nominal logic LO level)
$V_T = -5.2$ (pull-down voltage)

$$P_{quies} = \frac{4.91}{R} \quad [2.23]$$

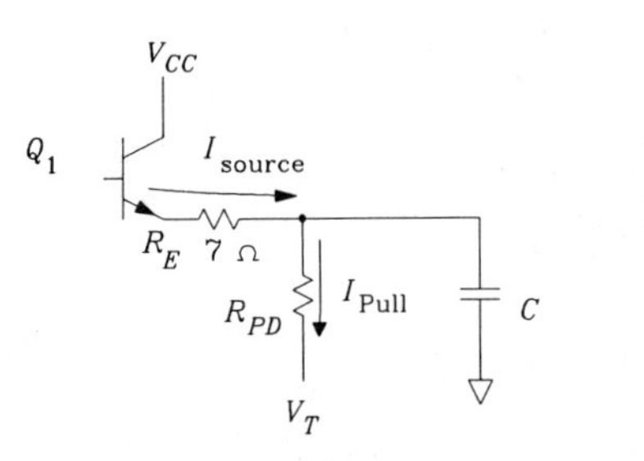

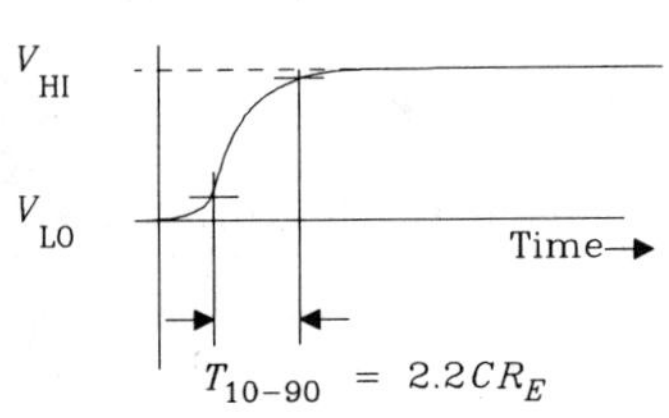

Rise time is determined by the equivalent emitter series resistance R_E and the capacitive load C

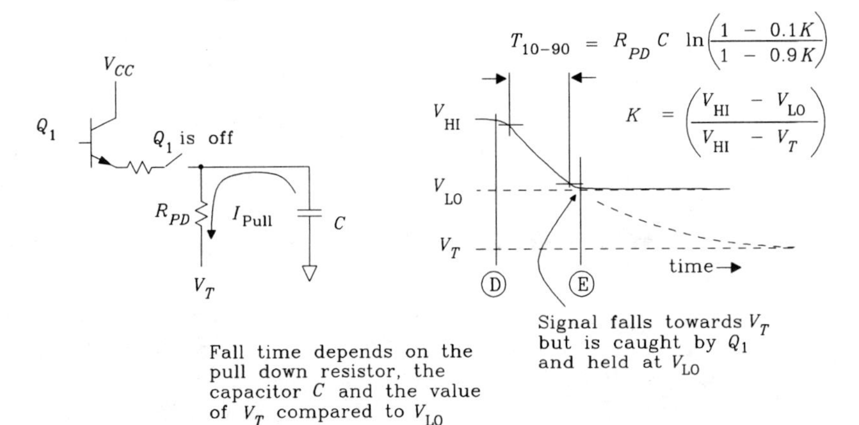

Fall time depends on the pull down resistor, the capacitor C and the value of V_T compared to V_{LO}

Figure 2.9 Rise and fall times of an emitter follower circuit.

When the same circuit is pulled down by resistor R to –2.0 V, Equation 2.22 reduces to

$V_{CC} = 0$ (positive supply voltage)
$V_{HI} = -0.9$ (nominal logic HI level)
$V_{LO} = -1.7$ (nominal logic LO level)
$V_T = -2.0$ (pull-down voltage)

$$P_{quies} = \frac{0.75}{R} \quad [2.24]$$

For the same resistor value, there appears to be a tremendous power advantage to using the –2.0-V termination. This happens because when pulled down to only –2.0 V, the pull-down resistor draws less current. Less current means less power. Less current also means a slower fall time when the circuit transitions from HI to LO.

Because the output circuit is an emitter follower, the rise time is independent of the pull-down current. Figure 2.9 shows the equivalent series resistance R_E of emitter Q_1 having a value of about 7 Ω for 10KH ECL logic. When charging up a load capacitance C, the source current is much greater than the pull-down current, and so the charging time constant equals the product:

$$T_{RC} = R_E C \tag{2.25}$$

The number T_{RC} is the time required for the output to go from a low state to 63% high. It takes more than twice as long to get to 90% high. The 10–90% rise time of a simple RC circuit is

$$T_{10-90} = 2.2T_{RC} = 2.2R_E C \tag{2.26}$$

The time constant (Equation 2.26) is usually shorter than the turn-on time of transistor Q_1, and so the output rise time usually equals the turn-on time of Q_1. Appendix B explains more precisely how to combine several independent rise time effects into a composite rise or fall time.

On falling edges, transistor Q_1 cuts off, and no current flows through the emitter. Only the pull-down resistor discharges capacitor C. This is where the relationship between power and rise time comes into play. The fall time is directly proportional to how fast we pull current out of capacitor C. The power dissipation is proportional to the quiescent pull-down current. Whether our pull-down resistor is connected to –5.2 or –2.0 V, we need a lot of current to discharge capacitor C quickly.

Figure 2.9 shows the decay waveform. At time D, transistor Q_1 cuts off. The output voltage decays with time constant $R_{PD}C$ toward voltage V_T. At point E the voltage has decayed down to V_{LO} and transistor Q_1 turns back on, arresting further decay. The output voltage then sticks at V_{LO}.

If Q_1 snapped off perfectly, the 10–90% fall time would be

$$T_{10-90} = R_{PD}C \cdot \ln\left(\frac{1-0.1K}{1-0.9K}\right) \tag{2.27}$$

Where the constant K is equal to

$$K = \frac{V_{HI} - V_{LO}}{V_{HI} - V_T} \tag{2.28}$$

When the time constant (Equation 2.27) is shorter than the turn-off time of transistor Q_1, the fall time closely approximates the turn-off time of Q_1. Appendix B explains more precisely how to combine several independent rise-time effects into a composite rise or fall time.

When ECL powered by –5.2 V is pulled down by resistor R_{PD} to –5.2 V, Equation 2.27 for fall time may be reduced by substituting numeric values:

$$V_{HI} = -0.9 \text{ (nominal logic HI level)}$$
$$V_{LO} = -1.7 \text{ (nominal logic LO level)}$$
$$V_T = -5.2 \text{ (pull-down voltage)}$$
$$K = 0.186 \text{ (constant } K\text{)}$$
$$\ln[(1-0.1K)/(1-0.9K)] = 0.164$$

$$T_{10-90} = 0.164 R_{PD} C \quad [2.29]$$

When the same circuit is pulled down by resistor R_{PD} to –2.0 V, Equation 2.27 for fall time reduces to

$$V_{HI} = -0.9 \text{ (nominal logic HI level)}$$
$$V_{LO} = -1.7 \text{ (nominal logic LO level)}$$
$$V_T = -2.0 \text{ (pull-down voltage)}$$
$$K = 0.727 \text{ (constant } K\text{)}$$
$$\ln[(1-0.1K)/(1-0.9K)] = 0.987$$

$$T_{10-90} = 0.987 R_{PD} C \quad [2.30]$$

To get the same fall time, the –2.0-V pull-down circuit requires a much smaller resistor than the –5.2-V pull-down circuit. Once the resistors are selected to equalize rise times, the power dissipation numbers in Equations 2.23 and 2.24 are about the same.

There is no great advantage in power or speed to either the –5.2- or –2.0-V termination. The resistor values are just different.

The –5.2-V pull-down does have the advantage of not requiring a separate power supply. On the other hand, the –2.0-V pull-down has the advantage of working well as a terminator at the end of a transmission line. Reasonable values for pull-down resistors to –2.0 V in ECL logic range from 50 to 100 Ω, corresponding roughly to the range of practical transmission line impedances. Reasonable ranges for terminating resistors to –5.2 V in ECL logic range from 330 to 680 Ω, a factor of 6 higher. These higher resistances do not work as well as terminators.

With either circuit, reducing the resistor value dissipates more power while reducing the fall time. For the same fall times, both circuits dissipate similar amounts of power.

2.2.6.4 Split Pull-Down Terminations

ECL circuits are sometimes terminated with the so-called *split termination* shown in Figure 2.10. Formulas for synthesizing the actual split termination resistance values from the desired composite impedance and terminating voltage are

$$R_1 = R_3\left[\frac{V_{CC} - V_{EE}}{V_T - V_{EE}}\right]$$
$$R_2 = R_3\left[\frac{V_{CC} - V_{EE}}{V_{CC} - V_T}\right] \quad [2.31]$$

where R_3 = desired composite impedance
V_T = desired effective terminating voltage
R_1 = top resistor (goes to V_{CC})
R_2 = bottom resistor (goes to V_{EE})

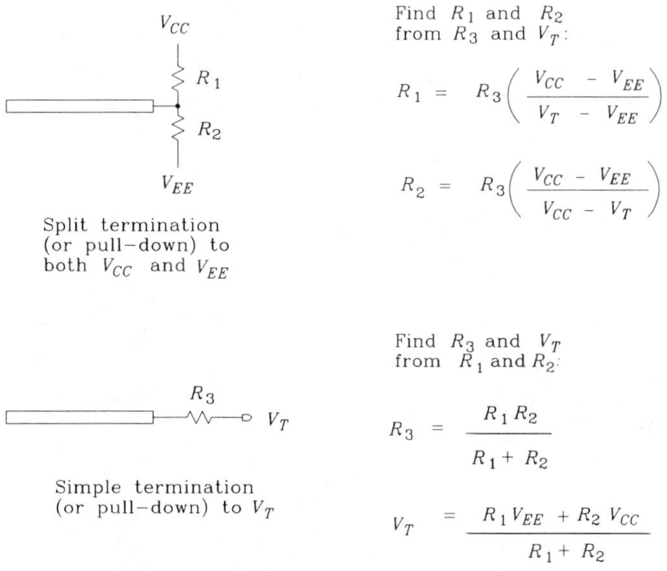

Figure 2.10 Split termination equivalent circuits.

2.2.6.5 Active Power Dissipated in Emitter Follower Output Circuit

This is a small effect in ECL system designs. Dissipation due to the pull-down resistor (which is also responsible for ensuring a quick discharge of any load capacitance) is usually much larger than the active power required to charge up any load capacitance.

The same is true for open collector circuits and for current source output circuits. Any capacitive loads present cause fall-time problems well before contributing appreciably to drive circuit dissipation.

2.2.6.6 Power Dissipated in TTL or CMOS Open Collector Output

A formula similar to Equation 2.22 calculates the quiescent power dissipation in a TTL open collector (or CMOS open drain) output. When pulled up to V_T volts by a Thevenin equivalent resistance R:

$$P_{\text{quies}} = \frac{1}{2}\,\frac{(V_T - V_{HI})(V_{HI} - V_{EE}) + (V_T - V_{LO})(V_{LO} - V_{EE})}{R} \qquad [2.32]$$

where V_T = effective terminating voltage for pull-up resistor
R = effective value of terminating resistor
V_{HI} = HI-level output (often equals V_T)

V_{LO} = LO-level output
V_{EE} = supply voltage to emitter (or source) of output transistor
P_{quies} = power dissipated in output driver

The BTL family of transceivers[8] uses open collector drivers with the pull-up resistors tied to +2.0 V. Operating logic levels are +2.0 and +1.0 V. The BTL drive circuit incorporates a Schottky diode, shown in Figure 2.11 as D_1, in series with its output pin. This diode is reversed biased when Q_1 turns off, resulting in a very low output capacitance of 6.5 pF typical. Low-output capacitance is the primary benefit of BTL technology.

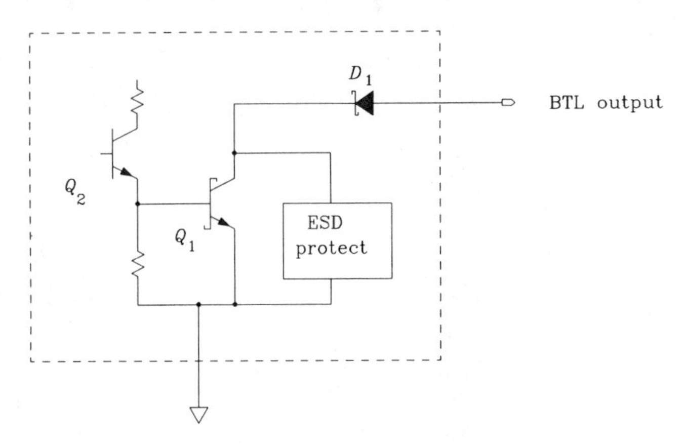

Figure 2.11 BTL drive circuit.

Totem-pole output circuits always leave a reversed biased base-to-emitter (or disabled gate-to-source) junction connected to the line when tristated. The capacitance of this junction, which must be physically big to support large output drive currents, is much greater than an ordinary input capacitance. In comparison, the BTL drive circuit has a much lower capacitance when switched to its OFF state.

2.2.6.7 Power Dissipated in Current Source Output Circuit

Current source outputs, used for some specialized bus applications, have the advantage of linearity. Their current outputs naturally superimpose on each other when driving a long bus, in contrast to voltage source outputs which interact in a nonlinear fashion.

Because these circuits are designed as linear class-A amplifiers, the driving transistor does not saturate and the output circuit consequently dissipates a lot of power.

Open collector drive circuits either draw a lot of current with little voltage drop or have a big voltage drop with no current. Both states consume little power. Current source drives, on the other hand, sometimes draw a lot of current with a big voltage drop in one or both states. Despite their power inefficiency, current source outputs have big advantages in long bus structures.

[8]BTL is a trademark of National Semiconductor Corporation. These transceivers are recommended for use with IEEE Standard 896.1 1987 Futurebus.

Example 2.3 shows off one big benefit of using current source drivers.

EXAMPLE 2.3: Use of Current Source Drivers

Some systems use a current source driver configured as a one-way bus (Figure 2.12). The clock driver demarks a succession of clock intervals beginning at times t_1, t_4, and t_8. These clock signals propagate from left to right parallel to the data bus. Each bus driver, named alpha, beta, or gamma, is responsible for inserting data on the bus at predetermined time slots. The timing of bus transmissions coincides with the arrival, at the transmitting device, of a clock signal. This arrangement guarantees that the each data cell arrives at the right-hand end of the bus properly framed within the clock interval, regardless of the physical position on the bus at which it was transmitted. The one and only receiver is located at the right end of the bus, and it clocks data off the bus synchronously with arriving clock signals.

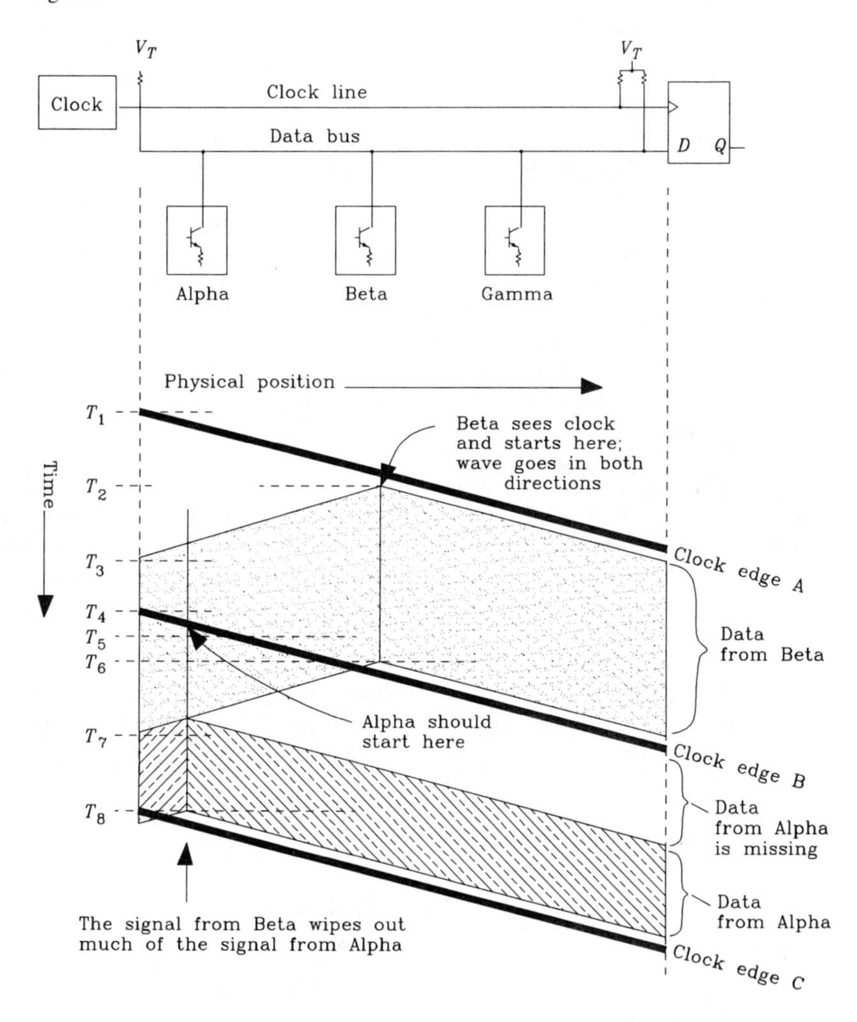

Figure 2.12 Current source drive current used on a long unidirectional bus.

Figure 2.12 shows two data cells, the first transmitted by Beta, and the second by Alpha. Beta begins sending at time t_2 and stops one clock interval later at time t_6. This time corresponds to the instant when clock signals A and B arrive at position Beta. Alpha starts sending at time t_5, when clock signals arrive there, and stops one clock interval later.

As each device transmits, its signal propagates both to the right and to the left along the data bus. At the right end of the data bus, both signals end up properly framed in their assigned positions.

So far, there is no theoretical limitation to the speed at which we may operate this bus. Because we do not have to wait for the clock to propagate down and back to remote devices, we can raise the clock rate to any arbitrary frequency, limited by the operating speed of the associated devices but not by the physical propagation velocity and length of the bus. The bus can carry several data cells simultaneously, all propagating to the right toward the receiver.

The disadvantage to this one-way arrangement is that while we care only about signals at the right end of the bus, the transmitted signals actually propagate in both directions. Examine what happens at point t_2 when Beta transmits. Its transmitted signal goes to the right (where it is received) and also backward toward Alpha. At point t_5, when Alpha should start transmitting, the tail end of datum A from Beta is passing by on the data bus.

If transmitter Alpha is a totem-pole driver, and if the values of datum A and datum B match, *no current flows from device Alpha onto the data bus until datum B has completely passed.* The data bus is already at the correct logic state, and so transmitter Alpha has no effect. It is as if Alpha were never connected. As soon as datum B passes, at time t_7, current flows from totem-pole device Alpha to hold the bus in a particular state. At the receiving end, because no transmitted wave emanated from Alpha until well into its assigned time slot, the front end of received datum B is missing.

Similarly, if the bit polarities are opposite, Alpha must work twice as hard to drive the bus output, and the front end of datum B, as seen at the D input, will show a larger-than-normal pulse.

The solution to this problem is to use a linear mode drive circuit which superimposes its own signal onto whatever happens to be present on the line. The correct circuit configuration is a current source, usually implemented as either an open collector or open drain current regulator circuit. This circuit injects a predetermined amount of current onto the data bus. The data bus, responding like a resistive load,[9] produces a corresponding shift in voltage. One of the two logic states is usually defined as zero current, and inactive devices switch to that state.

Depending on the length of the bus, each driver might operate in the presence of several simultaneous datum passing by, each from a different driver located at different positions along the cable. This requirement implies that the driver must supply a linear current over a wide range of voltages. The combination of constant drive current over a large range of voltages wastes a lot of power.

2.2.7 Output Dissipation

Power dissipated in terminating resistors, pull-down resistors, or other bias resistors adds to the overall power load on the power supply, and to the cooling requirements.

[9]The data bus must be terminated resistively at both ends.

Section 2.2.3 dealt with power dissipated in an output circuit which drives an external load. This section computes the power dissipated in the load itself.

First, remember that ideal capacitors dissipate no power. Power is dissipated in the driving circuit that charges and discharges the capacitor, but not in the capacitor itself.

For an individual resistor R connected between a data line and a fixed power supply voltage V_T, the power dissipation in the HI state is

$$P_{HI} = \frac{(V_{HI} - V_T)^2}{R} \qquad [2.33]$$

Power dissipated in the same resistor during the logic LO state equals

$$P_{LO} = \frac{(V_{LO} - V_T)^2}{R} \qquad [2.34]$$

Bias resistors should always be sized to handle the worst-case power dissipation, in case the data gets stuck in one state or another. Dissipation in bias resistors is often higher than in the driving circuits, so one sometimes worries more about burning resistors than about burning logic gates.

The power supply can be sized to handle the expected average dissipation, plus a moderate safety factor. Power supplies have fuses and thermoelectric cutoff circuits to protect them in case of overload; bias resistors do not. Don't underestimate the power dissipation in bias and terminating resistors.

> POINT TO REMEMBER:
>
> Always include active power dissipation and power dissipated driving heavy loads in device power calculations.

2.3 SPEED

Theoretical digital logic design focuses on the propagation delay of logic gates. In contrast, practical problems in high-frequency engineering often depend solely on a more subtle specification: the minimum output switching time. Figure 2.13 illustrates the difference.

Faster switching times cause proportional increases in problems with return currents, crosstalk, and ringing that are independent of propagation delay. Logic families having minimum switching times much faster than the propagation delay suffer an unnecessary penalty in system design because the device packaging, board layout, and connectors must accommodate fast switching times while the logic timing benefits only from propagation

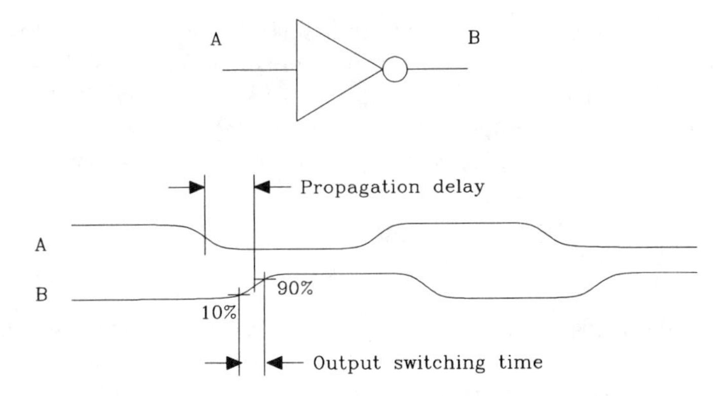

Figure 2.13 Output switching time versus propagation delay.

delay. *Given two logic families with identical maximum propagation delay statistics, the family with the slowest output switching time will be cheaper and easier to use.*

Many logic families are available at a variety of speed-power combinations. The TTL family includes the LS (low-power Schottky) and S (regular Schottky) varieties. All CMOS families exhibit a dramatic power-versus-speed relationship linking the power dissipation of any CMOS system proportionally to the clock speed. The ECL family includes MECL III, at nearly twice the speed of MECL 10KH but using more than twice the power.

Manufacturers emphasize the tradeoff of speed and power because it is easy to see on a data sheet. What they often don't list is the minimum device switching time. This parameter is difficult to control unless the manufacturer inserts special circuitry to slow down the output switching rate.

Circuits that limit switching time have slowly crept into a few logic families. ECL families have incorporated edge-slowing circuits since the introduction of the MECL 10K family in 1971. The FCT family, introduced in 1990, was the first CMOS circuit to incorporate edge-slowing mechanisms. Since that time, other manufacturers have been getting the idea.

Excessively fast switching times cause problems through two distinct mechanisms: effects created by sudden changes in voltage and effects created by sudden changes in current.

2.3.1 Effects of Sudden Change in Voltage, *dV/dt*

Referring back to Equation 1.1, recall that most of the frequency content of a digital signal lies below its knee frequency. The knee frequency F_{knee} is related to pulse rise time T_r but not to the propagation delay, clock rate, or switching frequency:[10]

$$F_{\text{knee}} = \frac{0.5}{T_r} \qquad [2.35]$$

[10]Remember from Chapter 1 that F_{knee} is an imprecise, but useful, measure of spectral content for which we do not formally specify the method of rise-time measurement.

Propagating pathways, including device packaging, board layout, and connectors, must have a flat frequency response up to at least F_{knee} if they are to distribute faithfully digital signals with switching times as fast as T_r. If the frequency response of a pathway is not flat up to F_{knee}, signals received at the far end of the pathway may have impaired rise times, lumps, overshoot, or ringing.

Too short a rise time (high dV/dt) pushes up the value of F_{knee}, making the signal propagation problem that much harder. This is the primary disadvantage of too short a rise time (too high a dV/dt).

The dV/dt on one circuit may also affect signals on another nearby circuit. This crosstalk occurs through the mechanism of mutual capacitance (see Section 1.9). Two circuit elements placed nearby will always interact capacitively. As hinted in Section 1.10.3, crosstalk due to mutual capacitance is much less of a problem in digital systems than crosstalk due to mutual inductance.

We can relate the maximum dV/dt in a circuit to its 10–90% rise time and the voltage swing ΔV:

$$\frac{dV}{dt} = \frac{\Delta V}{T_{10-90}} \qquad [2.36]$$

2.3.2 Effects of Sudden Change in Current, *dI/dt*

Sudden changes in current may affect signals on another nearby circuit. This crosstalk occurs through the mechanism of mutual inductance (see Section 1.10). Two circuit elements placed nearby will always interact inductively. To calculate the amount of inductive coupling, we must first estimate the rate of change of current in the source network. It makes sense that circuits having a high rate of change in current will have greater problems with inductive coupling. This is the primary disadvantage of too high a dI/dt.

Since our primary probing instrument, the oscilloscope, reads out voltage, not current, we need a way to translate from voltage rise-time readings to rates of change in current. Figure 2.14 illustrates the general situation. The rising voltage waveform $V(t)$ causes currents to flow in the load resistance and load capacitance equal to[11]

$$I_{\text{resistor}} = \frac{V(t)}{R} \qquad [2.37]$$

$$I_{\text{capacitor}} = \frac{dV(t)}{dt}C \qquad [2.38]$$

Differentiating both waveforms to find the *rate of change in current*, we get

$$\frac{dI}{dt}(\text{resistor}) = \frac{dV(t)}{dt}\frac{1}{R} \qquad [2.39]$$

$$\frac{dI}{dt}(\text{capacitor}) = \frac{d^2V(t)}{dt^2}C \qquad [2.40]$$

[11]We assume here that the voltage waveform is an integrated gaussian pulse.

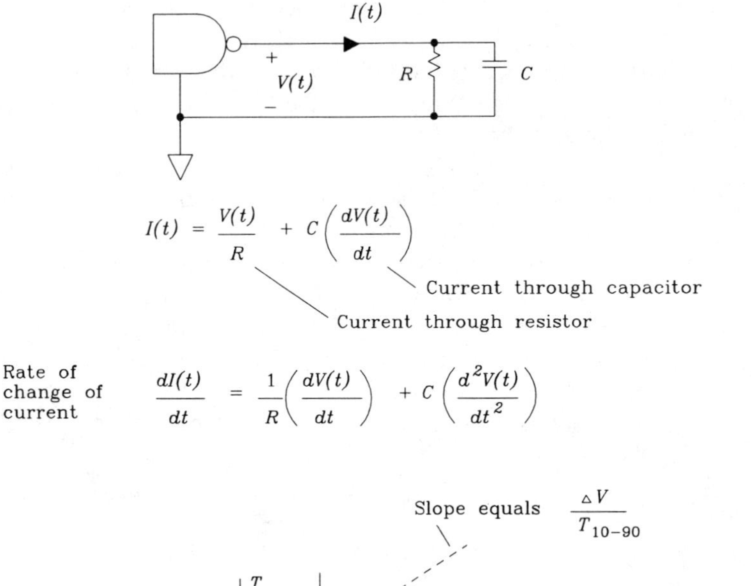

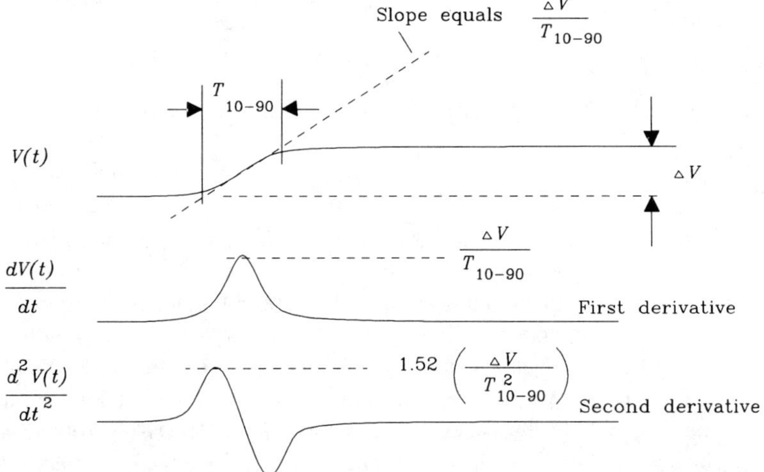

Figure 2.14 Relation of maximum current slew rate to voltage rise time.

The maximum value of the rate of change in current will be useful for determining the peak value of inductive coupling. For the resistor and capacitor individually, these maximum values are

$$\text{Maximum}\frac{dI}{dt}(\text{resistor}) = \frac{\Delta V}{T_{10-90}}\frac{1}{R} \tag{2.41}$$

$$\text{Maximum}\frac{dI}{dt}(\text{capacitor}) = \frac{1.52\ \Delta V}{T_{10-90}^{\ 2}}C \tag{2.42}$$

When driving a combination load having both resistive and capacitive parts, just sum the maximum numbers from Equations 2.41 and 2.42. Summing will somewhat overestimate the actual peak value but is close enough for our purposes. Figure 2.14 shows that the peaks in the first and second derivatives of $V(t)$ do not quite line up, so the peak rates of change in current for the resistor and capacitor occur at slightly different times. Sure, the sum isn't exactly accurate, but it's easy to remember and it comes close.

Equation 2.42 gives us a hint as to why mutual inductance is such a problem. The driving factor for mutual inductance, rate of change in current, is proportional to 1 over the square of the 10–90% rise time. *When we halve the rise time, we quadruple the amount of dI/dt flowing into capacitive loads.*

Let's work two examples comparing the rate of change of current in TTL and ECL systems. These examples show that ECL systems do not necessarily generate higher current-switching transients than TTL systems. The ECL system is faster but makes less noise.

EXAMPLE 2.4: Rate of Change of TTL Output Current

Assume a TTL gate is loaded with a capacitive load of 50 pF. Let $\Delta V = 3.7$ V, $C_L = 50$ pF, and $T_r = 2$ ns.

$$\frac{dI}{dt} = \frac{1.52 C_L \Delta V}{T_r^2} = 7.0 \times 10^7 \text{ A/s} \qquad [2.43]$$

EXAMPLE 2.5: Rate of Change of ECL Output Current

Assume an ECL gate is driving a 50-Ω resistive load. $\Delta V = 1.0$ V, $R_L = 50$ Ω, and $T_r = 0.7$ ns.

$$\frac{dI}{dt} = \frac{\Delta V}{R_L T_r} = 2.8 \times 10^7 \text{ A/s} \qquad [2.44]$$

2.3.3 The Bottom Line—Voltage Margins

Voltage margin is the difference between the guaranteed output of a logic driver and the worst-case sensitivity of a logic receiver. Logic families which operate based on received voltage have voltage margins, as opposed to optical logic which might have photon margins, or mechanical devices in the Babbage engine which undoubtedly had margins for mechanical linkage.

Figure 2.15 illustrates the margin setup for Motorola 10KH emitter-coupled logic gates at an ambient temperature of 25°C. These gates have voltage-sensitive inputs whose guaranteed 0 and 1 switching thresholds appear in the figure as V_{IL} min and V_{IH} max, respectively. Voltages received below V_{IL} min guarantee a logical 0 response, and above V_{IH} max guarantee a logical 1 response. Received voltages that fall between the two thresholds may be interpreted by the receiving circuit as a 1, as a 0, or as an *indeterminate state.*

The term V_{IL} min implies that, across all gates, it is the minimum value of V_{IL} required to guarantee a low input level. Most gates will switch low at voltages greater than V_{IL} min. These gates have an extra switching margin. Manufacturers do not state how great the low-side switching threshold might be; they only specify its minimum value. The principle applies, but reversed, for V_{IH} max.

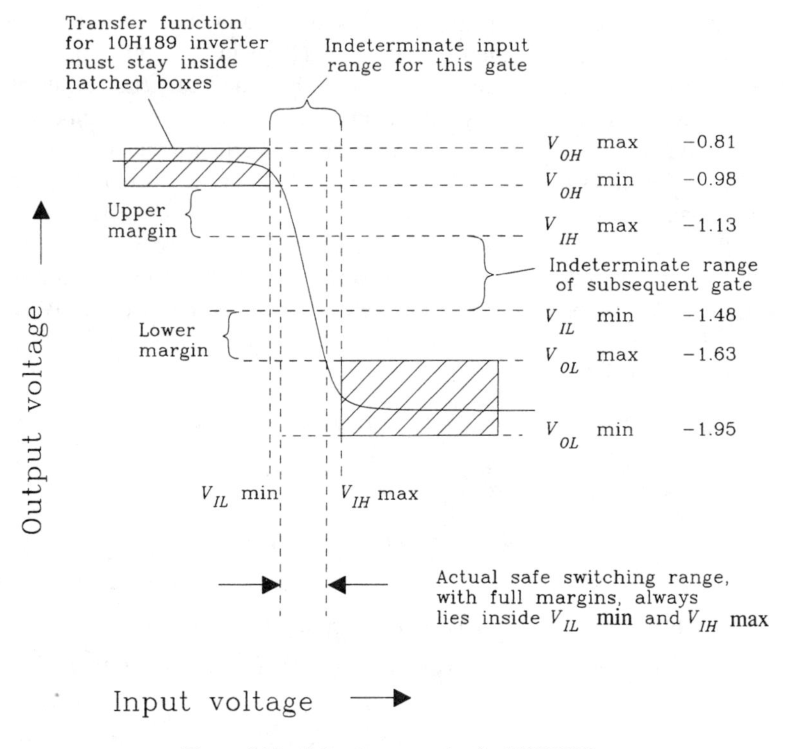

Figure 2.15 DC voltage margins for 10KH ECL.

Guaranteed 0 and 1 output voltages for the 10KH family are marked as V_{OL} (min and max) and V_{OH} (min and max), respectively. When transmitting a low level, the output is guaranteed to fall between V_{OL} min and V_{OL} max. When transmitting a high level, the output is guaranteed to fall between V_{OH} min and V_{OH} max.

There is no overlap between the guaranteed output range and the indeterminate input range. The absence of overlap means that static transmitted values, whether 0 or 1, will always be properly received.

Actual transmitted values for a particular gate are a function of the ambient temperature, the power supply, and the manufacturing process variables used to make that gate. The specification sheet says only that the output values for 0 and 1 will exceed V_{OL} and V_{OH},[12] respectively. Typical output levels are marked in Figure 2.15.

The term voltage margin refers to the difference between V_{OH} and V_{IH}, or the difference between V_{OL} and V_{IL}, whichever is least.

Just as logic outputs often exceed the worst-case specification, logic inputs are often more precisely discriminating than the worst-case switching thresholds. A typical 10KH inverter switching function is shown in Figure 2.15. As you can see, the circuit exhibits a gain of –4 in the switching region, saturating outside the required switching

[12]This means that the transmitted logic 0 voltage will be lower (more negative) than V_{OL} and the transmitted logic 1 voltage will be higher (more positive) than V_{OH}.

zone at values within the hatched output specifications. Just because this one gate switches well within the worst-case switching zone does not mean that every gate will act the same way. The next gate off the assembly line might have a different input DC offset and therefore switch close to one side or the other of the worst-case range. Military manufacturers screen parts at the factory to ensure that none ship with switching thresholds outside the acceptable range. Commercial manufacturers only statistically spot-check these parameters and hope for the best.

Why do we need margins? Margins compensate for imperfect transmission and reception of digital signals in real systems. Systems without adequate margins will not work in the presence of signal corruption such as

(1) DC power supply currents, flowing through the DC resistance of the ground path, cause ground voltage differentials between logic devices. A signal transmitted from one gate at a fixed potential above local ground will be received at a different potential if the ground reference between the transmitting and receiving gates is shifted.

(2) Fast changing return signal currents, flowing through the inductance of a ground path, cause ground voltage differentials between logic devices. These differentials affect the voltage potential of received signals just like DC ground differentials. This is a form of inductive crosstalk.

(3) Signals on adjacent lines may couple through either mutual capacitance or (more likely) mutual inductance to a given line, generating crosstalk. The crosstalk adds to the intended received signal, potentially moving a good signal closer to the switching threshold.

(4) Ringing, or reflection, on long lines distorts the appearance of binary signals. Signal transitions may appear smaller (or larger) at the receiver than at the transmitter. Margins allow some tolerance for signal distortion.

(5) The threshold levels on some logic families are a function of temperature.[13] A cold gate transmitting to a hot gate (or vice versa) may suffer reduced or negative margins.

Items (1) and (5) apply to all systems regardless of operating speed and must always be taken into account. Items (2) through (4) are peculiar to high-speed systems.

All three high-speed effects vary with the size of the transmitted signal: More return current induces higher ground differentials, more signal voltage (or current) induces more crosstalk, and larger transmitted signals exhibit more ringing and reflections. These proportional relations lead us to conclude that an important measure of tolerance to effects (2) through (4) in high-speed systems is the ratio of voltage margin to output voltage swing. Percentage calculations are easier to visualize, and more trans-

[13] 10KH threshold levels vary almost 100 mV over the temperature range of 0–70°C. ECL system designers must ensure that the entire system is at a uniform temperature, or else derate the switching margins to account for the difference.

portable among different logic families, than direct voltage readings. The *noise margin percentage* is equal to the lesser of

$$\frac{V_{OH}\,\text{min} - V_{IH}}{V_{OH}\,\text{max} - V_{OL}\,\text{min}} \quad \text{or} \quad \frac{V_{IL} - V_{OL}\,\text{max}}{V_{OH}\,\text{max} - V_{OL}\,\text{min}} \qquad [2.45]$$

The noise margin percentage of 10KH ECL logic is 17.8%, while 74AS TTL logic has a margin of only 9.1%. This difference is the basis for claims that ECL logic has better noise immunity than TTL. While the actual margin voltages for ECL are smaller than for TTL, they are a bigger percentage of the ECL voltage swing.

Of course, the 10KH family switches two to three times faster than the 74AS family. Faster switching time increases return-current problems, crosstalk, and ringing. Overall it is more difficult to control return current, crosstalk, and ringing with the MC10KH family than with the 74AS, but not two to three times harder.

POINTS TO REMEMBER:

Given two logic families with identical maximum propagation delay statistics, the family with the slowest output switching time will be cheaper and easier to use.

We can figure the *dI/dt* in an output circuit given the voltage rise time and the load.

When we halve the rise time, we quadruple the amount of *dI/dt* flowing into capacitive loads.

A complete voltage margin budget in a system accounts for the effects of power supply variations, ground shifts, signal crosstalk, ringing, and thermal differences.

2.4 PACKAGING

Variety in packaging springs from the same well that sustains so many different circuit configurations. The number of logic packaging schemes is enormous and growing daily.

Almost all packages, when used at high speeds, suffer from problems with lead inductance, lead capacitance, and heat dissipation.

2.4.1 Lead Inductance

The inductance of individual leads in a device package creates a problem called *ground bounce*. This phenomenon causes glitches in the logic inputs whenever the device outputs switch from one state to another. The magnitude of these glitches, and the effects they cause, are the subject of this section.

2.4.1.1 Unwanted Voltages on Ground Wires—Why Ground Bounce Occurs

Figure 2.16 represents an idealized logic die, wire-bonded to four pins of a DIP package. One transmit circuit and one receive circuit are shown. The transmit circuit shown is a totem-pole output stage, although any configuration exhibits the same problem at high speeds.

Suppose switch B of the output driver has just closed, discharging load capacitor C to ground. As the voltage across capacitor C falls, its stored charge flows back to ground, causing a massive current surge around the ground loop shown as $I_{\text{discharge}}$.

As the discharge current builds and then recedes, changes in that current, working across the inductance of the ground pin, induce a voltage V_{GND} between the system ground plane underneath the device and the ground internal to the package. The magnitude of this voltage is equal to

$$V_{GND} = L_{GND} \frac{d}{dt} I_{\text{discharge}} \qquad [2.46]$$

Shifts in the internal ground reference voltage due to output switching are called *ground bounce*.

The ground bounce voltage V_{GND} is usually small compared to the full-swing output voltage. It does not act to significantly impair the transmitted signal, but it interferes in a major way with reception.

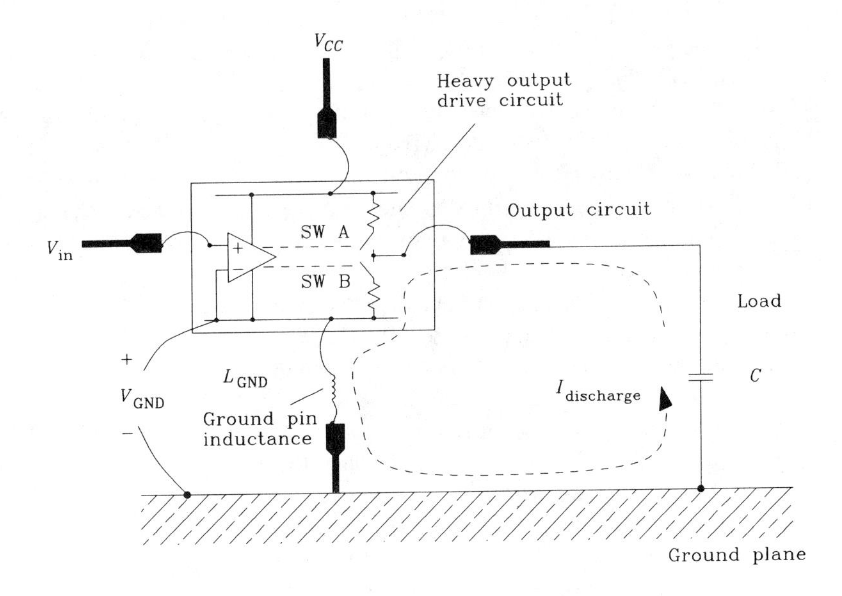

Figure 2.16 Lead inductance of a logic device package.

Consider the receiver section of the same die. This receiver differentially compares the input voltage V_{in} against its local internal ground reference.[14] This differencing operation appears in Figure 2.16 as a plus (+) input connected to V_{in} and a minus (–) input connected to the internal ground. Because the internal ground carries the V_{GND} noise pulse, the actual differential voltage seen by the input circuit is equal to

$$\text{Input circuit sees}: V_{\text{in}} - V_{GND} \tag{2.47}$$

Because the input circuit responds to the difference between its plus (+) and minus (–) inputs, it has no way of knowing whether the noise pulse V_{GND} has been added to the minus (–) input or subtracted from the plus (+) input. In other words, the V_{GND} pulse looks to the inputs circuit like noise directly superimposed on the input signal.

If we simultaneously switch N outputs from a chip into N corresponding capacitive loads, we get N times as much ground current and the pulse V_{GND} looms N times larger.

Ground bounce voltages are proportional to the rate of change in current through the ground pin. When driving capacitive loads, we expect the rate of change in current to look like the second derivative of the voltage. Referring to Figure 2.14, the second derivative of the voltage is a double-humped waveform, first bumping up, and then bumping down.

2.4.1.2 How Ground Bounce Affects Your Circuit

Figure 2.17 illustrates a ground bounce situation. Imagine a TTL octal D flip-flop, with a single clock input, driving a bank of 32 memory chips. At 5 pF per input, each address line is loaded with 160 pF.

Suppose data comes into the D input with plenty of setup time but with little hold time. Figure 2.17 illustrates data arriving with a 3-ns setup time and a 1-ns hold time. Assume this timing meets the requirements of our octal TTL flip-flop.

On clock edge A the flip-flop latches in data word FF hex. At clock edge B the flip-flop latches in data 00 hex. In both cases, the flip-flop propagation delay of 3 ns is slightly longer than the required hold time.

At point C, let the input data change to any pattern XX. Point C follows 1 ns after clock pulse B. At this point, the flip-flop has internally latched in the 00 data word, but the Q outputs have not yet switched from FF to 00.

The next to the bottom trace shows V_{GND}. After point A, when the Q outputs switch positive, the load-charging current flows in the V_{CC} pin, not the ground pin, so we get little noise on V_{GND}. At point D all eight outputs swing LO, and we get a big V_{GND} noise pulse. This noise pulse causes an error called *double-clocking*.

Double-clocking results from differential input action in the clock circuit. Internal to the flip-flop, the clock input measures the difference between the chip's clock pin and its ground pin. The bottom trace of Figure 2.17 shows this difference. The difference

[14]This is representative of TTL circuits. CMOS circuits tend to compare inputs against a weighted average of V_{CC} and ground. ECL and the 10G GaAs family compare inputs against V_{CC}. In all cases, the same effect applies, with slightly different topology.

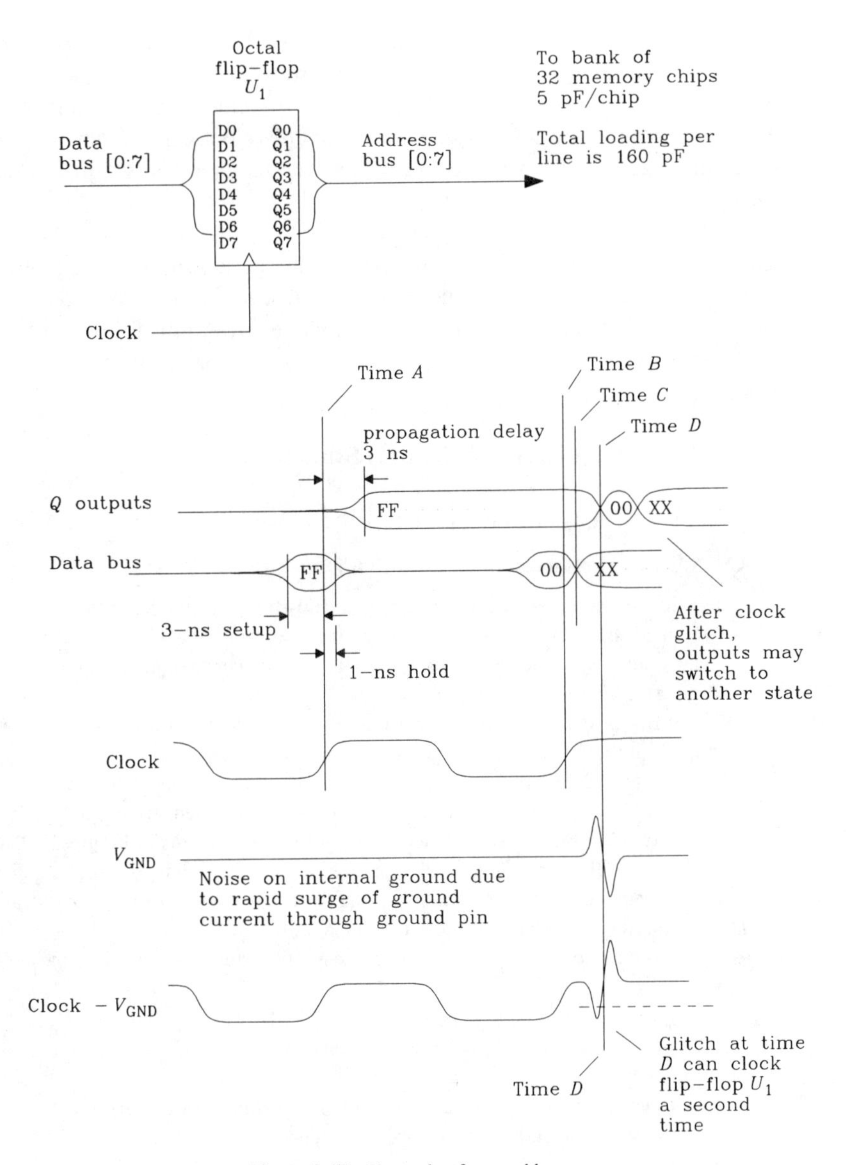

Figure 2.17 Example of ground bounce.

waveform has a clean clock edge at *B*, followed by a big glitch induced by signal currents flowing in the ground pin. *The flip-flop will reclock itself on this pulse.*

If the data input has changed by the time the second clock at *D* happens, the flip-flop will proceed ignorantly to state XX. The *Q* outputs at point *D* momentarily flip to the correct state and then mysteriously flip to some wrong condition.

External observations of the clock input show a perfectly clean signal; it is only internal to the logic package that anything is amiss.

The double-clocking error happens on DIP flip-flop packages which have very fast output drivers connected to heavy capacitive loads. Large latches in the FCT family, sold in a DIP package, have exhibited this problem. Surface-mounted packages, with their shorter pins, are less susceptible to double-clocking. As new generations of flip-flops get faster, we will need new packages with less and less ground inductance to house them.

Providing special power pins for the output drivers separate from those used for referencing input signals elegantly circumvents the ground bounce problem. Since little current flows in the input ground reference pins, no ground bounce effect occurs. Most ECL families, and many gate arrays, use separate power pins for this purpose.

Edge-sensitive input lines such as resets and interrupt service lines are particularly susceptible to ground bounce glitches.

2.4.1.3 Magnitude of Ground Bounce

Let's look at a concrete example to see how big the ground bounce pulse can be.

EXAMPLE 2.6: Measurement of Ground Bounce

For this measurement we will use a quad flip-flop, configured so that three of its outputs are toggling while the fourth output stays fixed at zero. We have the ability to switch 20-pF loads onto any of the three active outputs. This experimental arrangement can show ground bounce with no load or with heavy loads.

Because the inactive fourth output stays at logic LO, it serves as a window into the chip through which we may measure the internal ground voltage.

Figure 2.18 shows the arrangement. The clock and asynchronous reset lines alternately set and reset the three active outputs. For this experiment we use a 74HC174 flip-flop.

With all three loads connected we get the waveforms in Figure 2.19. When the Q outputs switch HI, there is a small V_{GND} glitch. This corresponds to switching currents internal to the device (see Section 2.2.2). When the Q outputs switch LO, the big ground bounce pulse appears. In this example it is about 150 mV tall.

A pulse of only 150 mV may not seem like much, but consider these facts:

(1) The low-side voltage margin on HCT logic is only 470 mV.
(2) If we had eight simultaneous outputs switching, the pulse would be eight-thirds times bigger.
(3) Ground bounce reduces the available residual noise margin used to compensate for other noise and signal distortion effects.

Identical measurements carried out on a 74F174 flip-flop result in a ground bounce of 400 mV.

2.4.1.4 Predicting Ground Bounce Magnitude

To make useful predictions about ground bounce we need to know four facts: The 10–90% switching time of the logic device, the load capacitance or resistance, the lead inductance, and the switching voltage.

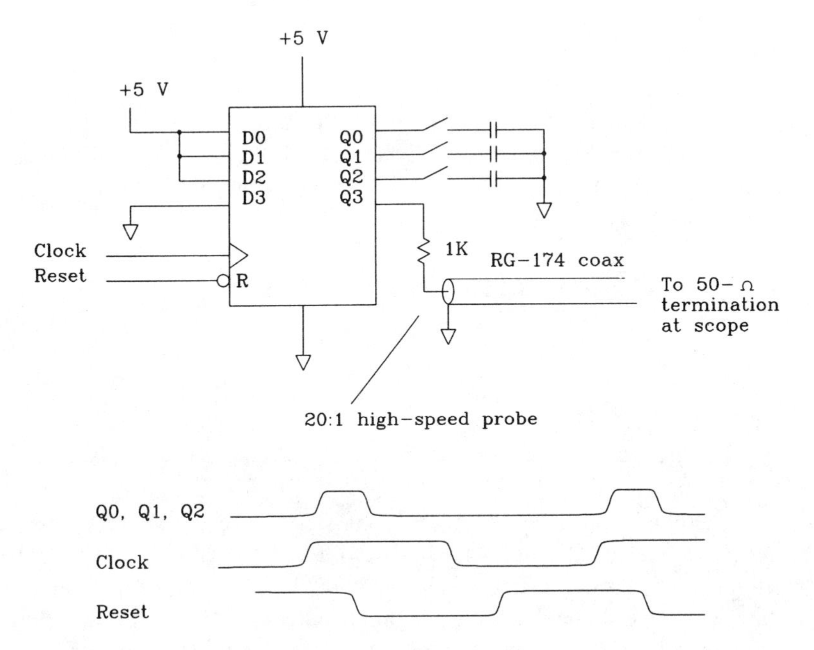

Figure 2.18 Measuring ground bounce.

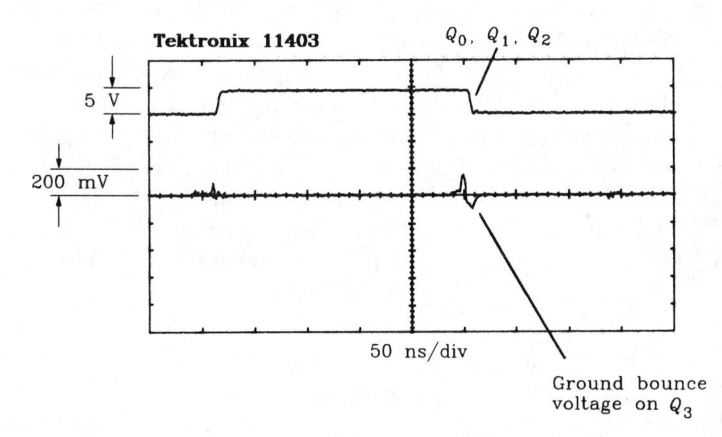

Figure 2.19 Ground bounce on 74HC174 with three loads of 20 pF.

For a resistive load R, we can use Equation 2.41 to find the rate of change in current and then the definition of inductance (Equation 1.17) to compute the ground bounce amplitude:

$$|V_{GND}| = L \frac{\Delta V}{T_{10-90}} \frac{1}{R} \quad [2.48]$$

For a capacitive load C, we can use Equation 2.42 to find the rate of change in current and then the definition of inductance (Equation 1.17) to compute the ground bounce amplitude:

$$|V_{GND}| = L\frac{1.52\,\Delta V}{T_{10-90}^{\ 2}}C \qquad [2.49]$$

The factors ΔV and T_{10-90} depend on the logic family. Here are typical figures.

Table 2.2 compares the switching characteristics of five logic families: Signetics 74HCT CMOS, Texas Instruments 74AS TTL, Motorola 10KH ECL, GigaBit Logic 10G GaAs, and NEL GaAs.[15]

TABLE 2.2 SWITCHING CHARACTERISTICS OF FIVE LOGIC FAMILIES

	74HCT CMOS	74AS TTL	10KH ECL	10G GaAs	NEL GaAs
ΔV_{max} (V)	5	3.7	1.1	1.5	1.0
T10–90 (ns)	4.7	1.7	0.7	0.15	0.05

Ground lead inductance is a strong function of the package type. Larger packages have more lead inductance. Packages with internal ground planes do better but do not eliminate the ground bounce problem. Wide, low-inductance internal ground plane structures still have skinny leads connecting the internal ground plane to external ground.

The most promising techniques for dramatically reducing lead inductance are wire bond, tape automated bonding (TAB), and flip-chip. All three techniques shorten the ground wire connections between the chip and its printed circuit board. See Figure 2.20. For an excellent overview of modern packaging techniques, see footnote.[16]

The wire bond method places an unsealed die on its back on a printed circuit board and welds tiny bonding wires between the chip pads and the printed circuit board. The chip and its wire bonds are then sealed with a blob of coating material or covered with a hermetically sealed lid over the entire circuit board.

Wire bonding is a mechanically simple method with plenty of tolerance for changes in either the chip bonding pad locations or printed circuit board wiring. Wire bonding can be done by hand for very low-volume applications.

Tape automated bonding replaces the wire bonds with a mass termination technique. Interconnecting wiring used to connect the chip to a printed circuit board is first printed on a very thin flexible substrate (a flex circuit). This substrate may have more than one layer, including a ground layer for impedance control. Solder bumps are then placed on the chip bonding pads, and the chip reflow-soldered to the flex circuit. The chip now has the flex circuit bonded to its face. As a second step, the combination chip and flex circuit is reflow-soldered to the printed circuit board. The result is then sealed with a blob of coating material or covered with a hermetically sealed lid over the entire circuit board.

[15]At the time of writing, NEL holds the record among commercially available digital logic families for switching speed. For more information contact KBK, Inc., New York, New York.

[16]H. B. Bakoglu, *Circuits, Interconnections, and Packaging for VLSI,* Addison Wesley, Reading, Mass., 1990.

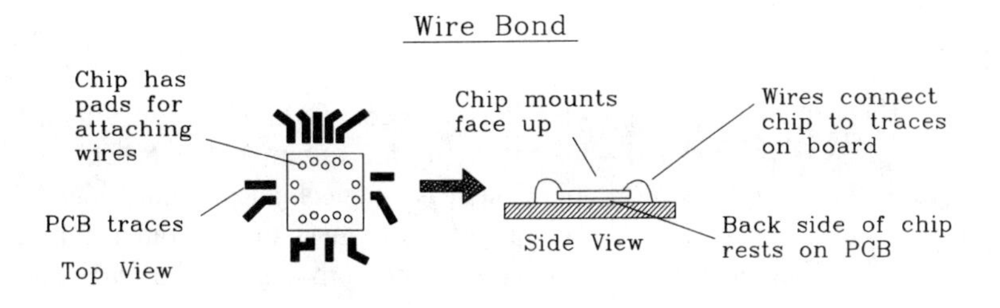

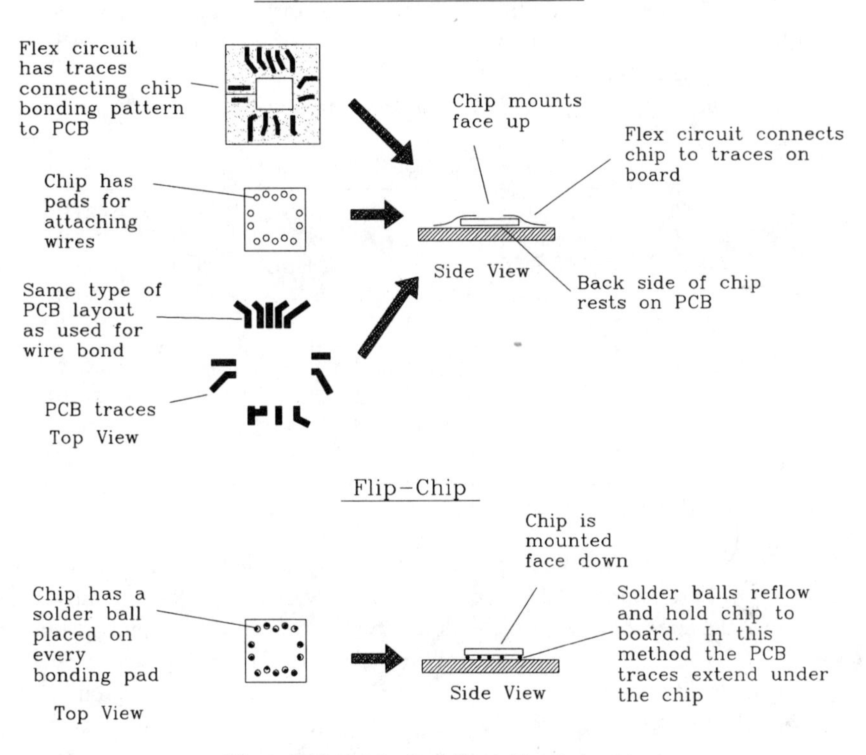

Figure 2.20 Methods of direct chip attachment.

Tape automated bonding, being a mass attachment technique, is very quick. It has the advantages of providing a continuous ground plane for all signals and also providing some mechanical compliance between the chip and printed circuit board. TAB can accommodate lead spacings as small as 0.08 mm (300 leads/in.). The disadvantages of TAB are that a different flex circuit is needed for each chip and that the flex circuit must change if either the printed circuit board or chip bonding patterns change.

Flip-chip technology first places solder balls on each chip attachment pad. The chip is then turned face down onto the printed circuit board and directly reflow-soldered in place. Flip-chip mounting is often used on ceramic multichip modules which incorporate advanced cooling structures and an overall hermetic seal around the entire enclosure.

Electrically, flip-chip technology is ideal. The bonding lengths are miniscule, and so all parasitics associated with packaging are minimized. Mechanically and thermally, flip-chip technology is miserable. There is no mechanical compliance between chip and printed circuit board except the limited springiness of the solder balls themselves. The thermal coefficient of expansion between the chip and printed circuit board must match extremely closely.

Cooling difficulties are exacerbated with the flip-chip method because the chip substrate is held up off the printed circuit board. In both the wire bond and TAB methods, the chip mounts with its back side touching (often glued to) the printed circuit board which provides a good conduit for heat dissipation.

Table 2.3 lists typical lead inductance figures for various packages.

TABLE 2.3 LEAD INDUCTANCE OF LOGIC PACKAGES*

14-pin plastic dual in-line package (DIP)	8 nH
68-pin plastic DIP	35 nH
68-pin surface-mount plastic leaded chip carrier (PLCC)	7 nH
Wire bonded to hybrid substrate	1 nH
Solder bump to hybrid substrate	0.1 nH

*Much of this data is taken from H. B. Bakoglu, *Circuits, Interconnections, and Packaging for VLSI,* Addison Wesley, Reading, Mass., 1990, Table 6.2. Reprinted by permission of Addison-Wesley Publishing Co., Inc., Reading, MA.

2.4.1.5 Factors that Reduce Ground Bounce

Slowing down the output switching time is a good idea. The 10K ECL family, the CMOS FCT family, and a few newer bus drivers incorporate circuitry designed to slow down the edge transition time with minimal impact on overall propagation delay.

Some manufacturers put multiple ground wires on their packages. This is a also good idea if the grounds are spaced evenly around the die. If the grounds are all near each other, going from one to two grounds nearly halves the ground inductance, but increasing the number of nearby grounds beyond two has a diminishing effect. Spreading the grounds out evenly around the chip is much better than lumping them together.

Components which bring out a separate ground reference pin for the input circuitry attack the problem in a more subtle way. These circuits, like the 10K family, provide a separate sense wire for the internal reference voltage generator which has a direct path to the external ground. This pin does not carry large ground currents and subsequently acquires no ground bounce. This is an excellent method of attacking ground bounce problems. For chips with separated grounds, make sure each ground wire has a direct path to the ground plane. Connecting the two grounds together and then running them through a trace to ground defeats the purpose of having independent ground leads.

Differential inputs are a similar and even more effective means of achieving the same end.

2.4.2 Lead Capacitance

Stray capacitance between adjacent pins of a logic device can couple noise voltages onto sensitive inputs. Figure 2.21 depicts a situation where mutual capacitance C_M couples together pins 1 and 2 of a logic device.

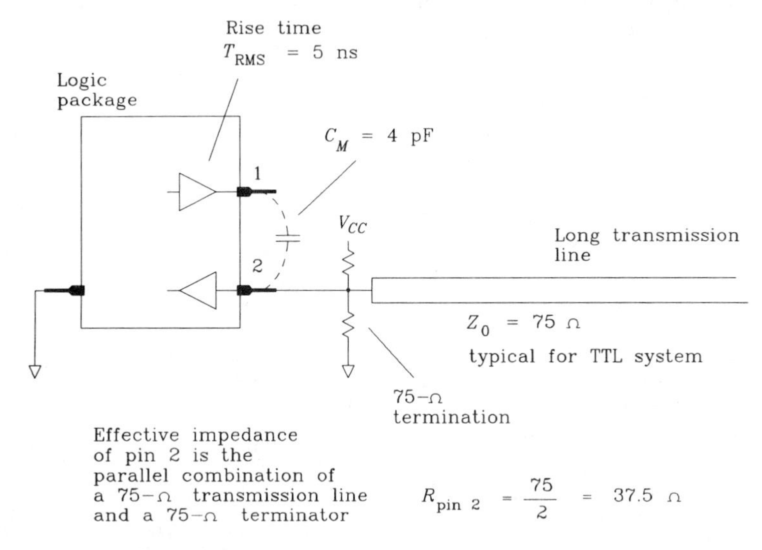

Figure 2.21 Stray capacitance between pins on a logic package.

We may compute the percentage crosstalk introduced in circuit 2 by circuit 1 using Equation 1.30:

$$\text{Crosstalk} = \frac{R_2 C_M}{T_{10-90}} \qquad [2.50]$$

where $C_M = 4\ pF$ (mutual capacitance of circuits 1 and 2)
$R_2 = 37.5\ \Omega$ (parallel impedance of 75-Ω long transmission line and 75-Ω terminator)
$T_{10-90} = 5$ ns (voltage rise time of signal on pin 1)

In this example the crosstalk is 0.03 (3%).

The capacitive crosstalk problem becomes more serious as rise times get shorter. It also grows worse with higher impedance input connections.

Figure 2.22 illustrates the high-impedance input problem. The ASIC in Figure 2.22 generates a clock and also debounces a switch input. Without C_1 and C_2 the impedance of R_1 and R_2 are so high that we expect capacitive crosstalk to be a problem. Using Equation 2.50 we obtain a ridiculous crosstalk factor of 8. This means practically all the clock signal from pin 1 will appear on pin 2.

Capacitors C_1 and C_2 reduce the impedance of the receiving circuit at high frequencies, heading off capacitive crosstalk problems. When dealing with capacitive loads on a receiving circuit, the percentage of crosstalk is just equal to the ratio of the capacitances:

$$\text{Crosstalk} = \frac{C_M}{C_1} \qquad [2.51]$$

With C_1 set to 0.01 μF, we get crosstalk of only 0.0004. This amount of crosstalk is immaterial. Checking the time response of R_1C_1 we get a time constant of 0.1 ms. No one operating the switch will know the difference.

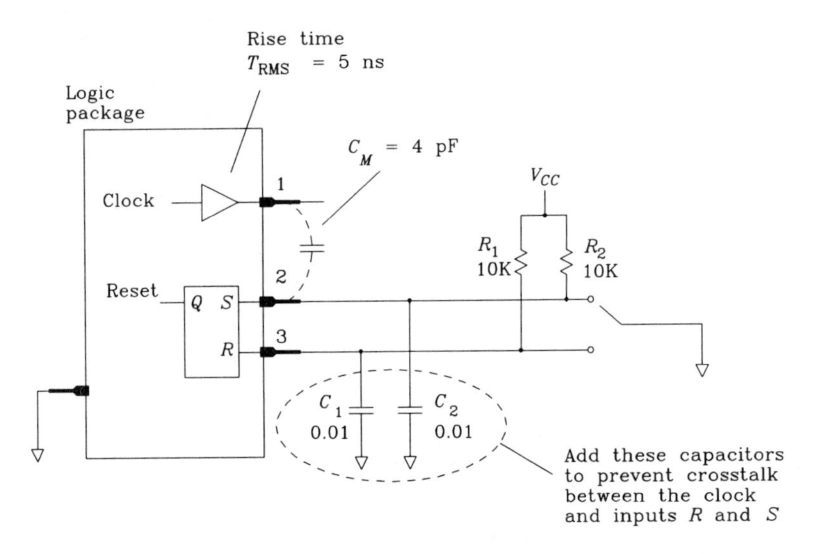

Figure 2.22 Fixing a stray capacitance problem on a debouncing circuit.

Table 2.4 lists order-of-magnitude figures for the adjacent lead capacitance of various packages.

TABLE 2.4 ADJACENT LEAD CAPACITANCE OF LOGIC PACKAGES*

14-pin plastic dual in-line package (DIP)	4 pF
68-pin surface-mount plastic leaded chip carrier (PLCC)	7 pF
Wire bonded to hybrid substrate	1 pF
Solder bump to hybrid substrate	0.5 pF

*Much of this data is from H. B. Bakoglu, *Circuits, Interconnections, and Packaging for VLSI,* Addison Wesley, Reading, Mass., 1990, Table 6.2.

2.4.3 Heat Transfer—Θ_{JC} and Θ_{CA}

Instead of getting too theoretical, let's start with an experiment to see how power dissipation and temperature interrelate. Rig up a 14-pin DIP package with nothing but a 1-Ω resistor body inside.[17] Wire the resistor to pins 7 and 14. Also wire up a temperature sen-

[17]Slice the package in half, dig out the internal circuitry to make room for your resistor and temperature sensor, and glue the thing back together.

sor to pins 1 and 2 so we can see the temperature inside the package. By varying the voltage across pins 7 and 14, we can control the power dissipation internal to the package.

Next mount the package in still air inside a temperature-controlled chamber, set the ambient temperature to 86°F (30°C), and set the power dissipation to zero. A few minutes later, the temperature inside the package will level out to 30°C.

Next make a plot of internal temperature versus power dissipation. Allow the circuit time to come to thermal equilibrium between each reading. Figure 2.23 shows the result.

The data points make a straight line, indicating we get a temperature rise of 83°C per watt of dissipation. This linear relation between temperature and power is typical of all logic device packages.

Figure 2.24 shows the results starting at different ambient temperatures of 30, 70, and 110°C. The slope of the temperature curves are the same in all cases—we have just

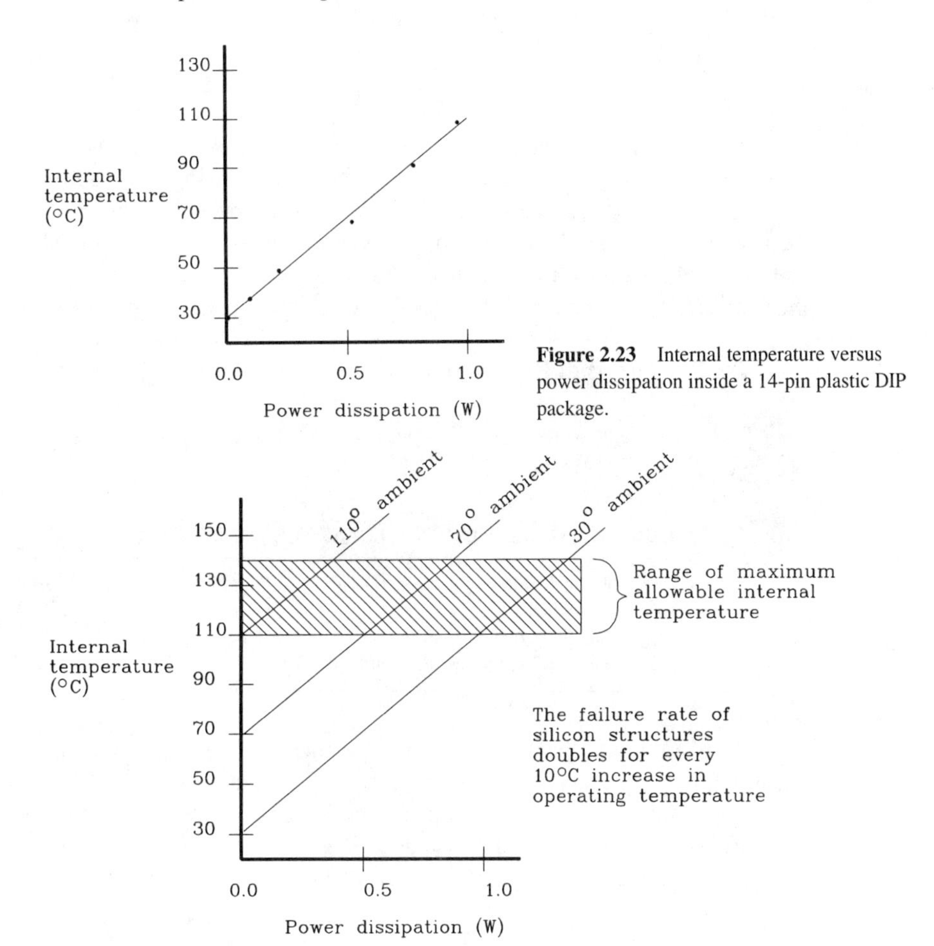

Figure 2.23 Internal temperature versus power dissipation inside a 14-pin plastic DIP package.

Figure 2.24 Internal temperature versus power dissipation and ambient temperature (14-pin plastic DIP package).

offset their starting points. From these observations we may synthesize a general formula for predicting the internal temperature of a logic device. The internal temperature (called the junction temperature) is equal to the ambient temperature plus an offset proportional to the internal power dissipation P.

$$T_{\text{junction}} = T_{\text{ambient}} + \Theta_{JA} P \qquad [2.52]$$

The constant of proportionality Θ_{JA} is called the thermal resistance, from junction to ambient, of the package. The constant Θ_{JA} is a property of the die attachment method internal to the package, the package material, the package size, and any special heat-dissipating features attached to the package such as fins or wings.

Sometimes a manufacturer will separate the thermal resistance into component pieces relating to the internal workings of the package and its method of mounting. The most common partition accounts separately for the temperature rise from junction to case and from the package case to the outside ambient environment.

$$\Theta_{JA} = \Theta_{JC} + \Theta_{CA} \qquad [2.53]$$

Manufacturers make this partition for us because we can usually do nothing to affect Θ_{JC}, but we can do plenty to affect Θ_{CA}. Manufacturers of add-on heat sinks provide detailed literature and technical reports on improvements in Θ_{CA} achievable with their products. To predict the maximum internal junction temperature when using a special heat sink, we must find Θ_{JC} from the manufacturer of the device package, Θ_{CA} from the heat sink manufacturer, and the total device power dissipation by our own calculation.

2.4.3.1 Thermal Resistance—Junction to Case

Table 2.5 lists some typical figures for Θ_{JC} (junction-to-case resistance) of various packages.[18,19]

TABLE 2.5 Θ_{JC} JUNCTION TO CASE THERMAL RESISTANCE

Package	Θ_{JC}
16-pin plastic dual in-line package (DIP)	34°C/W
16-pin ceramic DIP	25°C/W
40-pin ceramic leaded chip carrier (LCC) with 10K square mil die	5.5°C/W
132-pin ceramic LCC with 50K square mil die	1.4°C/W

The bigger the package, the smaller the thermal resistance. This makes sense because a larger package has a greater surface area and conducts heat more effectively to its surface. Note that for the larger packages, we specify the die size. These are about the largest dies that will fit in each listed package. Smaller dies in the same package have greater thermal resistances, as they have a smaller area of surface contact between die and package.

[18] *Motorola MECL System Design Handbook,* Motorola Inc., Phoenix, Ariz., 1988, p. 111.

[19] *GigaBit Logic Standard Cell Array Design Manual,* GigaBit Logic, Newbury Park, Calif., 1989, pp. 5–7.

Engineers specifying packaging for logic devices become interested in the thermal resistance of the die itself, the bond between die and package, any conductive or heat-spreading materials embedded in the package, the package material (ceramic conducts heat better than plastic), and the package physical configuration (flat, thin packages are better than thick, squat ones).

2.4.3.2 Thermal Resistance—Case to Ambient

Table 2.6 lists some typical figures for Θ_{CA} (case-to-ambient resistance) for various packages.[20,21] Air speed near the package has a strong impact on heat conduction. Air speed assumptions are listed along with the entry for each package type.

TABLE 2.6 Θ_{CA} CASE TO AMBIENT THERMAL RESISTANCE

16-pin dual in-line package (DIP) in still air	80°C/W
16-pin dual in-line package (DIP) in 400 ft/min air flow	35°C/W
72-pin ceramic pin grid array (PGA) in still air	34°C/W
72-pin ceramic pin grid array (PGA) in 400 ft/min air flow	18°C/W
72-pin ceramic pin grid array (PGA) in 400 ft/min air flow with heat sink	10°C/W

Figure 2.25 plots the composite junction-to-ambient thermal resistance of a Motorola 72-pin grid array package as a function of ai: flow. The composite thermal resistance from junction to ambient Θ_{JA} decreases with increasing air flow both with and without the Motorola recommended heat sink. Both curves include a fixed value Θ_{JC} of 4°C/W. Most heat sinks show a similar increase in efficiency versus air flow velocity.

Figure 2.26 is a scatter plot showing the thermal resistance from case to ambient Θ_{CA} for all heat sinks manufactured by Thermalloy, Inc., of Dallas, Texas. The two curves show how their heat sinks work in still air and with an air flow of 1000 ft/min. The inescapable conclusion of this scatter plot is that it takes a large volume of material

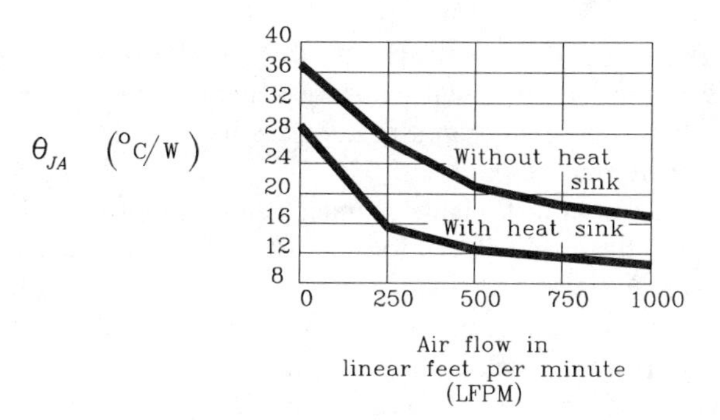

Figure 2.25 Typical thermal resistance versus air flow for a Motorola 72-lead pin grid array (PGA package). (Data courtesy of Motorola Inc.)

[20] *Advanced Low Power Schottky, Advanced Schottky Logic Data Book,* Texas Instruments, Dallas, Tex., 1986, pp. 1–14.

[21] *MCA800ECL/MCA2500ECL Macrocell Array Design Manual,* Motorola, Phoenix, Ariz., 1986, p. 36.

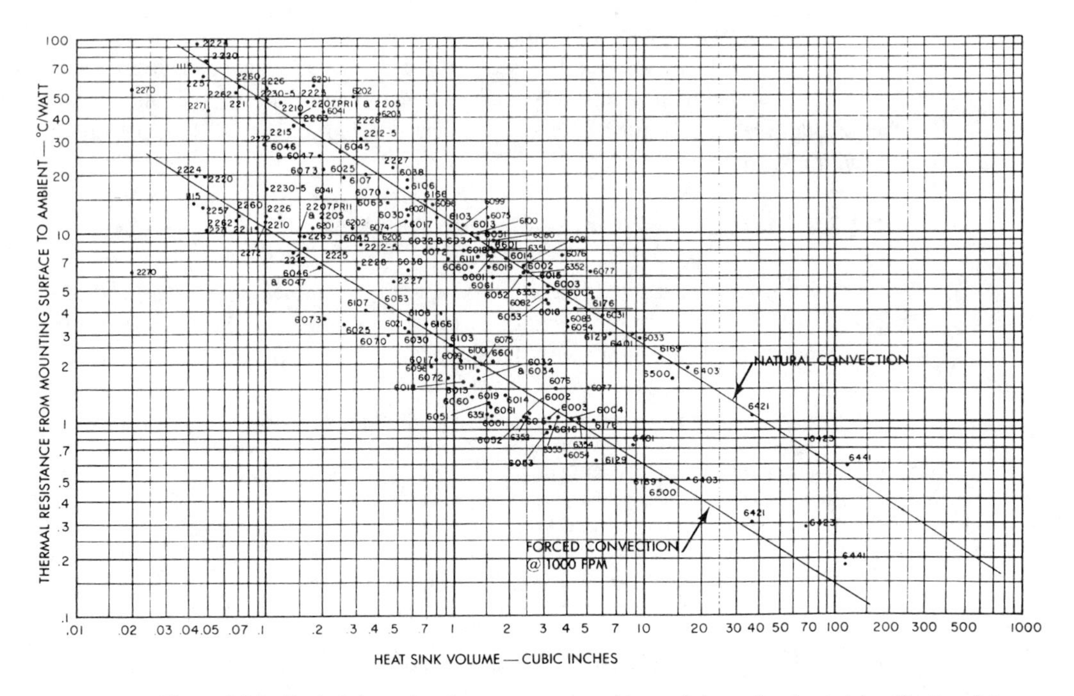

Figure 2.26 Typical thermal resistance, case to ambient, of thermalloy heat sinks. (Scatter plot courtesy of Thermalloy, Inc.)

in contact with the air to dissipate large amounts of heat. Increased air flow lowers the thermal resistance, but not as much as using a bigger heat sink.

The slope of the scatter plot is about –2/3. This means that the heat sink efficiency goes up as the 2/3 power of volume, which is the same as the square of linear size. Making a heat sink 40% larger in every dimension halves the thermal resistance.

2.4.3.3 How Much Air is 400 ft/min?

Manufacturers often glibly quote heat sink specifications at 400 ft/min, or above, as evidence of their product's superior performance under standard operating conditions. Unfortunately, 400 ft/min is a lot more air than a heat sink is likely to receive without meticulous attention to air flow design.

To convert air flow expressed in feet/minute to miles/hour, multiply by 0.0114, or about 1/100. An air flow of 400 ft/min works out to 4.5 mi/hr. On land, this is a gentle breeze, but in the restricted air space inside a computer chassis it takes a sizeable fan to move this much air. A large fan is required because as air spreads out from the fan, it slows down. Also, air flow in confined spaces develops turbulent whirlpools and dead zones. We must blow extra air into a cavity to ensure that at every point inside we attain some minimum acceptable air speed.

The fans inside a typical personal computer chassis generate about 150 ft/min.

POINTS TO REMEMBER:

At high speeds, the inductance of logic device packaging is critical.

Output switching currents flowing through a ground pin cause ground bounce, which can cause double-clocking of flip-flops.

Thermal resistance is the ratio of temperature rise to power dissipation.

Heat flows from a silicon die to its package, and from the package to the ambient surroundings: $\Theta_{JA} = \Theta_{JC} + \Theta_{CA}$

400 ft/min is a lot of air.

3

Measurement Techniques

All scientific instruments have limitations. When using an oscilloscope to probe inside a digital machine, as with any instrument, we must learn to live with its limitations and to account for them in our analysis of the results.

3.1 RISE TIME AND BANDWIDTH OF OSCILLOSCOPE PROBES

Three primary limitations of oscilloscope systems are inadequate sensitivity, insufficient range of input voltage, and limited bandwidth.

For all but the most sensitive digital work, we are well above the minimum sensitivity level of any reasonable oscilloscope. On the high side, digital signals being less than 5 V,[1] we are well within the maximum voltage range of most oscilloscope inputs. The most serious remaining limitation is bandwidth.

Your oscilloscope vertical amplifier undoubtedly has a bandwidth rating, as does your oscilloscope probe. What do these numbers mean? Few engineers would try using a 100-MHz scope on 200-MHz digital signals, but how about using it on 99-MHz signals? What precisely does *bandwidth* mean, and how does it affect digital signals?

Figure 3.1 gives us a clue. The two traces in Figure 3.1 depict precisely the same signal viewed with two probes having widely different bandwidths. The top trace rises quickly, while the bottom trace rises much more slowly. The top trace was recorded with a very fast rise-time probe, and the other with a probe having a limited bandwidth of 6-MHz. The 6-MHz probe, originally manufactured as a noise-filtering, very-high-impedance input probe, exaggerates the differences you will see among practical digital probes. The lower bandwidth probe slurs out and slows down both rising and falling edges of digital signals. In signal processing terminology, the lower bandwidth probe filters out high-frequency components of the signal under test.

[1]This will remain true as long as vacuum tubes stay out of favor.

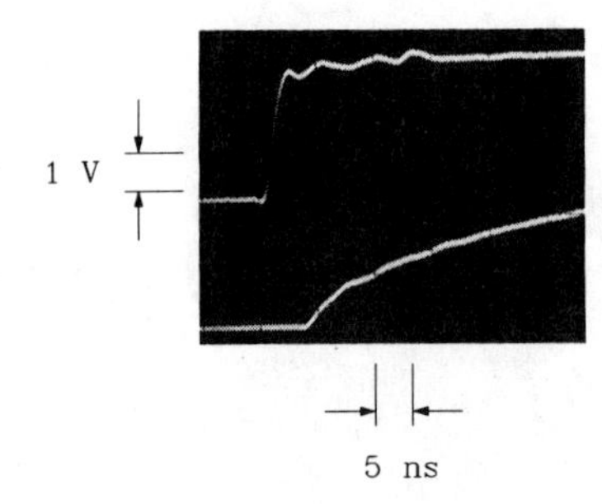

Figure 3.1 Two probes having different bandwidth display the same signal differently.

Figure 3.2 breaks apart the oscilloscope system into component pieces, separately showing the input signal, the probe, and the vertical amplifier. In Figure 3.2, a perfect signal with a razor-thin rising edge separately feeds each stage, so that we may directly observe the distortion caused by each piece of the system. Both the probe and the vertical amplifier do the same thing: They degrade the rise time of their input signals.

Figure 3.2 quantifies the rise-time degradation of each processing stage separately.

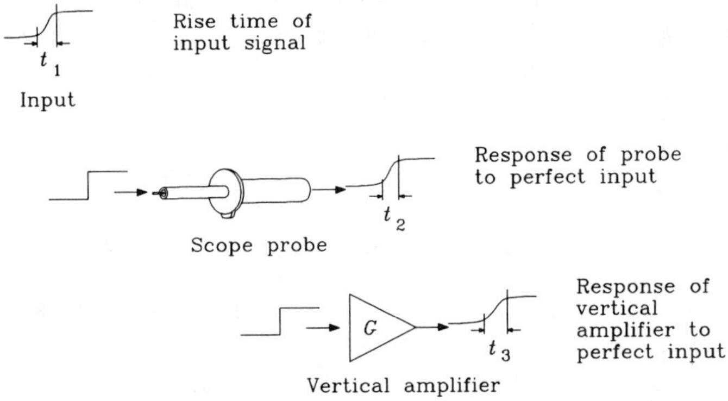

t_1
Input
Rise time of input signal

t_2
Scope probe
Response of probe to perfect input

G
t_3
Vertical amplifier
Response of vertical amplifier to perfect input

Figure 3.2 Rise time of oscilloscope components.

When a realistic input feeds the combination of probe and vertical amplifier, as in Figure 3.3, the rise time of the result is equal to the square root of the sum of the squares of the rise times of each component.

$$T_{\text{rise composite}} = \left(T_1^{\,2} + T_2^{\,2} + \cdots + T_N^{\,2}\right)^{1/2} \qquad [3.1]$$

Whenever processing stages connect in series, the squares of their rise times add. For this example, an appropriate measure of rise time would be the 10–90% rise time.[2]

Oscilloscope manufacturers commonly quote the *3-dB bandwidth,* $F_{3\text{dB}}$, of probes and vertical amplifiers instead of rise time. The conversions between 3-dB bandwidth and 10–90% rise time are as follows (see also Equation 1.6). *These approximations assume*

[2]Equation 3.1 holds strictly true only when each impulse response in Figure 3.2 is gaussian. For other impulse response shapes Equation 3.1 is very close to true, but not exact. See Appendix B for more information about the exact calculation of rise time in cascaded systems.

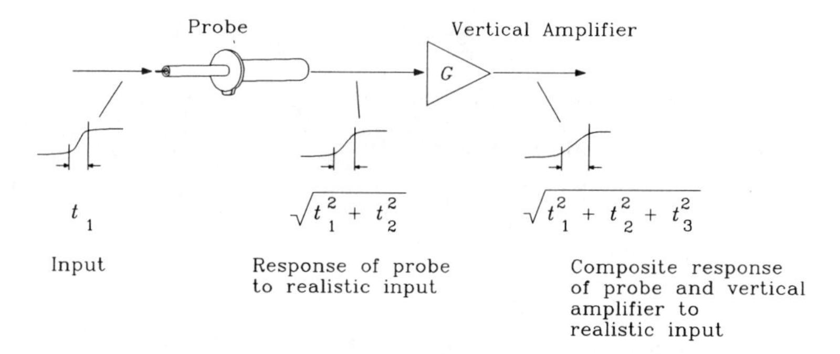

Figure 3.3 Composite rise time of an oscilloscope system.

the frequency response of a probe, being a combination of several random filter poles near each other in frequency, is gaussian.

$$T_{10-90} = \frac{0.338}{F_{3\text{dB}}} \qquad [3.2]$$

Specifications from instrument manufacturers quoting RMS bandwidth, or equivalent noise bandwidth F_{RMS}, should be converted according to (see also Equation 1.7)

$$T_{10-90} = \frac{0.361}{F_{\text{RMS}}} \qquad [3.3]$$

In some of our work with shop-built probes, we will analyze the performance of simple low-pass filters. These filters do not have gaussian frequency response curves. In these cases, the 10–90% rise time of the circuit is related to the filter time constant in this way:

For a *LR* low-pass filter:

$$T_{10-90} = 2.2\frac{L}{R} \qquad [3.4]$$

For a *RC* low-pass filter:

$$T_{10-90} = 2.2RC \qquad [3.5]$$

For a two-pole *RLC* filter near critically damped:

$$T_{10-90} = 3.4(LC)^{1/2} \qquad [3.6]$$

EXAMPLE 3.1: Rise Time Degradation

Bob buys an oscilloscope rated at 300 MHz and a probe also rated at 300 MHz. Both specifications are 3-dB bandwidths. How will this combination affect signals having 2-ns rise times?

$$T_{r\ \text{scope}} = 0.338/300\ \text{MHz} = 1.1\ \text{ns}$$
$$T_{r\ \text{probe}} = 0.338/300\ \text{MHz} = 1.1\ \text{ns}$$
$$T_{r\ \text{signal}} = 2.0\ \text{ns}$$

$$T_{\text{displayed}} = (1.1^2 + 1.1^2 + 2.0^2)^{1/2} = 2.5 \text{ ns} \quad [3.7]$$

Bob's scope will display the 2.0-ns actual rise time with a 2.5 ns apparent rise time.

EXAMPLE 3.2: Calculating Input Rise Time

If Bob's scope displays a 2.2-ns rising edge, can you guess the actual input rise time?

Invert the rise-time formula (Equation 3.1) to find what rise time would cause a 2.2-ns displayed result.

$$T_{\text{actual}} = (2.2^2 - 1.1^2 - 1.1^2)^{1/2} = 1.6 \text{ ns} \quad [3.8]$$

The actual 10–90% rise time of the displayed 2.2-ns signal must be 1.6 ns.

Please do not take this example too seriously. It is accurate only if the input waveform has no overshoot, if the 10–90% rise time of the equipment is accurately known, and if measurements are made precisely under conditions of no noise. A better approach to finding the rise time would be to use a faster probe or a faster scope.

In a pinch, this technique can extend the usefulness of Bob's scope by a factor of 2 or 3.

POINT TO REMEMBER:

When figuring a composite rise time, the squares of 10–90% rise times add.

3.2 SELF-INDUCTANCE OF A PROBE GROUND LOOP

The primary factor degrading the performance of ordinary 10:1 oscilloscope probes, when used for digital electronics, is the inductance of their ground wire. Manufacturers report probe performance as measured using a test jig connected directly to the probe tip and outer probe shield. *No ground wire is used for the probe bandwidth measurement.* Since digital engineers commonly use probes with a plastic clip covering the probe tip, and a ground wire connected to the middle of the probe barrel, we should investigate how these modifications affect probe performance.

Figure 3.4 shows a typical probe arrangement. The probe tip connects to the circuit under test, and the ground lead connects the probe barrel to a convenient local ground reference point. Note that the ground connection is made using a thin wire several inches in length.

Examine the equivalent circuit for this probe arrangement, drawn at the bottom of Figure 3.4. Here we assume the probe has an input impedance of 10 pF shunted by 10 MΩ.[3] As you can see, currents flowing into the probe must traverse the ground loop

[3]A typical value for oscilloscopes.

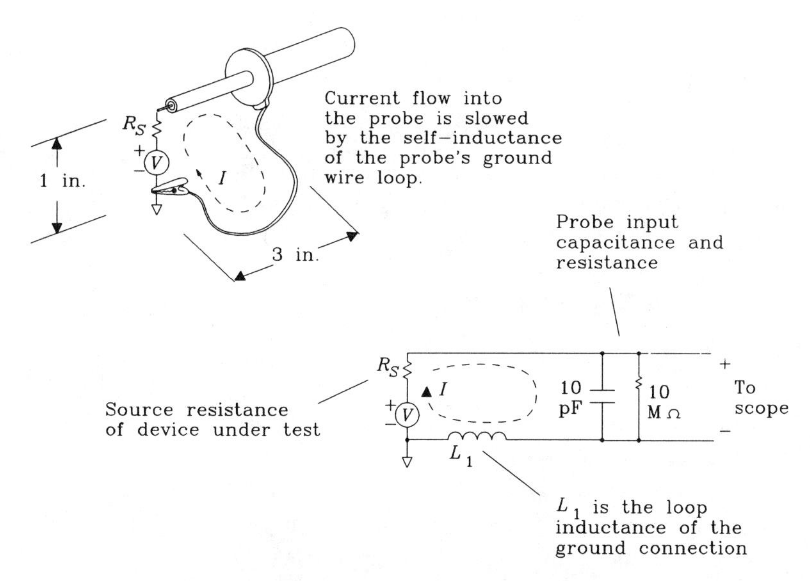

Figure 3.4 Electrical model of an oscilloscope probe.

on the way back to their source. The self-inductance of the ground loop, represented in our schematic by series inductance L_1, impedes these currents.

How does the inductance L_1 affect our measurements? The reactance of L_1, working into the load impedance of the probe input, has a finite rise time. We will calculate the value of L_1, find the 10–90% rise time, and then discuss its significance.

3.2.1 Calculating Ground Loop Inductance

The dimensions of the ground loop in Figure 3.4 are 1 in. × 3 in. A typical ground wire size for this type of probe is American Wire Gauge (AWG) 24, having a diameter of 0.02 in. Using the inductance formula from Appendix C for the case of a rectangular loop, the resulting inductance is

$$\begin{aligned} L &\approx 10.16\left[1\ln\left(\frac{2\times3}{0.02}\right)+3\ln\left(\frac{2\times1}{0.02}\right)\right]\text{nH} \\ &\approx 200\text{ nH} \end{aligned} \quad [3.9]$$

3.2.2 Finding the 10–90% Rise Time

The LC time constant for this circuit is

$$C = 10\text{ pF}$$
$$L = 200\text{ nH}$$
$$T_{LC} = (LC)^{1/2} = 1.4\text{ ns} \quad [3.10]$$

The 10–90% rise time for a critically damped two-pole circuit[4] of this type is 3.4 times the LC time constant:

$$T_{10-90} = 3.4T_{LC} = 4.8 \text{ ns} \qquad [3.11]$$

This 4.8-ns rise time is an indication of trouble. Note that in Example 3.1 we found that a 300-MHz-rated probe should have a 10–90% rise time of 1.1 ns. Here we see that the 3-in. ground wire has already degraded our 10–90% rise time to 4.8 ns.

3.2.3 Estimating the Circuit *Q*

Figure 3.4 includes a resistor in series with the signal source. This resistor models the output impedance of whatever gate is driving the signal under test. For TTL or high-performance CMOS drivers, this source impedance is about 30 Ω. For ECL systems (silicon or GaAs), the output impedance is about 10 Ω.

The Q, or resonance, of this LC circuit is seriously affected by the source resistance of the signal under test. The series combination of L, C, and the source resistor R_S makes a series resonant circuit having a Q of approximately

$$Q \approx \frac{(L/C)^{1/2}}{R_S} \qquad [3.12]$$

In Equation 3.12, Q is the ratio of energy stored in the loop to energy lost per radian during resonant decay. A high-Q circuit rings for a long period after excitation from an outside source. This resonance shows up as a large peak in the frequency response of the circuit.

In the circuit in Figure 3.4, as we reduce the source resistance R_S, the LC filter develops a big resonance near 100 MHz. The frequency response plots in Figure 3.5 illustrate this effect for source resistances of 5, 25, and 125 Ω.

A 5-Ω source resistance gives a 29-dB resonance. Digital signals having cutoff frequencies higher than 100 MHz will be very much distorted by this probe circuit.

A 25-Ω source resistance gives a 15-dB resonance. Digital signals having cutoff frequencies higher than 100 MHz will be distorted by this probe circuit.

The 125-Ω plot shows critical damping (Q = 1). Source resistances near 125 Ω yield the best frequency response with this probe.

Digital signals having knee frequencies lower than 100 MHz exhibit no artificial ringing or overshoot when displayed with the probe illustrated in Figure 3.4. Equation 1.1 tells us we need a rise time longer than 5 ns to guarantee a cutoff frequency below 100 MHz:

$$\text{Rise time} > \frac{0.5}{100 \text{ MHz}} = 5 \text{ ns} \qquad [3.13]$$

[4] An underdamped circuit with the same L and C has an even slower rise time.

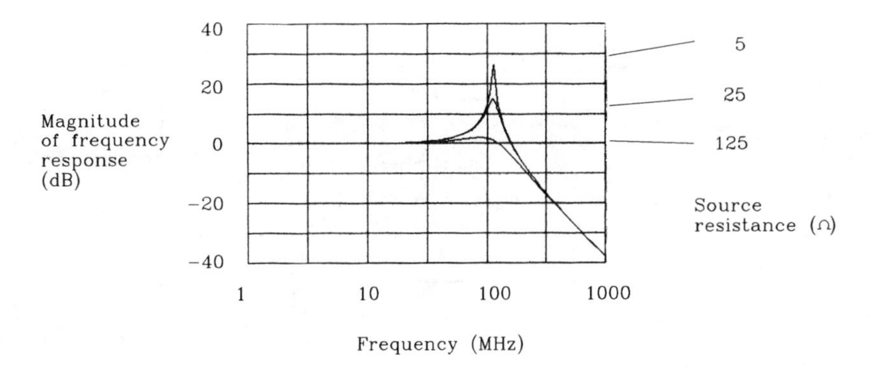

Figure 3.5 Frequency response of a probe with a ground wire.

The 5-ns rise-time limit is a function of this particular probe arrangement having 200 nH of loop inductance and a 10-pF shunt capacitance.

3.2.4 Significance of Results

We predict that probes having ground wires, when used to view very fast signals from low-impedance sources, will display artificial ringing and overshoot.

Figures 3.6 and 3.7 compare our prediction to actual measurements. These experiments were conducted using a very-low-capacitance FET input probe, rated at 1.7-pF

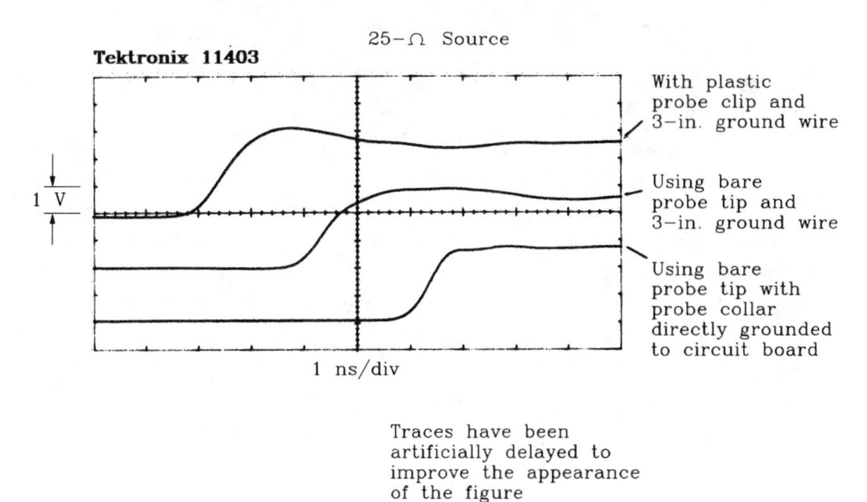

Figure 3.6 Ringing induced in a 1.7-pF probe by a 3-in. ground wire when viewing a 25-Ω source.

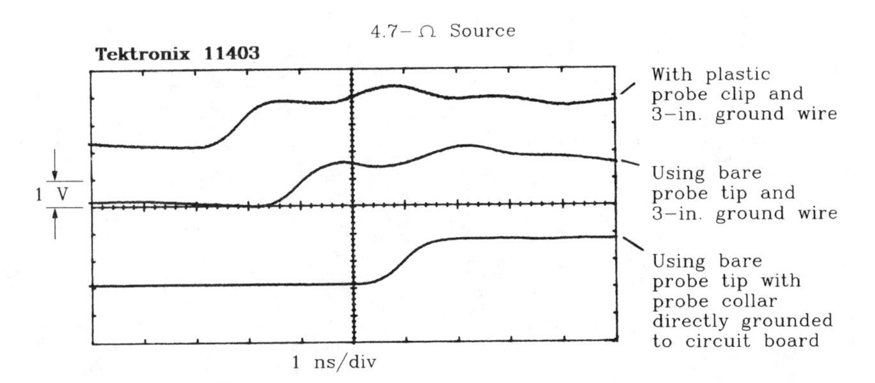

Figure 3.7 Ringing induced in a 1.7-pF probe by a 3-in. ground wire when viewing a 4.7-Ω source.

shunt capacitance and 1 GHz 3-dB bandwidth, connected to a Tektronix 11403 digital sampling scope. In Figure 3.6 the source impedance is 25 Ω, while in Figure 3.7 the source impedance is 4.7 Ω. The top trace in each photo is recorded using a standard plastic probe clip covering the probe tip, and a 3-in. ground wire. The middle trace is recorded with the bare probe tip touching the signal under investigation, and a 3-in. ground wire. Evidently, taking off the plastic probe clip has little effect. These traces show approximately 15% overshoot in the 25-Ω case and a much larger 29% overshoot in the 5-Ω case.

The ringing periods evident in the figures are in the range 2–6 ns. A quick check of the expected *LC* time constant reveals:

$$T_{LC} = (LC)^{1/2} = (200 \text{ nH} \times 2 \text{ pF})^{1/2} = 0.63 \text{ ns} \qquad [3.14]$$

The ringing period predicted by an *LC* time constant of 0.63 ns is

$$\text{Expected period} = 2\pi T_{LC} = 4.0 \text{ ns} \qquad [3.15]$$

So far, the measured results and the theory line up pretty well. What about the bottom trace in each figure? Why is it better?

The bottom trace in each figure gives us a good hint as to how to solve the overshoot problem. In the bottom trace, we have removed the plastic barrel which holds together the ground wire assembly and removed the ground wire, exposing the metal shield which covers the probe all the way out to the bare probe tip. Then, using a small knife blade, we connect the metal probe shield directly to the circuit ground as near the signal sense point as possible (see Figure 3.8). This shorts the metal probe shield to ground with as little inductance as practical. Traces for both 25- and 5-Ω sources show much improvement in the overshoot when using this direct ground attachment method.

Why does grounding the probe close to the signal source help? The basic reason is that we have radically reduced the ground loop inductance of the probe assembly. Reducing the inductance reduces the probe rise time (Equations 3.10 and 3.11) and lowers the Q (Equation 3.12).

How small must we make the probe ground wire inductance to guarantee a low Q and a fast rise time? Is it possible to just use a shorter ground wire instead of both-

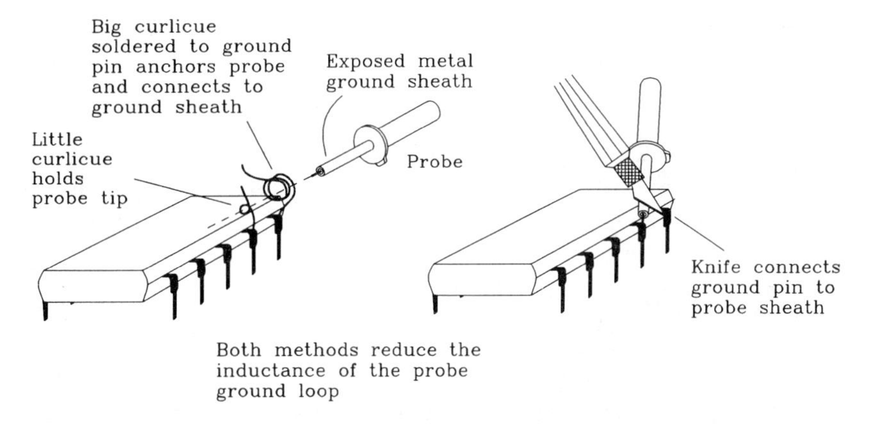

Figure 3.8 Methods for grounding a probe tip near a signal under test.

ering with the knife blade method? Table 3.1 lists the 10–90% rise time in nanoseconds and Q as a function of the ground loop inductance for both TTL (30-Ω) and ECL (10-Ω) measurements.

TABLE 3.1 EFFECT OF GROUND LOOP INDUCTANCE ON 10- AND 2-pF PROBE PERFORMANCE

Ground loop inductance (nH)	10-pF probe			2-pF probe		
	$T_{10\text{–}90}$	Q_{TTL}	Q_{ECL}	$T_{10\text{–}90}$	Q_{TTL}	Q_{ECL}
200	2.8	4.7	14.1	1.3	10.5	32.0
100	2.0	3.3	9.9	0.89	7.4	22.0
30	1.1	1.8	5.4	0.49	4.1	12.0
10	0.6	1.1	3.2	0.28	2.4	7.1
3	0.3	0.6	1.7	0.15	1.3	3.9
1	0.2	0.3	1.0	0.09	0.7	2.2

For a 10-pF probe, we would have to get the loop inductance down below 10 nH to obtain a low overshoot performance on TTL rise times of 1 ns. For ECL circuits, we would need even lower inductance.

For a reduced loop inductance, let's try replacing the ground wire in Figure 3.4 with a bigger wire. If the original wire was AWG 24, we can try AWG 18, which has twice the diameter. Reworking Equation 3.9 for this new ground lead,

$$L \approx 10.16\,[1\ \ln(3/0.02) + 3\ \ln(1/0.02)]\ \text{nH} \tag{3.16}$$

$$\approx 170\ \text{nH} \tag{3.17}$$

See how slowly inductance changes as a function of wire diameter? Doubling the wire diameter in this case makes only a 15% change in inductance. The slow variation in inductance as a function of wire diameter is a result of the logarithms which appear in the formula for inductance. To make a big difference in inductance (a factor of 10) we must increase the wire diameter until the two sides of the loop almost come in contact with each other.

Wire stiffness, on the other hand, is proportional to the cube of wire diameter and so goes up markedly with increases in diameter. The stiffness and inductance factors work against each other. There is no reasonable way to cure the probe inductance problem by just using a bigger wire.

Inductance is roughly proportional to loop area and to wire length. Common fixes for inductance problems involve shortening wires or reducing loop sizes rather than increasing wire diameter.

Table 3.1 shows that users of 2-pF probes obtain better rise times than with a 10-pF probe but encounter even higher Q problems when probing low-impedance signals.

POINTS TO REMEMBER:

A 3-in. ground wire used with a 10-pF probe induces a 2.8-ns 10–90% rise time. In addition, the response will ring when driven from a low-impedance source.

Fattening the ground wire hardly helps with ringing.

Radically shortening the ground loop improves ringing and reduces rise time.

3.3 SPURIOUS SIGNAL PICKUP FROM PROBE GROUND LOOPS

Any ground wire loop, in addition to lengthening the 10–90% rise time of a probe, picks up noise. Additive noise coupled through the probe ground loop masquerades as noise that is naturally present at the signal node under test. This additive noise, if synchronous with the signal under test, is difficult to separate from real features of the signal.

Figure 3.9 shows an integrated circuit in a dual in-line (DIP) package sending a digital signal to a load of 50 pF. The signal current loop is shown in heavy black lines. The changing current in loop A, acting through the mutual inductance of loops A and B, induces voltages in loop B.

We will first estimate the changing currents in loop A and then calculate the mutual inductance of A and B. Finally, we will use the definition of mutual inductance to find the noise voltage received by an oscilloscope under these conditions.

3.3.1 Changing Currents in Loop *A*

Assume the IC drive circuit conforms to Example 2.4. The maximum dI/dt is 7.0×10^7 A/s.

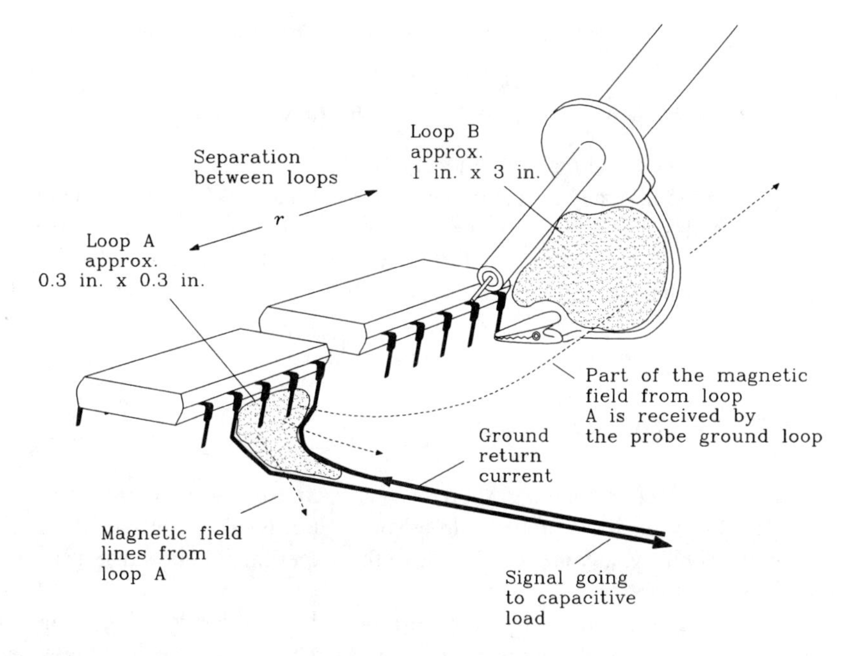

Figure 3.9 A probe ground loop picks up spurious noise voltages.

3.3.2 Mutual Inductance of Loops *A* and *B*

Dimensions for loops *A* and *B* appear in Figure 3.9, and so we need only apply the formula from Appendix C for the mutual inductance of two loops.

$$L_M = 5.08 \frac{A_1 A_2}{r^3} \quad [3.18]$$

$$= 5.08 \frac{(0.3 \times 0.3)(1 \times 3)}{2^3} \quad [3.19]$$

$$= 0.17 \text{ nH} \quad [3.20]$$

where A_1 = area of loop 1, in.^{2}
A_2 = area of loop 2, in.^{2}
r = separation of loops, in.
L_M = mutual inductance between loops 1 and 2, H

3.3.3 Apply the Definition of Mutual Inductance

The noise voltage induced in loop *B* is the product of the rate of change in current in loop *A* and the mutual inductance of loops *A* and *B*:

$$V_{\text{noise}} = L_M \frac{dI}{dt} = (0.17 \text{ nH})(7.0 \times 10^7 \text{ V/s}) = 12 \text{ mV} \quad [3.21]$$

where L_M = mutual inductance of loops A and B, H

dI/dt = rate of change of current in loop A, A/s

V_{noise} = noise voltage induced in loop B, V

Current switching transients in loop A induce noise pulses of only 12 mV in loop B. By itself, 12 mV may be safely ignored. What happens, though, if the probe ground loop moves near a 32-bit bus? It is highly likely that the noise voltages from each bus line will add, resulting in a nettlesome disturbance of 0.384 V. This disturbance is comparable to the total voltage margin in TTL systems and represents a serious source of measurement error.

Faster logic compounds the noise-pickup problem.

3.3.4 A Magnetic Field Detector

To view inductive coupling, short an oscilloscope probe tip to ground as shown in Figure 3.10. Don't touch the probe tip to anything else. Ideally, you should see no signal at all. On the contrary, anyone who has done this experiment knows that near fast digital logic, you see plenty.

The combination of probe and ground loop responds to changing magnetic fields, which induce voltages around the loop. As the loop moves near high-speed digital circuits, it picks up noise coupled through mutual inductance. This same noise adds to any measurement made using a similar-sized ground loop.

If you crush the ground wire tightly against the body of the scope probe, the loop area decreases and you will see less coupling. The amount of noise coupling is directly proportional to the ground loop area. If lots of coupling comes from a particular area

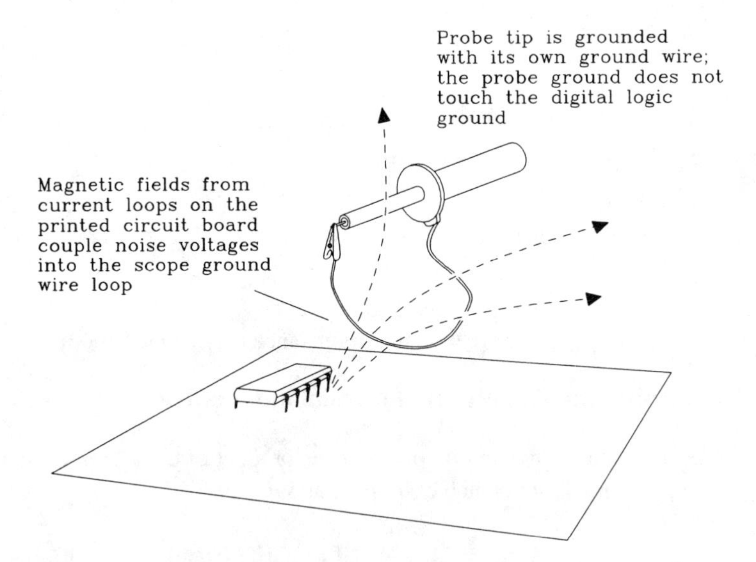

Figure 3.10 Magnetic field detector.

(such as a connector), turning the loop perpendicular to the magnetic field lines can partially cancel the coupled signal.

The exposed probe tip area is usually so small that significant mutual capacitances do not exist between it and digital circuitry. Try holding the probe near a high-speed digital circuit, with no ground wire at all, to see if mutual capacitance between the circuit and the probe tip induces measurable noise current in the probe. Probes are very well electrostatically shielded.

> POINTS TO REMEMBER:
>
> Ground the probe near the signal of interest to reduce the ground wire pickup loop area.
>
> Keep the probe ground wire as short as possible or use a knife blade to short the probe shield directly to the circuit board ground.
>
> Make a magnetic field detector to test for noise induced by mutual inductive coupling.

3.4 HOW PROBES LOAD DOWN A CIRCUIT

The act of using a probe changes the behavior of the circuit under test. Surely, we have all encountered circuits that work when we are probing them but malfunction as soon as we withdraw the probe. This is a common occurrence, due simply to the loading effect of the probe upon the circuit in question.

When a probe loads down a circuit, what changes in waveform do we expect? The nature of the change induced in a circuit depends mainly on three factors:

- The knee frequency of the digital signal under test (Equation 1.1)
- The source impedance of the circuit under test at the knee frequency
- The input impedance of the scope probe at the knee frequency.

Believing for the moment that typical digital source impedances range from 10 to 75 Ω, we need only study the behavior of scope probes versus frequency. Figure 3.11 shows the input impedance of three popular probe styles:

(1) 10 × passive probe with 0.5-pF, 1000-Ω input
(2) 10 × FET active input probe with 1.7-pF, 10-MΩ input
(3) 10 × passive probe with 10-pF, 10-MΩ input.

Referring to Figure 3.11, over the range of interesting rise times, the probes with higher shunt capacitance have much lower impedances. At high frequencies, only the shunt capacitance matters.

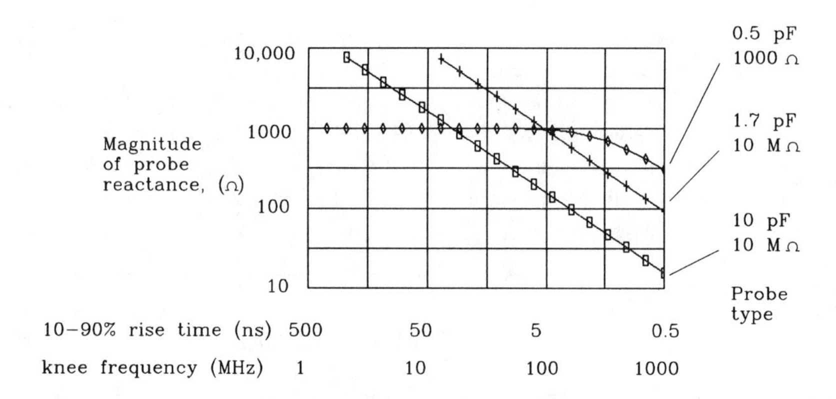

Figure 3.11 Probe input impedance.

If we want the probe to have no more than a 10% effect on the circuit under test, then the probe impedance should be at least 10 times higher than the source impedance of the circuit under test. For any rise time less than 5 ns, the 10-pF probe fails to pass muster.

EXAMPLE 3.3: Probe Loading

Referring to Figure 3.12, we connect a signal source through a long transmission line of 50-Ω impedance to a 50-Ω terminator. At the terminating location connect a sensing probe consisting of a 1000-Ω resistor feeding a short length of RG174 50-Ω coax. The other end of the sense coax leads to a 50-Ω terminated input on a high-speed sampling scope.

We may now connect various loading probes to the test point, and watch their effects.

Figure 3.13 reveals the results when loading the circuit with a Tektronix P6137 probe. The P6137 probe is a 10×, 10-pF, 10-MΩ style connected to a portable 400-MHz scope. The first trace was recorded with no loading probe connected, the second with the loading probe connected using a 6-in. ground wire, and the third with the bare tip of the loading probe on node A and the probe body directly grounded with a small knife blade.

The first trace has the best rise time of 600 ps, with a moderate amount of ringing. The second trace, while exhibiting little rise-time degradation, has a bigger undershoot ripple following the initial rising edge. The first trace ripples also but stays within half a division of the asymptote. The final trace displays a rise time of 800 ps, with very little ripple.

Let's compute the expected rise-time degradation and compare it to these experimental results.

When connected, as in the third trace, with little series inductance, the probe behaves as a simple capacitive load. The source impedance of the test point in Figure 3.12 is 25 Ω.[5] When coupled to a capacitive load of 10 pF, the *RC* rise time is

$$T_{RC} = (25\ \Omega)(10\ \text{pF}) = 250\ \text{ps} \qquad [3.22]$$

[5]A 50-Ω terminator in parallel with a 50-Ω driving cable.

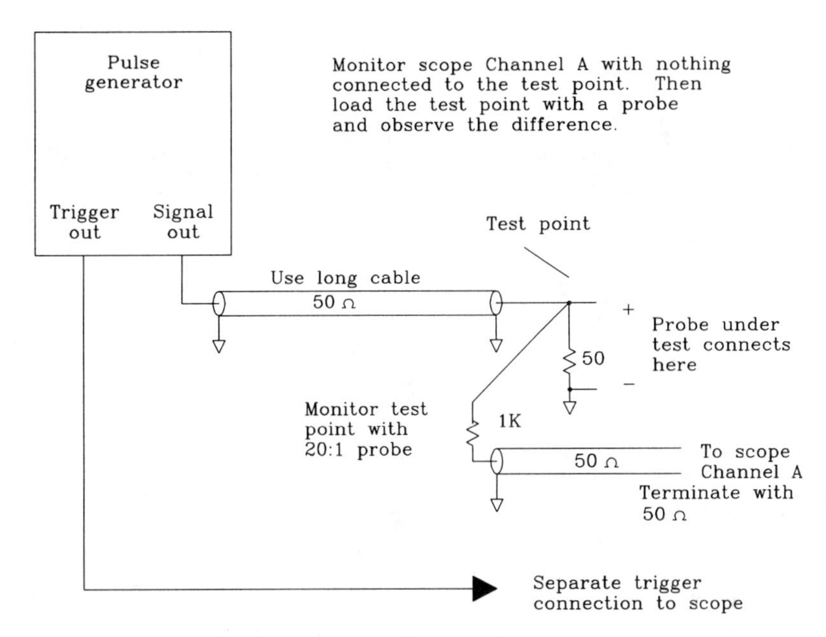

Figure 3.12 Fixture for testing probe loading.

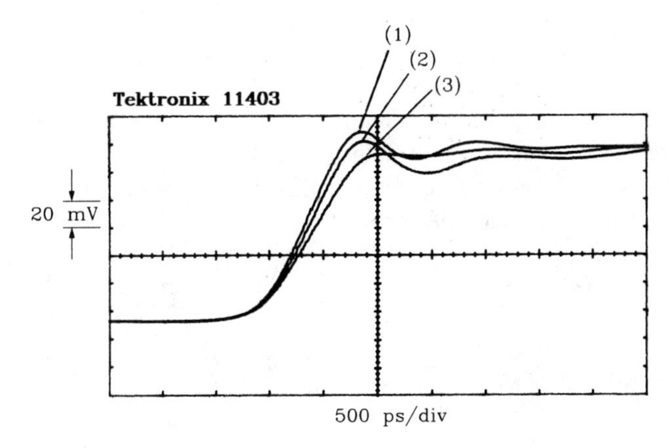

(1) No load

(2) P6137 10−pF probe using 6−in. ground wire

(3) P6137 10−pF probe with collar directly grounded using small knife blade

Figure 3.13 A 10-pF probe loading a 25-Ω circuit.

The 10–90% rise time of an *RC* circuit is 2.2 times larger:

$$T_{10-90} = 2.2T_{RC} = 550 \text{ ps} \qquad [3.23]$$

When combined with the native 600-ps rise time of the test signal, we get a composite result of:

$$T_{10-90\ \text{composite}} = [(600\ \text{ps})^2 + (550\ \text{ps})^2]^{1/2} \\ = 814\ \text{ps} \qquad [3.24]$$

This figure favorably compares to the measured result of 800 ps. That's as accurate as we can expect to get.

While the probe loading increased the rise time by 200 ps, it increased the delay by only 100 ps. That's because most gates switch near the center of the rising edge, not at the 10 or 90% points.

POINTS TO REMEMBER:

A 10-pF probe looks like 100 Ω to a 3-ns rising edge.

Less probe capacitance means less circuit loading and better measurements.

3.5 SPECIAL PROBING FIXTURES

Most portable oscilloscopes in digital development labs use probes having a 10-pF input capacitance and a 3- to 6-in. ground wire. With this type of probe, there is not much hope of accurately viewing, say, 2-ns rising edges. Compounding this problem, the probe itself will significantly alter the signal rise time and pulse shape when attached to a circuit.

This section presents three measurement techniques that directly attack the problems of ground loop inductance and shunt capacitance.

3.5.1 Shop-Built 21:1 Probe

Figure 3.14 illustrates a typical shop-built 21:1 probe. This probe is made from ordinary 50-Ω coaxial cable (RG-174, RG-58, or RG-8) soldered to both the signal under test and the local circuit ground at the point of usage. The probe terminates at the scope into a 50-Ω input jack.[6] The sensing end of the coaxial cable, terminated at its far end by the oscilloscope, looks entirely resistive. The total input impedance of this probe is therefore 1050 Ω. The combination of a 1000-Ω feed resistor and the 50-Ω resistance of the coax forms a resistive divider with a voltage division ratio:

$$\text{Division ratio} = \frac{50}{50 + 1000} = 0.048 \qquad [3.25]$$

With your scope set to the 50-mV/division setting, the displayed vertical sensitivity will be

[6]If your scope does not have a 50-Ω input jack, use an in-line 50-ohm terminator at the scope front panel.

$$\text{Vertical sensitivity} = \frac{0.050 \text{ V/div}}{0.048} \qquad [3.26]$$
$$= 1.04 \text{ V/div}$$

A little adjustment on the vertical sensitivity vernier can trim this to 1.00 V/division if necessary.

The advantages of the 21:1 probing arrangement are threefold:

- The DC input impedance is 1050 Ω (as opposed to just using the coax, which would be a 50-Ω load).
- The shunt capacitance of a 1/4-W 1000-Ω resistor is about 1/2 pF. This is very favorable.
- The probe rise time is very fast (see below).

Three factors dominate the calculation of the 10–90% rise time for this shop-built probe: the rise time of the BNC connector, the rise time of the coax, and the rise time of the sense loop.

Assuming a 50-Ω BNC input jack is used on your oscilloscope, the BNC connector introduces a series inductance in the 50-Ω cable at the point where the shield spreads out away from the center conductor to connect with the BNC fitting. Table 3.2 shows the series inductance of several types of coax connectors and the 10–90% time constant that goes with them.[7,8] If your scope does not have a 50-Ω internal terminator, use an external terminating plug to terminate the coax. This connection can add significant parasitics to the setup, especially if it uses a BNC "T" fitting with a separate terminator plug. Get a good-quality in-line terminator for this setup.

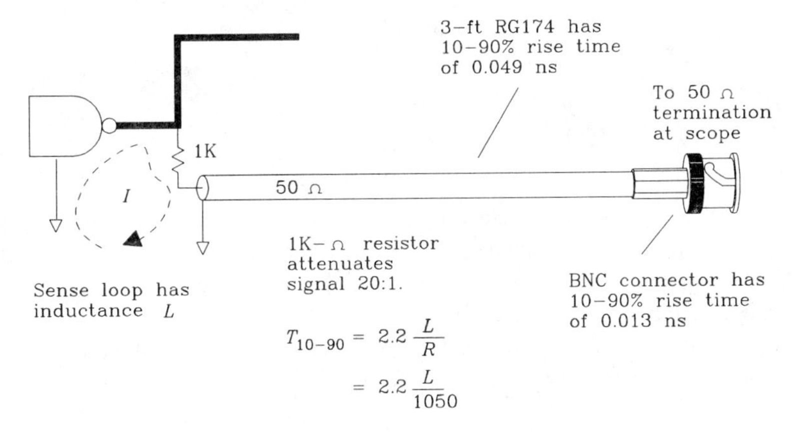

Figure 3.14 Shop-built 20:1 probe.

[7]The time constant due to relaxation of the series L, terminated in 50 Ω at both sides, is $L/100$. The 10–90% rise time is 2.2 times this value.

[8]Computed from VSWR measurements quoted in *Cambridge Products UHF and RF Coaxial Connectors* (1987 catalog), Bloomfield, Conn.

TABLE 3.2 10–90% RISE TIME OF MALE COAX CONNECTORS

Type	$L_{connector}$	t_{10-90} (ns)
RG-58 BNC twist-on	1.	0.022
RG-58 BNC double-crimp	0.5	0.011
RG-174 BNC double-crimp	0.5	0.011
RG-8 N-type	0.2	0.004

Table 3.3 reports the 10–90% rise time of various lengths and types of cables. The cable rise time is proportional to the square of the distance. There is a fixed constant of proportionality between the results for each cable type.

You can approximate the rise time of a coaxial cable by first finding the frequency at which the attenuation equals 3.3 dB. This value, for coaxial cable, is the knee frequency. Find the rise time from the knee frequency in the ordinary way: $T = 0.5/F_{knee}$. This formula works only for short lengths of cable (total attenuation less than a few decibels).

Note that the attenuation at high speeds is proportional to the square root of frequency. This fact can help interpolate between attenuation specification points in the cable manufacturer's catalog. Attenuation is directly proportional to length.

TABLE 3.3 10–90% RISE TIME OF COAX CABLE

Feet	$T_{RG\text{-}174}$ (ns)	$T_{RG\text{-}58}$ (ns)	$T_{RG\text{-}8}$ (ns)
1	0.004	0.002	0.0002
2	0.014	0.006	0.001
3	0.032	0.014	0.002
4	0.056	0.024	0.004
5	0.088	0.038	0.006
10	0.35	0.15	0.025
20	1.4	0.61	0.10
50	8.8	3.8	0.64

The 21:1 probe's sense loop starts at the signal source and includes the 1000-Ω sensing resistor, its attachment to the coax, the coax ground connection to the printed circuit board, and the ground path back to the signal source. Keep this path very tight for good results.

As a function of sense loop diameter, Table 3.4 lists values for the sense loop inductance and the 10–90% rise time. Table 3.4 assumes the sense loop is made mostly of AWG 24 wire.

Because the shop-built probe incorporates a 1K-Ω input resistor, the rise-time degradation, *L/R*, due to the inductance of the sense loop is much smaller than when working with a 50-Ω coax or with a 10-pF input probe. The shop-built probe has a terrific rise time. With bigger resistors, the rise time gets even better.

One factor limiting the use of attenuating probes is the end-to-end shunt capacitance of the attenuating resistor. A 1/4 W resistor body normally has 1/2 pF from end to

TABLE 3.4 10–90% RISE TIME OF PROBE SENSE LOOP

Loop diameter (in.)	L_{sense} (nH)	t_r (ns)
0.1	3.9	0.01
0.2	11.4	0.02
0.5	31.0	0.06
1.0	80.0	0.17
2.0	200.0	0.42
5.0	500.0	1.1
10.0	1220.0	2.6

end. At very high frequencies the shunt capacitance feeds additional unwanted power into the coax, further loading the circuit under test.

Using a smaller resistor body (1/8 W) reduces problems with shunt capacitance. Watch out for the power limitation on 1/8-W 1000-Ω resistors, which can be used only up to ±11 V.

Another way to combat shunt capacitance is to intentionally place a capacitor in shunt with the sensing end of the coax. The correct-value capacitor, working with the shunt capacitance of the sense resistor, forms a matched 21:1 divider network. This network has a flat frequency response up to exceptionally high frequencies. Commercial oscilloscope probes use this technique. It's hard to fabricate an exact capacitance divider arrangement in the lab.

Attenuating probes have a low Q. You will experience few overshoot and ringing problems with a properly made 21:1 probe.

Tektronix manufactures a variety of low-impedance, passive attenuating probes built basically like the shop-built model. This product line includes the P6156, P6150, and P6231. The P6156 may be used with any vertical amplifier provided it has a BNC input and 50-Ω internal termination.

EXAMPLE 3.3: 10–90% Rise Time of Shop-Built Probe

Build a 21:1 probe using 6 ft of RG-174, a BNC dual-crimp connector, and a probe loop of 0.5-in. diameter.

From Table 3.2, $t_{BNC} = 0.013$ ns
From Table 3.3, $t_{cable} = 0.19$ ns
From Table 3.4, $t_{loop} = 0.08$ ns

$$t_{composite} = (t_{BNC}^2 + t_{cable}^2 + t_{loop}^2)^{1/2} = 0.206 \text{ ns} \quad [3.27]$$

POINT TO REMEMBER:

A shop-built 21:1 probe has a terrific rise time.

3.5.2 Fixtures for a Low-inductance Ground Loop

Most scope probes have a removable plastic *IC* grabber clip covering the probe tip. Remove the plastic clip, revealing the probe barrel. If necessary, then disassemble the apparatus holding the ground wire in place, exposing the low-inductance probe ground sheath. This metal sheath, or ground collar, extends out almost to the end of the probe tip. It serves two purposes: electrostatically shielding the probe tip and providing a good ground point near the tip for implementing a low-inductance sensing loop.

Figure 3.8 shows two methods for achieving a low-inductance sense loop using the metal probe sheath.

The big curlicue in Figure 3.8 is just a piece of resistor wire wrapped around the metal ground sheath and then soldered to a convenient ground pin. It mechanically holds the probe in place and simultaneously grounds it. The little curlicue holds the probe tip. These fixtures can be applied to a circuit board anywhere. Curlicues work well for engineering purposes but are not permanent and not easily manufactured.

The ground pad method simply places a small square of exposed ground material (no solder mask) near each probe point. With the probe held against the sense point, use a small knife blade[9] to connect the probe ground sheath to the ground pad. A 0.035-in. exposed ground pad is big enough. If the top surface of the board is not otherwise grounded, then use a 0.020-in. via to bring the ground to the surface and place an exposed 0.035-in. pad around it.

Ground pads are useful for probing and serve as attachment points for repair circuitry should a design require rework. Some analog engineers make a practice of leaving the entire top ground surface exposed during prototype runs so it will be easy to access.

The sense loop inductance for both methods falls between 3 and 30 nH, depending on the craftsmanship exercised.

Tektronix manufactures a probe fixture specifically designed to connect a probe tip to a circuit under test with very little ground loop inductance. This fixture, shown in Figure 3.15, works well if the circuit's mechanical layout allows the probe to project perpendicular to the printed circuit board. Circuit cards which must be embedded in card cages can't use this fixture.

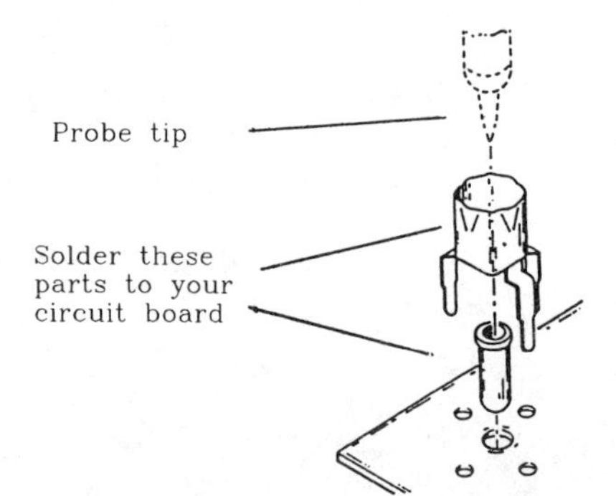

Figure 3.15 Low-inductance connector for Tektronix probes. (Figure courtesy of Tektronix, Inc.)

[9]Paper clips will work, but the sharp knife edge better penetrates oxide layers, establishing a low-resistance contact.

Some oscilloscope probes come with tiny clips that attach to the probe barrel, facilitating a direct connection between the barrel and ground.

3.5.3 Embedded Fixtures for Probing

Removable probes disturb a circuit under test while it is being serviced and then leave the circuit in a different condition when the probe is removed. Consider how a 10-pF load[10] from a scope probe affects a high-speed signal. Embedded probe arrangements can leave the circuit in the same state all the time.

In addition, the embedded probing fixtures recommended below have parasitic capacitances on the order of 1 pF, much less than those of a 10-pF probe.

The embedded fixture in Figure 3.16 accomplishes a 21:1 probe function, provides a convenient test connection point, and leaves the circuit connected the same way all the

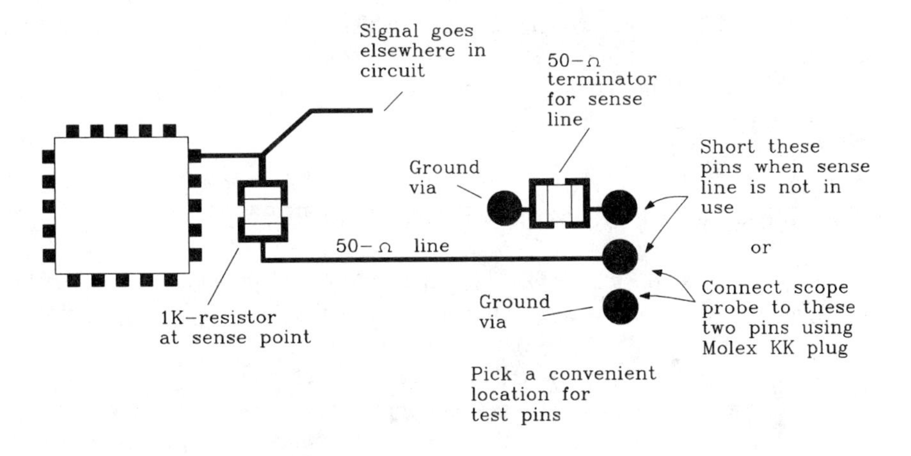

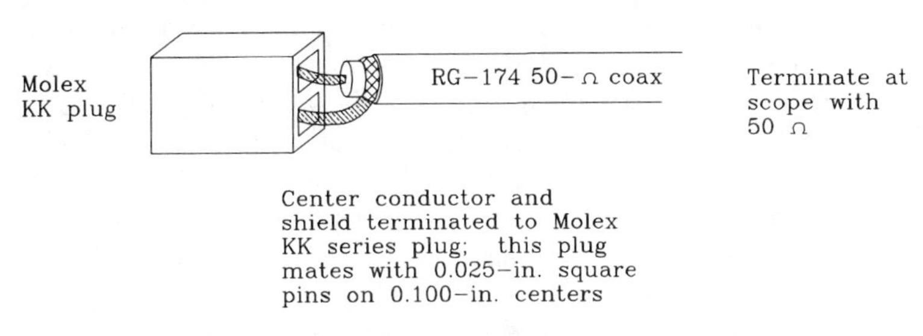

Figure 3.16 Embedded probe fixture.

[10]Ten pF looks like 100 Ω to a 3-ns rising edge.

time. A 1000-Ω sense resistor connects the circuit under test to a 50-Ω test trace. The 50-Ω test trace routes to a convenient test point on the board. In Figure 3.16, means are provided for shorting the test point through 50 Ω to ground when the scope is not connected.

Many options exist for connecting a test point to test equipment. PC-mounted BNCs are one readily available alternative, but they consume a lot of board area.

Figure 3.16 shows 0.025 in. square posts on 0.1 in. centers used as test point connectors. These parts are cheap, and they mate with a variety of female connector housings. The authors prefer a MOLEX/WALDOM KK series terminal housing. RG-174 coax directly crimps into the female pins of the MOLEX/WALDOM KK housing, which then mates with the male pins on the circuit board. Expect a series loop inductance of 10 nH with this connector. When used in series with 50 Ω cable, the resulting T_{10-90} is 0.22 ns. If you mount the MOLEX pins very close to the 1000-Ω sense resistor, the effective resistance in series with the connector loop inductance is 1000 Ω, and the T_{10-90} drops to 0.025 ns.

Whatever connection method you choose, provide a way to terminate the test trace when it is not in use or provide a way to disconnect, or not remove, the sense resistors. Figure 3.16 indicates a shorting jumper which engages a 50-Ω terminating resistor when the test trace is not in use.

Terminating the test trace provides a constant resistive 1050-Ω load to the circuit under test, even when the probe is removed.

3.6 AVOIDING PICKUP FROM PROBE SHIELD CURRENTS

Oscilloscope probes have two wires, one connecting the circuit under test to the vertical amplifier (the sense wire), and the other connecting local digital logic ground to the scope chassis ground (a shield wire). Normally, we think about a scope responding to voltages on its sense wire. This section explains how a scope responds to signals on its shield wire, also.

Any voltage difference between logic ground and the scope chassis causes current to flow in the shield wire. Shield current, coursing through the resistance R_{shield} of the shield wire in Figure 3.17, generates a voltage drop across V_{shield}. The center conductor of the probe cable, the sense wire, carries no shield current and therefore has no voltage drop across it.

When the shield and sense wires both touch ground in a working circuit, the differing voltage drops across the two wires cause the vertical amplifier to experience a voltage difference between them. It has no way of knowing whether this difference is due to an actual signal voltage at the far end of the probe cable or due to shield current. Although you would like the scope to show no voltage, what it shows is the shield voltage.

The scope responds to shield voltage as if it were a real signal.

Shield voltage is proportional to shield resistance but not to shield inductance. This happens because the shield and center conductor are magnetically coupled. Any changing magnetic field generated by current flowing in the shield encircles both the shield and the center conductor, inducing identical voltages in both wires. Inductive voltages appear on both wires, unlike the resistive voltage drop which appears only across the shield.

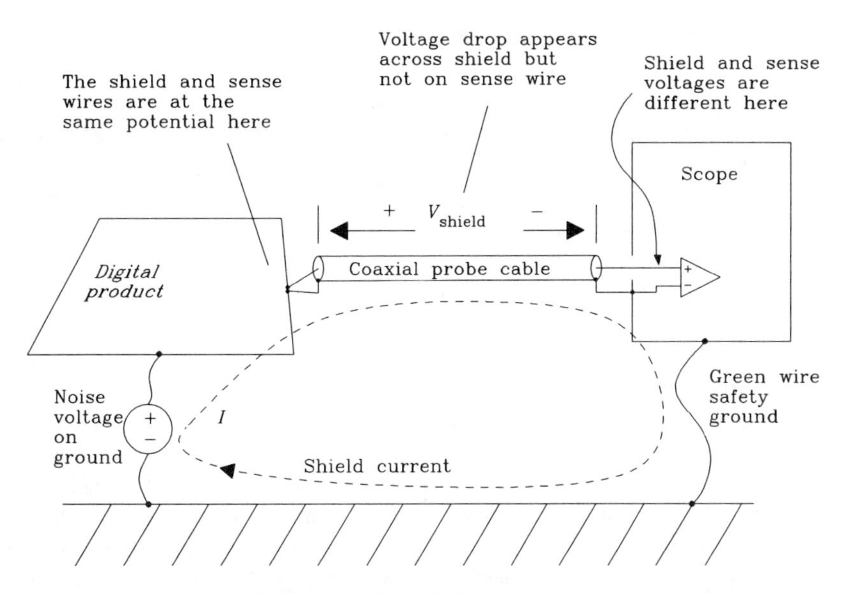

Figure 3.17 Noise pickup from probe shield currents.

Shield voltage is easy to observe:

(1) Connect your scope ground and tip together.

(2) Move the probe near a working circuit without touching anything. At this point you see only the magnetic pickup from your probe sense loop.

(3) Cover the end of the probe with aluminum foil, shorting the tip directly to the probe's metallic ground sheath. This reduces the magnetic pickup to near zero.

(4) Now touch the shorted probe to the logic ground. You should see only the shield voltage. If the shield voltage appears very small, ignore it.

Shield noise bedevils digital systems that control large power equipment. Huge 60-Hz AC currents flowing elsewhere in the equipment can induce voltages on the digital logic ground, which in turn create shield noise. If shield noise causes trouble, there are nine ways to attack it.

(1) Lower the shield resistance. This is hard to accomplish if you bought the probe. If you are using shop-built coax probes, try a bigger coax. Change from RG-174 to RG-58, or from RG-58 to RG-8. The stiffness of a larger coax makes this method impractical for all but fixed instrumentation setups.

(2) Add a shunt impedance between the scope and logic ground. This causes more of the noise current to flow through the shunt and less through the shield. This is usually not practical, especially at high frequencies. Finding a good ground on the circuit board and attaching it to the scope with a low enough inductance to make any difference is almost impossible.

If the shunt is as long as the probe wires, there will not exist any sufficiently large-diameter object to make any difference (inductance varies as the logarithm of diameter). If the shunt is much shorter than the probe length, it might work.

(3) Turn off the circuit board. Or, turn off sections of it. This method works only when observing just one part of the circuit. If you suspect problems with shield current noise, this is a good test. It will determine if the noise is really emanating from your circuit or from some other source.

(4) Put a big inductance in series with the shield. Using a big high-frequency magnetic core, make 5 to 10 turns through it with the probe. This raises the inductance of the probe shield, lowering the current. This method works well in the range 100 kHz–10 MHz. Below 100 kHz you will need a very large inductor to make any difference. Above 10 MHz the effectiveness of magnetic cores deteriorates.

(5) Redesign the board to reduce radiated fields. Change a two-layer board to a four-layer board with solid ground planes. Reducing the radiated fields lowers the tendency of your ground plane to develop noise voltages in the first place.

(6) Disconnect the scope safety ground. Disconnecting the scope safety ground defeats a crucial safety feature of your AC power system. Should any energized part of the scope power supply contact its outer case, the scope chassis becomes energized with 110 V. This is deadly. Normally, if such a failure occurs, the safety ground shunts massive amounts of AC power current to ground, tripping your local circuit breaker. This disconnects power to the unit, possibly saving your life.

Nevertheless, you should know what effect disconnecting the scope safety ground has on high-frequency signals.

Perfect isolation of the scope chassis from the safety ground breaks the probe shield ground loop at the oscilloscope, reducing probe shield currents. Unfortunately, disconnecting the safety ground wire does not achieve perfect isolation.

Most scopes have, internal to their power supply, a 0.01-μF capacitor connected between the chassis and each of the AC power wires, which in turn lead back to ground. Even without the capacitor, there is enough parasitic capacitance in the power transformer to form a high frequency path between the chassis and the AC power wires.

At frequencies above 10 MHz, the scope has sufficient natural capacitance to earth anyway, so that isolating the safety ground is futile.

This method works at audio frequencies, but not for high-speed digital logic.

(7) Use a triaxial shield on the probe. Connect the triaxial shield to the scope frame on one end and the circuit board ground at the other. The scope probe must run inside the triaxial shield for its entire length. Connect the triaxial shield to exactly the same point as the probe ground. At high frequencies, most of the shield current diverts, because of the skin effect, to the outer shield. Because the inner probe shield carries no current, it exhibits no resistive voltage drop and no noise voltage develops. This sounds counterintuitive, but it really works. Make the triaxial shield from aluminum foil, or slit open an old RG-8 shield and wrap it around the probe. Minimize the exposed probe length between the triaxial shield ground and the probe point to reduce magnetic noise coupling into that loop.

If you wish to make your own 21:1 triax probe, POMONA sells a BNC-to-triax adapter useful for this purpose. Plug the BNC male end of the adapter into the scope BNC jack. The other end of the POMONA adapter has a triax female fitting with both outer and middle grounds connected internally to the single BNC ground. Terminate one end of the triax in a normal triax male fitting and plug it into the adapter. On the circuit board end of the triax, just directly solder the outer and middle shields together.

(8) Use a 1:1 probe instead of a 10:1 probe. A 10:1 probe does not attenuate the shield voltage effect. Because 10:1 probes do attenuate actual logic signals, using a 10:1 probe makes shield voltages appear relatively 10 times larger.

(9) Use a differential probe arrangement. Figure 3.18 shows the proper arrangement for differential measurement. Probe 1 connects to the signal point, while probe 2 connects to the circuit ground. The shields of both probes tie together at point G_S, and *do not connect to the circuit board.* A separate ground strap connects the circuit board to the scope ground. This separate strap is necessary only if the circuit board has no suitable connection to true earth ground through some other mechanism.

Set the scope to subtract the probe-2 signal from the probe-1 signal. This operation is never perfect and usually benefits from minor adjustment. Tie both scope probes temporarily to a common signal point while adjusting the gain balance between probes to best cancel their waveforms. Next, always touch both probes temporarily to ground to see what residual noise pickup exists. That's what we set out to reduce, and so it is worth checking to see if we got it.

Shield currents do not appear when using differential probes because the shields do not touch anything. This is the major benefit of differential probing. For circuits having a floating ground, or a ground energized above a true earth ground, differential probing may be the only alternative.

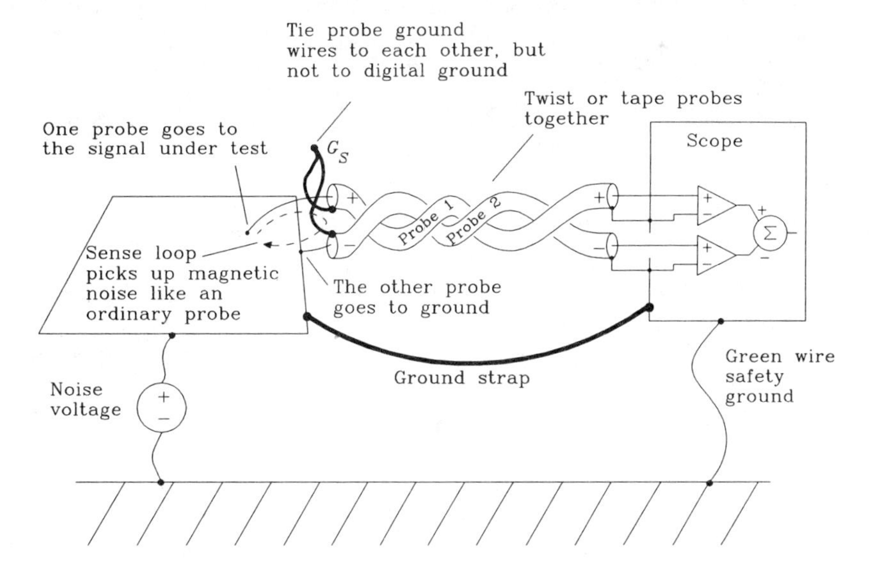

Figure 3.18 Using differential probing to eliminate shield current noise.

Keep the probes close together, minimizing the total size of the magnetic pickup loop between them. Any magnetic pickup in this loop will induce voltages between the two probes. Twisting or taping the probes keeps them close.

As with an ordinary probe, keep the ground sense point close to the measurement point. Noise couples by way of mutual inductance into the sense loop between the two probes just as with an ordinary single-ended probe.

The probes must be of the same type and length in order for differential sensing to work. Imbalances in the frequency response or delay of the two probes will result in common-mode signals appearing on the screen.

Some scopes come with special differential amplifier modules and matched probes having matched gain and frequency response characteristics. These modules have extraordinary common-mode rejection but usually a lower bandwidth than is useful for high-speed digital problems.

Beware of using 10× probes for differential measurements. The high frequency compensating adjustments, as well as the DC gain, must match perfectly in order to obtain useful common-mode cancellation. This rarely works at high speeds.

POINTS TO REMEMBER:

A single-ended scope probe responds to shield voltage as if it were a real signal.

To see if you are getting noise induced by shield currents shield the tip of your probe with foil and then touch the probe and its ground to the circuit board ground.

Tie both differential scope probes temporarily to a common signal point while adjusting the gain balance between probes to best cancel their waveforms.

3.7 VIEWING A SERIAL DATA TRANSMISSION SYSTEM

Figure 3.19 illustrates a 100-Mbit/s data transmission system. Due to intersymbol interference, and also to additive noise, this system exhibits more jitter in the output waveform *D* than in the transmitted signal *A*. This section shows how to properly characterize jitter in the output waveform.

Our first step is to connect channel 1 of our oscilloscope to signal *D*. Then we also select channel 1 for triggering and adjust the scope to trigger on positive-going transitions. We see a pattern like Figure 3.20.

Notice that the waveform shows no jitter at the trigger point. This is a key indicator that something is awry. The scope waits for a positive-going transition and then shifts the

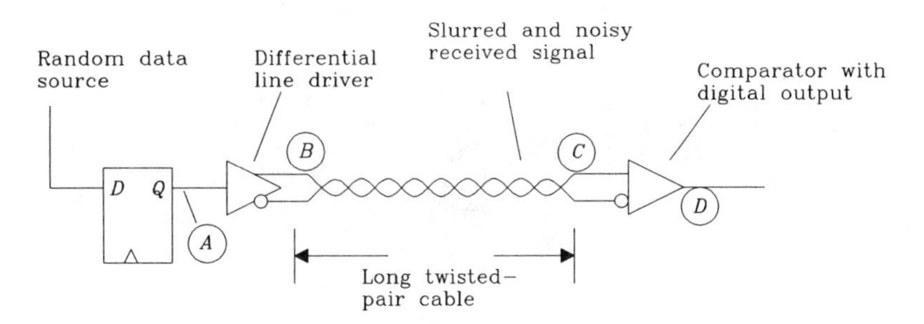

Figure 3.19 Typical data transmission circuit.

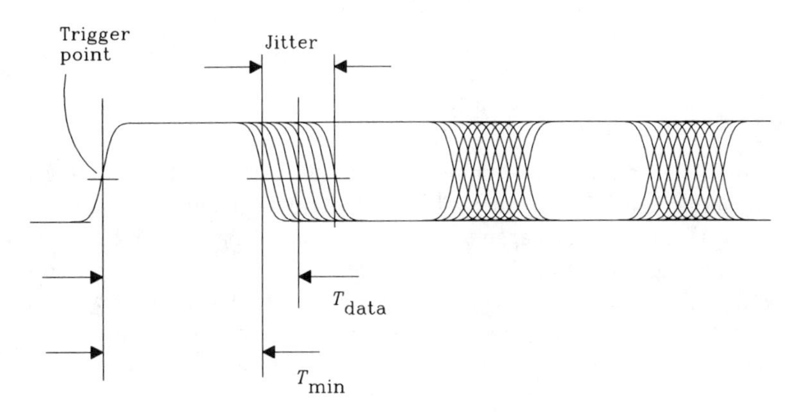

Figure 3.20 First attempt at eye pattern observation.

data waveform to line up the trigger point with our left cursor mark. The first pulse properly represents the minimum space between transitions, but the jitter error surrounding subsequent clocking points is twice the actual clock-to-data jitter.

Figure 3.21 shows the measurement properly displayed. The signal in Figure 3.21 is triggered using the source clock as an absolute reference. The apparent jitter here is one-half the previous value. The previous technique shifted each waveform, lining up all the rising edges at one point. That shifting *added* jitter to all the other transition zones. The source clock is a stable, nonjittery signal, a solid point of reference for all data measurements.

One student has asked, "Why can't we just use the technique in Figure 3.20 and divide the result in half?" The answer is, that while the eye pattern in Figure 3.20 was sufficiently wide open to permit jitter measurement, we are not always so fortunate. Sometimes the eye does not open at all until we apply the superior triggering technique of Figure 3.21.

When the source data clock is not available,[11] try triggering on the source data signal (position *B* or *A* in Figure 3.19). The data at the source has almost no jitter.

[11]The source clock is generated internally in high-speed serializing chips like the AMD7468.

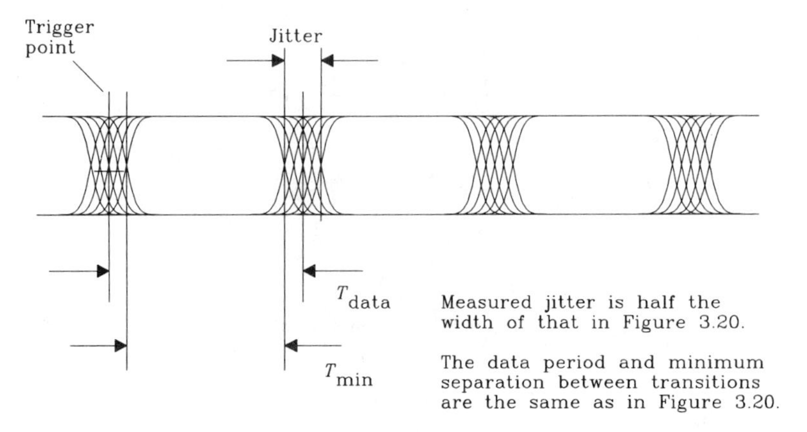

Figure 3.21 Eye pattern triggered by a source data clock.

Some oscilloscopes, most notably the newer digital sampling models, have poor triggering capabilities, especially on nonperiodic signals like data waveforms. While the vertical input may be capable of displaying signals at very high speeds, the triggering circuit may not trigger on them. When faced with a poor scope trigger circuit, first construct a digital circuit which divides down the system clock, and trigger from that secondary waveform. With the triggering stability improved, you may notice a decrease in the apparent rise time of signals displayed on the scope.

POINT TO REMEMBER:

View a serial data stream by triggering on the clock.

3.8 SLOWING DOWN THE SYSTEM CLOCK

High-speed digital signals usually include ringing, crosstalk, and other noise. At full clock speed, these various effects superimpose. This superposition makes it difficult to ferret out the true character of each individual effect. Slowing down the main system clock can help isolate effects.

A sufficiently slow clock allows all signal transients to decay before starting the next clock cycle. Reflections and ringing left over from cycle n will not appear in cycle $(n + 1)$. We can then separately see the entire response to each digital transition. Sometimes the response is much longer than you've anticipated. Better terminations usually cure this problem.

> **POINT TO REMEMBER:**
>
> During testing, a sufficiently slow clock allows all signal transients to decay before starting the next clock cycle.

3.9 OBSERVING CROSSTALK

Crosstalk problems, because they involve the interaction of several supposedly disconnected sections of logic, are difficult to corner. These problems often appear intermittently, may be data pattern-dependent, or may occur very rarely. That makes them difficult to observe. Errors due to crosstalk usually involve a confluence of several factors:

- Reduced logic margins due to ringing
- Marginal compliance with setup and hold requirements
- Multiple data lines coupling together.

If you suspect crosstalk is a problem, here are some ways to quantify the amount of crosstalk present without waiting for an error to occur.

First, rig up a coax 21:1 probe on the line you wish to monitor (the primary signal). Before connecting the probe to the primary signal, solder the sense resistor to a nearby ground, turn on the digital machine, and measure the amount of residual noise from magnetic pickup of the sense loop and shield current. This noise should be less than 2% of the digital signal size.[12] If you have more than 2% noise, the crosstalk may not show clearly. Work on the probe arrangement until it picks up less than 2% noise.

Next connect an external trigger to the scope. This trigger should be synchronous with the suspected source of crosstalk and must be present throughout the experiment. Using your external trigger, take another look at the noise coming from the 21:1 probe.

Now connect the 21:1 probe to the primary signal. You should see a combination of the primary signal, ringing due to that primary signal, crosstalk, and the noise present in our measurement system.

Our objective is to identify and quantify the crosstalk. Crosstalk, by its very nature, is often difficult to see. To amplify the visible effects of crosstalk, three avenues of attack are now open to us: turning off the primary signal, turning off the crosstalk, or generating artificial crosstalk.

3.9.3.1 Turning off the primary signal

Cut the primary signal at its origin and connect it to ground at that point. If its logic driver will take the punishment, just short the primary signal to ground at the driver. The

[12]Remember you are using a 21:1 probe. The scope should display 1/21 of 2%, or about 0.1% of a full digital signal.

short-to-ground is critical, as noise coupled through mutual inductance will disappear if the primary trace is open-circuited.

When shorting a logic gate, the short circuit must be accomplished with a broad, flat, very-low-inductance object such as a knife blade or a piece of copper foil. A wire $\frac{1}{2}$ in. long used to short out a logic driver has enough inductance to let through a sizable pulse. We want the output at this point to be zero.

With the output turned off, crosstalk should stand out clearly.

If we are dealing with a bus, now is a good time to start varying the bus patterns. Run a series of experiments where exactly one data line at a time changes while the others remain fixed. Depending on the bus layout, some lines will influence the primary signal in a positive sense, and others may induce negative crosstalk. When you flip the polarity of any individual data transition, its crosstalk should flip polarity also. For each data line, find the polarity that causes positive interference.

As a last experiment, arrange a data pattern with a transition on every line at the same time, each oriented to generate positive interference. This shows off the worst possible interference. The crosstalk levels attainable on a 32-bit bus are striking.

3.9.3.2 Turning off the crosstalk

Arrange a data pattern which you believe will exhibit crosstalk. Make two photos of the primary signal, one with the system operating normally, and the other with the interfering line(s) disconnected.

The interfering line may simply be cut, or shorted to ground at its source. Either method works. Grounding the line is not particularly important, as long as we reduce its current to zero.

The difference between the two photos is the crosstalk. If your digital oscilloscope has a signal manipulation feature (the Tektronix 11403 does), then store both waveforms and subtract them numerically.

3.9.3.3 Generating artificial crosstalk

With the system turned off or disabled, short the driving end of the primary signal. Now induce a step edge of known rise time on the interfering trace and measure the induced voltage on the primary signal, as in Figure 3.22.

Crosstalk is proportional to the *dV/dt* of the signal driving the interfering trace. This technique is best applied to a bare board before stuffing it with chips.

> POINT TO REMEMBER:
>
> Amplify the visible effects of crosstalk by temporarily changing your system.

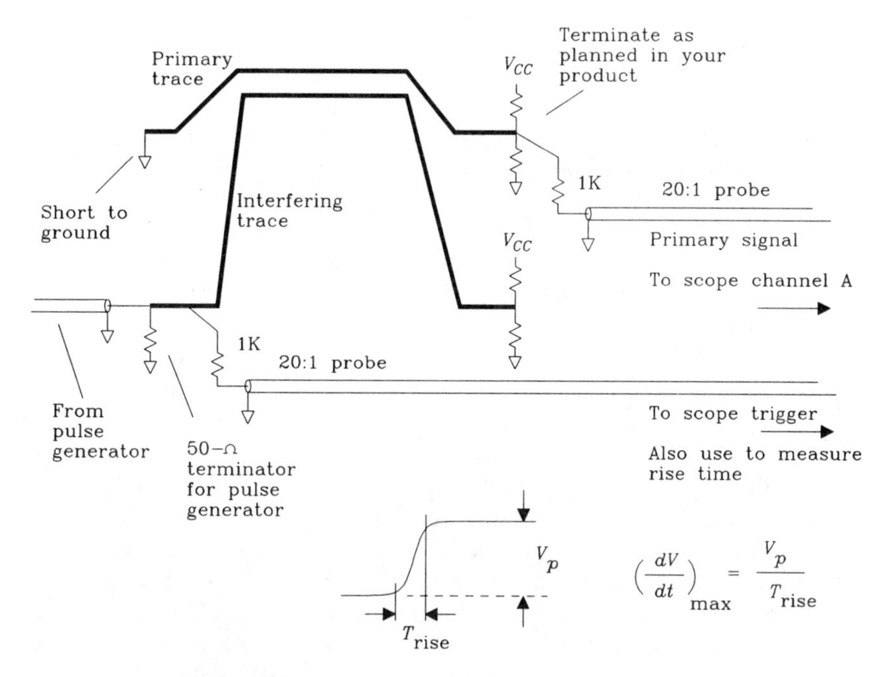

Figure 3.22 Measuring crosstalk between two signal traces.

3.10 MEASURING OPERATING MARGINS

Digital circuits often either work or don't work. Unlike analog circuitry, digital systems exhibit few marginal, or degraded, states of operation. Once a digital system is working, it is difficult to quantify how well it performs or how much operating margin it has left. This section documents techniques useful for quantifying the operating margins of digital systems.

Manufacturing engineers accustomed to statistical quality control recognize the relation between measuring quality and maintaining quality. That manufacturing principle applies directly to digital products.

These measurements are global, taking into account effects throughout a system. They assume you have a go-nogo test that shows whether the system is working or not. The go-nogo test should be as comprehensive as possible and designed so logic failures in any area of the system will generate a nogo response.

In each test, using a go-nogo tester, subject the system to one of the stresses listed below. We will measure how much stress the system absorbs before failing. This procedure converts the simple go-nogo test into a quantitative measure of product quality.

Make sure your test continues running even in the face of failure. We want it to report nogo, automatically restart, and keep going even when something goes wrong. This property makes it easy to vary the stress in and out of the error zone to make sure we are getting accurate readings. By setting the stress level to induce an error every few seconds, you can use a logic analyzer to catch the mistake. Designing around a mistake once it has been located is the easy part. If the go-nogo test halts on the first mistake, you can't establish a steady error rate and you may never debug the problem.

3.10.1 Additive Noise

Appropriate for small circuits having very high-speed processing elements, the additive noise test simply adds random noise to every node in the circuit. The best random sources for this test are sources with limited excursions like sine waves, square waves, and pseudorandom binary patterns.

Inject the noise at each circuit node using a series resistor calculated not to load the circuit. A series resistance of 1K Ω works for TTL, HCMOS, and ECL.

During development, inject noise into one node at a time to characterize its performance. This experimental evidence is useful later if you suspect a change in the layout introduces more ringing (lowering the tolerance to additive noise).

Once you know the relative sensitivities of each node, make up a set of resistors calibrated to inject into each node, one at a time, its critical noise current. Connect all the resistors with switches to a common noise source.[13] Now as we vary the source noise level, all the nodes should fail at about the same source noise level. Any node which has deteriorated will show up as a failure at a lower noise level. A scatter plot of the noise failure level for each product versus its manufacturing serial number readily reveals shifts in the manufacturing process.

The additive noise test is difficult to administer, as it requires either a bed-of-nails test fixture or special connectors on the printed circuit to accommodate the noise connection.

Additive noise testing is suitable for data receivers, clock recovery loops, phase-locked loops of all types, analog I/O interfaces, and buses. In short, any place where a lot of information passes through a limited number of test nodes.

3.10.2 Adjusting the Timing on a Large Bus

Most large bus systems are synchronized by a common clock, distributed along the bus. For these systems, the design engineer works out a detailed timing analysis showing the theoretical guaranteed setup and hold times for transfers taking place across the bus.

To test the setup and hold timing assumptions, we need a way to vary the transmitted data timing, either advancing or retarding it, until the bus fails. By recording the amount of timing adjustment the system will accept before failing, we arrive at a quantitative measure of the bus timing margin.

To apply this test, first set up a data transfer from device *A* to device *B* across the bus. Make sure there exists a way to tell when the system is making errors. It's best if the system reports an error rate, or flashes a light when errors occur, but keeps on going.

Now cut the clock distribution trace on the bus between the two devices. Feed each device with a different clock. We will arrange for the two clocks to have the same fre-

[13] A really fine test would use independent noise at each node. Several independent noise sources may be synthesized from a common generator using a coaxial delay line.

quency as the system clock, but with slightly different phases. By advancing (or retarding) the timing of one bus clock compared to the other we can determine how much timing margin remains on the bus.

For this test we will need a special circuit which produces two frequency-locked clocks with adjustable output phases. Any of the five ideas reported below can accomplish this task.

3.10.2.1 Clock adjustment by coax delay

For clocks up to 20 MHz, make up a coax delay selector box from segments of coax and ordinary switches. From a single clock source, run one clock (A) through the delay selector and the other clock (B) through a fixed length of coax. Use a coax impedance (50, 75, or 93 Ω) which matches the native impedance of the bus.

Pick the length of the fixed coax such that, with the delay selector box set to midrange, the two clocks come out with matched timing. This may take some fiddling with the fixed delay length.

Don't try to make up an elaborate binary selector because it is too difficult to get more than a few coax segments cut to matching lengths so the delay steps linearly. Instead, use two multiposition switches and install lengths of 1, 2, 3, ..., 10 and 10, 20, 30, ... units of delay.

3.10.2.2 Clock adjustment by pulse generator

A pulse generator with an adjustable delay, trigger-to-output, makes a fine clock adjuster. From a single clock source, run one clock (A) to the pulse generator trigger input, and the other clock (B) through a fixed length of coax directly to the bus. Use a coax impedance (50, 75, or 93 Ω) that matches the native impedance of the bus. Set the pulse generator to produce a pulse length equal to one-half the nominal clock period.

Pick the length of the fixed coax such that, with the pulse generator delay set to midrange, the two clocks come out with matched timing. This may take some fiddling with the fixed delay length.

Many pulse generators will not retrigger if the next trigger input comes before the existing pulse is complete, limiting the delay adjustment range to 0–180 degrees. With the fixed coax set to 90 degrees of delay, the effective adjustment range is –90 to +90 degrees.

3.10.2.3 Simple circuits for clock phase adjustment

The circuit in Figure 3.23A shows a hex inverter used to produce delays of from 30 to 160 nanoseconds. The delay of each section is from 5 to 35 ns, depending on the setting of the variable resistor.[14] The delay time in each section must be no more than 12% of the clock cycle time to guarantee reliable operation.

[14]Assuming a nominal 5-ns delay per gate.

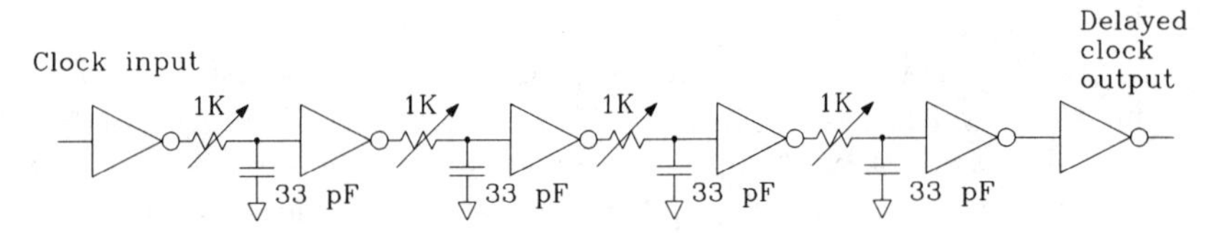

Figure 3.23A TTL or CMOS adjustable-delay network.

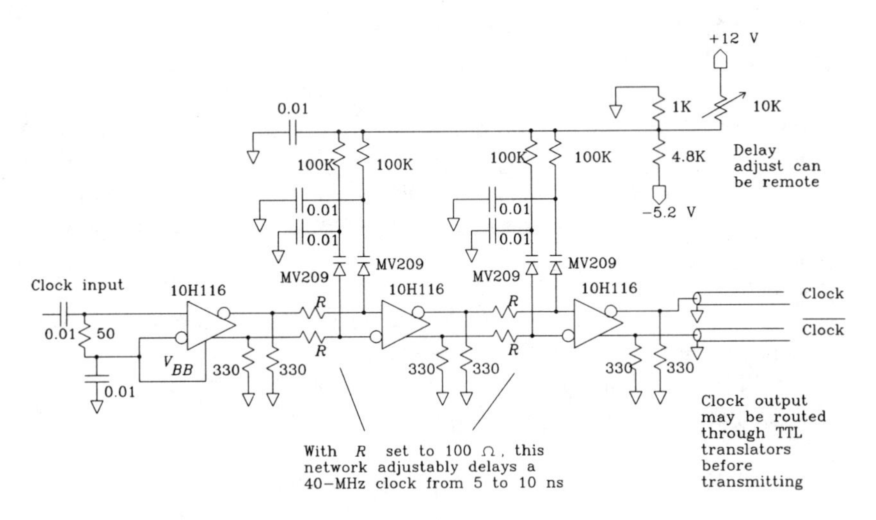

Figure 3.23B ECL remotely adjustable-delay network.

Balancing the number of delay sections (two or four) and adjusting the resistors equally among the sections (try using ganged potentiometers) will keep duty cycle distortion to a minimum. Use at least one extra inverter at the end of the delay chain to square up the output signal before it returns to the system.

The disadvantage of the circuit in Figure 3.23A is that the signal must traverse the physical potentiometers. For high-speed systems, this implies that the potentiometers must be unusually small and that they be located physically close to the active circuitry. The circuit in Figure 3.23B avoids those difficulties by using varactors. A *varactor* is a reverse biased diode whose capacitance varies as a function of the voltage applied across it. The circuit in Figure 3.23B works at higher speeds than the circuit in Figure 3.23A.

Each section of the variable phase shift network shown in Figure 3.23B delays its input from 2.5 to 5 ns. This network uses an *RC* phase shift network adjusted by MV209 varactor diodes.[15] Cascading several sections increases the total delay variation. Figure 3.23B uses two sections, giving a delay range from 5 to 10 ns.

This particular design works well with a clock frequency of 40 MHz. For use at other frequencies, pick a new value for *R:*

[15]The capacitance of this diode varies over the range 14–40 pF, as the DC control voltage across it varies from 10 to 1 V.

$$R = 100\ \Omega\ \frac{40\ \text{MHz}}{F_{\text{clock}}} \qquad [3.28]$$

Provide a separate regulated power supply for the variable delay unit and keep it at a constant temperature (outside your temperature test chamber) for greater stability.

With either circuit, from a single clock source, run one clock (A) through the adjustable delay, and the other clock (B) through a fixed length of coax directly to the bus. Use a coax impedance which matches the native impedance of the bus. Pick the length of the fixed coax such that, with the adjustable delay set to midrange, the two clocks come out with matched timing.

3.10.2.4 Clock adjustment by a phase-locked loop

Figure 3.24 outlines the Cadillac of clock phase adjustment circuits. For large-scale production testing, it may be worth building such a circuit. For general lab use, it is usually too much trouble.

The circuit divides the bus clock by N and compares it, using a phase-frequency-type comparator, with a local oscillator, also divided by N. The circuit locks the local oscillator onto the same frequency as the bus clock, but with a phase determined by the phase-shifting network.

Because phase locking occurs at a frequency N times lower than the clock oscillator, an adjustment by Y degrees in the phase shift network results in an adjustment of $N \times Y$ degrees in the high-frequency clock output. As a result, the phase shift network need only produce a trivial phase shift at the divided clock rate. A varactor-controlled RC shifter does the job handily.

This circuit can make adjustments greater than ±180 degrees. Large adjustments are useful for systems which distribute one high-speed clock and then divide it down locally to produce local control signals. Large adjustments are also useful for debugging

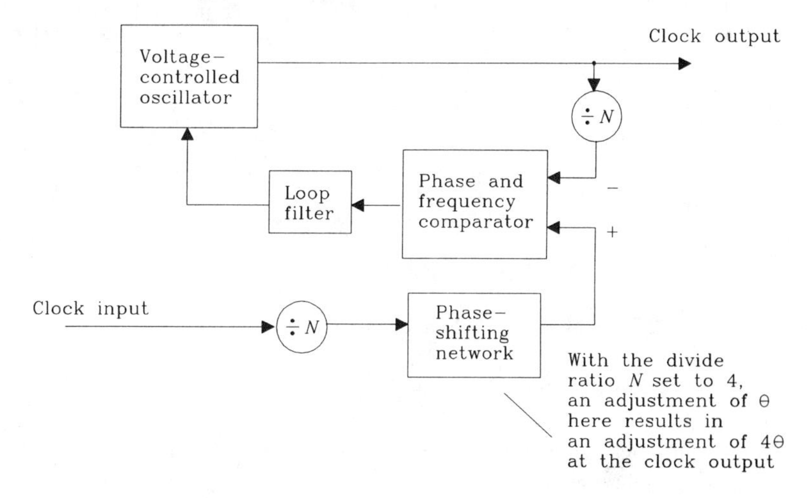

Figure 3.24 Adjustable-delay network using phase-locked loop.

asynchronous circuits intended to accommodate several clock cycles worth of jitter, such as T3 synchronizers used in telecommunications, and FIFO circuits.

The stability of the VCO and noise immunity of the phase-detecting network are critical in this circuit. Unless analog design is your forte, get some help building this circuit.

3.10.2.5 Clock adjustment by voltage variation

Varying the terminating voltage or adding a voltage to the clock lines through pull-up or pull-down resistors will induce minor changes in the clock receiver switching times, thus adjusting the effective clock phase. The same approach works for bus interfaces.

The disadvantage of this approach is that the reliable adjustment range is limited to only a small fraction of the signal rise time.

3.10.3 Power Supply

Adjusting the logic power supply over a ±10% range induces a small amount of delay variation. You may be able to modulate the failure rate of an extremely sensitive system by adjusting the power supply. More likely, a system with adequate margins will just work across the range.

The curves in Figure 3.25 of delay and setup time versus power supply voltage for CMOS and TTL flip-flops indicate the expected range of variation. The CMOS 74HC174 circuit is more than twice as sensitive as the TTL 74F174 circuit to power supply variations.

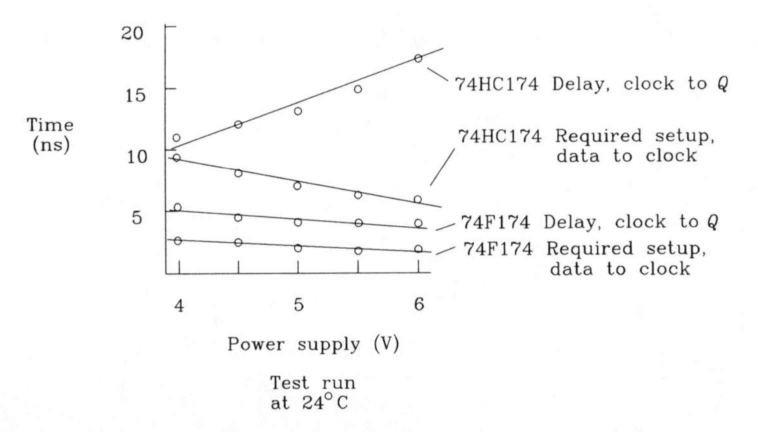

Figure 3.25 Delay and setup time versus power supply voltage.

3.10.4 Temperature

Temperature variations are similar to power supply variations; they induce small changes in delay characteristics.

Temperature variations are more difficult physically to implement than power variations. On the bench, engineers commonly use cans of cooling spray to lower the spot temperature on a circuit, or king-sized blow dryers to raise the temperature.

Remember that many cooling sprays contain dangerous chemicals that damage the earth's protective ozone layer. If you must use a cooling spray, first construct a small cardboard cage around the area of the circuit that needs cooling and then direct the spray into the cavity. The amount of cooling needed in such a small enclosed area will be greatly reduced, and the rate of temperature drift back toward room temperature will be slowed.

It takes considerable skill to induce consistent temperature variations by hand, modulating the duty cycle of the heated (or cooled) air to control the temperature. Always tape a temperature sensor on top of the circuit in question to make sure you do not violate its maximum operating temperature rating.

Systems with air intakes have a natural port for forcing hot or cold air throughout the entire system. A dryer vent hose and duct tape can permanently connect the system to sources of hot or cold air, or just point a hot air gun at the air intake.

Many companies invest in large hot or cold ovens or rooms for thermally cycling products as part of their manufacturing final test procedure. These rooms are uncomfortable for engineering development, restricting both the type and size of auxiliary equipment that will fit in them. Also, engineers do not like to spend time in temperature chambers, but they do provide a realistic test environment.

The curves in Figure 3.26 of delay and setup time versus temperature for CMOS and TTL flip-flops indicate the expected range of variation. The CMOS 74HC174 circuit is more than four times as sensitive as the TTL 74F174 circuit to variations in temperature.

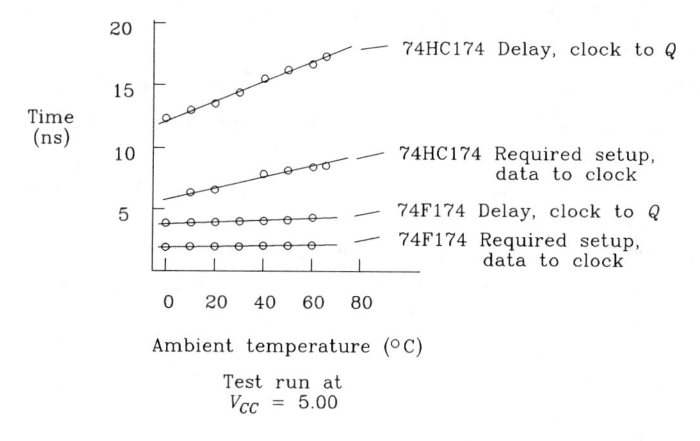

Figure 3.26 Delay and setup time versus temperature.

3.10.5 Data Throughput

Designers commonly use test suites to prove the correctness of logic inside digital machines. A designer might compose a suite of operations that exercises each individual logic connection inside a new machine. Seeing that the results are correct at each step, the natural conclusion is: "The machine works."

Unfortunately, real systems are more complex than that. Many real computers may pass a step-by-step logic test but fail at real operating speed or at real operating throughput. This statement may make little sense until you have accumulated lots of experience with complex systems, but it is true.

At high utilization, buses and other structures inside a high-speed digital machine generate lots of noise. The more data that runs through the machine, the more noise it generates. The best test plans step through progressively higher levels of data traffic intensity, building up to a final test using data patterns of enormous size, emphasizing patterns that stress pipeline logic, memory access, and other timing of critical sections of logic. Good data patterns uncover unexpected avenues of noise coupling which often break down normal operations.

POINT TO REMEMBER:

Measure how much stress a system absorbs before failing a go-nogo test. This procedure converts a simple go-nogo test into a quantitative measure of product quality.

3.11 OBSERVING METASTABLE STATES

Synchronous flip-flop circuits, when treated with respect, behave very predictably. As long as you obey the setup and hold time rules, the Q output faithfully matches the D input after every clock transition.

When using flip-flops to synchronize signals external to a digital machine, we cannot always guarantee the required setup and hold times. An external, asynchronous signal may change at any time without regard to internal synchronous clocks.

How can we resolve this difficulty? Is there a way to clock asynchronous signals into a synchronous digital system without sometimes violating the setup and hold time of the synchronous system? No, there is not. Therefore we must understand what happens to a flip-flop when we violate its setup and hold times.

The effect caused by setup and hold-time violations is called the *metastable state*. This section presents a measurement apparatus for experimenting with metastable states, an explanation of the results, and a few rules for dealing with the problem.

3.11.1 Measuring Metastability

Figure 3.27 illustrates a basic metastable observation fixture for discrete flip-flops. You'll need at least a two-channel oscilloscope to use this fixture.

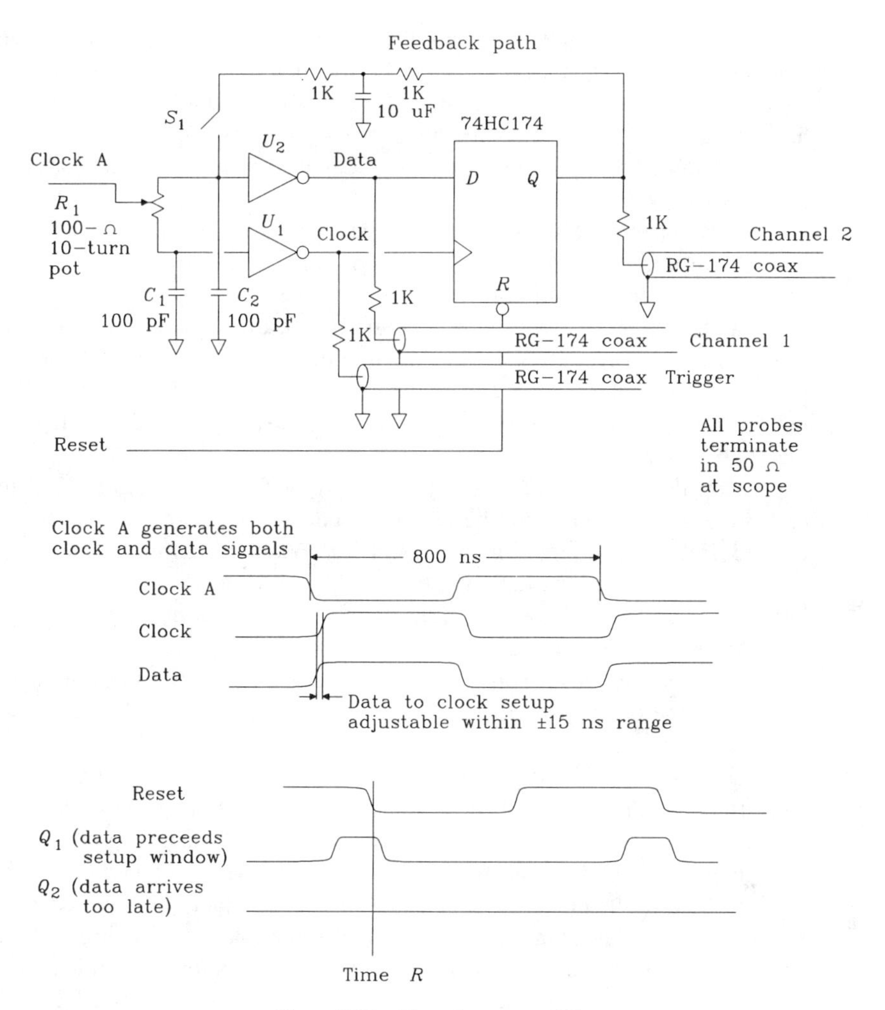

Figure 3.27 Observing metastability.

The CLKA waveform is a square wave, which is delayed through either arm of R_1 and the capacitors C_1 and C_2. With R_1 turned toward the DATA output, the CLK delay is maximized. With R_1 turned toward the CLK output, the DATA delay is maximized. The relative position of DATA versus CLK can be adjusted over a range of about ±15 ns.

The RESET waveform applies a negative-going reset pulse after each positive clock edge. This resets the flip-flop to a known state before each clock. You can make the RESET pulse from a delayed version of the clock.

All critical signals in Figure 3.27 have 21:1 probe taps using 1K-Ω resistors. Connect a scope first to the DATA and CLK signals.

Leaving the feedback switch S_1 open, turn the pot slowly from maximum data advance to maximum delay. Make a rough sketch of the relative DATA-CLK timing versus the potentiometer setting. Check that the adjustment spans a wide enough range. At maximum data advance, the data should arrive before the minimum required setup time. At maximum data retard the data should not turn on until after the minimum hold period has passed.

Calculate how many picoseconds of delay adjustment you get for each turn of the potentiometer.

Now connect the scope to the CLK and Q signals. Terminate the DATA coax with a 50-Ω resistor, so its response won't change. Trigger the scope on the CLK signal and adjust the potentiometer to maximum data advance.

At first, the D input clearly meets the setup requirements, and the Q output responds with a waveform like Q_1, shown in Figure 3.27. Every clock the Q output is set HI, and every cycle at time (R) the Q output is reset to LO. Don't reset the flip-flop with the inverse of the clock or else the transients caused by the reset will mix in with the metastable effect.

Turn the pot, retarding the data, until, after the data passes beyond the minimum setup window, at some point the Q output suddenly snaps off. The data is now arriving too late, and the Q output never switches HI, a condition drawn as waveform Q_2 in Figure 3.27. The position of the data, relative to the clock, at which the flip-flop just fails to latch the rising D input is called the *critical switching point*. The critical switching point lies between the minimum setup and hold times quoted by the flip-flop manufacturer. The manufacturer provides a spread between these two limits to ensure that the critical switching time on all parts, across extremes of temperature and voltage, stays between the limits.

Data arriving before the critical switching point seems to be always latched, and data after this time seems to be always not latched. Isn't that what we want? Yes, but we must look closer to understand the true nature of the metastability problem.

Figure 3.28 plots measurements made with this setup comparing the flip-flop delay, clock to Q, with the measured data setup time. In this plot, the time scale shows the difference between the actual data setup time and the critical switching time on a logarithmic scale. Whenever the data arrives more than 3 ns before the critical switching time, the clock-to-Q delay stays fixed at 13.5 ns. As the data moves toward the critical switch-

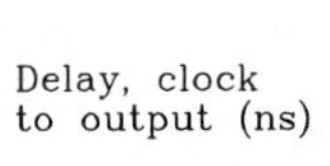

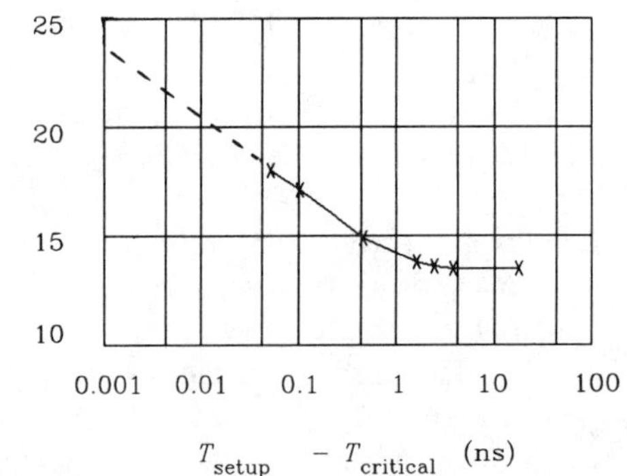

Figure 3.28 Output delay versus data setup time 74HC174.

ing boundary, the Q output still switches HI, but *the clock-to-Q delay gets longer.* For data arrivals very close to the critical switching point, the clock-to-Q delay is proportional to the logarithm of the difference between the data setup time and the critical switching point.

This increase in the clock-to-output delay as a function of the input setup time is the essence of metastability. You cannot work around it, all flip-flops do it, and it wreaks havoc with high-speed synchronizer designs. *The only possible cures make it less likely but never eliminate it.*

How long can the clock-to-output delay become? That depends on how close to the critical switching point the data waveform lies. It could be very long indeed. The next section explains why.

3.11.2 Understanding Metastable Behavior

Figure 3.29 is a simplified schematic of a digital flip-flop circuit. In this example, symmetric positive and negative voltages power the amplifier. The positive feedback drives any positive voltage on capacitor C toward the positive supply, or any negative voltage toward the negative supply.

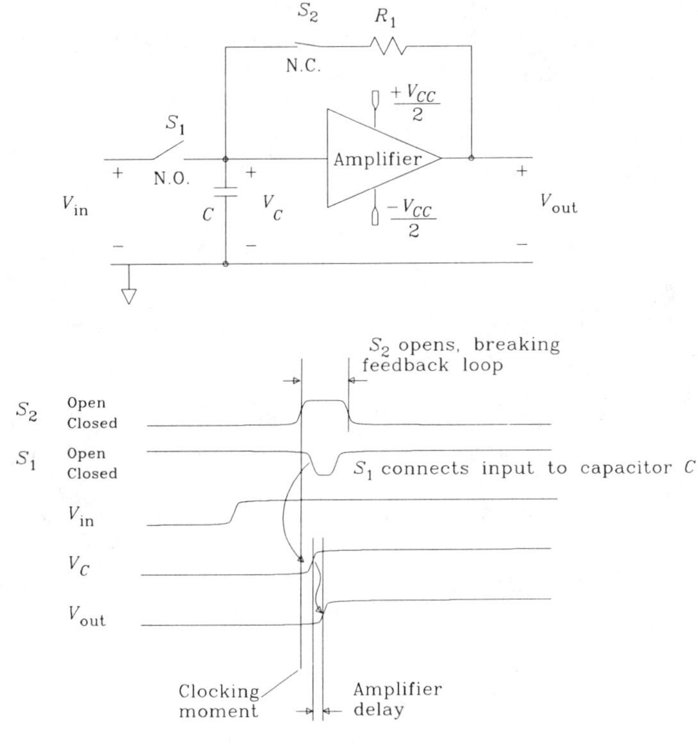

Figure 3.29 Simplified flip-flop circuit.

When clocked, this circuit will always stabilize to either a positive or negative state. All flip-flops operate on this or similar principles.

The lower portion of Figure 3.29 shows a timing diagram for the flip-flop. At the clocking moment, switch S_2 opens for a short period. While S_2 is open, switch S_1 briefly pulses closed, charging capacitor C to the input voltage V_{in}. When S_2 closes again, ending the cycle, positive feedback through resistor R_1 forces the amplifier to saturate either at the positive or negative state, holding the latched bit.

Chip manufacturers have tried all kinds of crazy circuits to get the sequencing of S_2 and S_1 just right. Regardless of what circuit they try, the flip-flop always exhibits a metastable effect.

If the input is a binary logic signal, it should be clearly positive or negative at all times. The flip-flop amplifier, once slewed in the correct direction while S_1 is closed, merely holds the circuit in one state or another.

What happens if the flip-flop clocks just as the input is changing? When S_1 closes, capacitor C charges to match the input. When S_1 opens, capacitor C is left with whatever charge was present at the input when the switch opened. If switch S_1 opens just as the data input is changing, we may latch onto capacitor C a voltage very near zero. That doesn't look very binary!

The setup and hold-time requirements on flip-flops ensure that data is never changing while switch S_1 is open. Inside a synchronous digital system, we can guarantee these requirements are always met. When interfacing to external asynchronous signals, we cannot prevent data from changing at the clocking moment.

How long the amplifier takes to slew to one power rail or the other depends on the value of V_C when S_2 closes. From that moment, the amplifier is exponentially unstable, having an output voltage equal to

$$V_{out}(t) = V_{in}\, e^{Kt} \qquad [3.29]$$

The value K is a time constant related to the amplifier bandwidth and the feedback component values.

If the input voltage at the sampling instant happens to be very close to zero, it could take a long time for the output to slam against one power rail or the other. This hesitation is called a *metastable state*.

Since the succeeding logic needs a 90% transition to satisfy its voltage margins, we must wait for the amplifier to respond completely before declaring its latching operation complete.

The metastable delay can be quite long if the input voltage is very near zero. How close to zero must the input be to generate a metastable delay of T seconds?

Solve Equation 3.29 for the case where the output just reaches the power rail at time T:

$$\left|V_{in} e^{KT}\right| = \frac{V_{CC}}{2} \qquad [3.30]$$

$$\left|V_{in}\right| = \frac{V_{CC}}{2e^{KT}} \qquad [3.31]$$

where V_{in} = how close the input must be to zero
T = to generate a metastable delay of T seconds
K = constant dependent on amplifier and switches
V_{CC} = power supply voltage

Equation 3.31 establishes the relation between the input voltage at the sampling moment and how long you must wait for an answer from the flip-flop, the *resolution time, T.*

Using the rise time of the input signal, we can convert the voltage V_{in} into a time offset. This translation is accomplished by observing that for voltages near zero, the signal waveform is linear with a slope proportional to the edge transition rate. If the input signal transition is located within T_w of the clocking moment, the input voltage will be within V_{in}.

$$T_w = V_{in}\frac{T_{10-90}}{V_{CC}} \qquad [3.32]$$

Equation 3.33 translates the same result into the time domain, telling us the relation between when an input arrives and how long we must wait for an answer.

Substituting Equation 3.31 for V_{in} in Equation 3.32:

$$|T_w| = \frac{T_{10-90}}{2}e^{-KT} \qquad [3.33]$$

If a rising data edge arrives outside the metastable window of $\pm T_w$, the output delays less than T seconds. If the data arrives inside the metastable window, the output data delays more than T seconds.

All flip-flops exhibit metastable behavior. Their metastable window width T_w can always be characterized as

$$|T_w| = Ce^{-KT} \qquad [3.34]$$

The constants C and K are properties of the particular flip-flop used, and T is the resolution time.

EXAMPLE 3.4: Metastable Error Rate

What is the chance that the circuit in Figure 3.30, implemented using the Actel ACT-1 gate array, will produce an output pulse when the input changes? Simplistic application of synchronous logic theory implies it never will, but we know better.

First check the worst-case setup times:

$$\begin{aligned} T_{PD} &= 9.3 \text{ ns (clock to } Q_1\text{, with good setup time)} \\ T_{PD} &= 9.3 \text{ ns (inverter-XOR combination)} \\ T_{SU} &= \frac{5.0 \text{ ns}}{23.6 \text{ ns}} \quad \text{(setup time for } D_2\text{)} \end{aligned}$$

Any clock slower than 42 MHz (23.6 ns) should satisfy all the propagation and setup times. Y_1 and Y_2 should always match, and the output Q_4 should never go high.

The only way the circuit can fail is if metastability causes Q_1 to change late, missing the setup window for D_2 (because of the propagation delay through gates G_1 and G_2), but not so late that D_3 misses it.

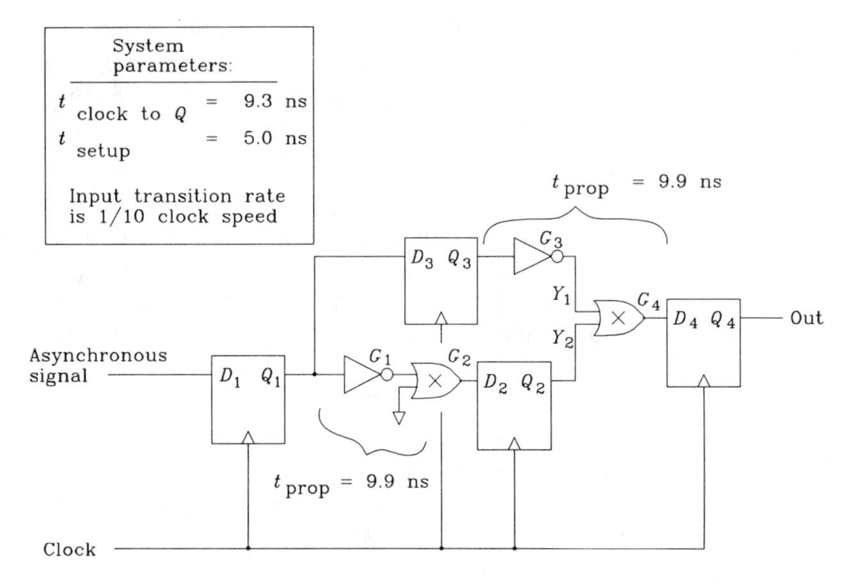

Figure 3.30 Analysis of a metastable circuit.

If the actual clock speed F is less than 42 MHz, then we can budget some metastable delay for Q_1 without missing the setup window for D_2. The permissible extra delay time allotted to metastability is

$$T_r = \frac{1}{F} - 23.6 \text{ ns} \tag{3.35}$$

This delay T_r is called the allowable *resolution time.*

The metastable window within which Q_1 takes longer than T_r to stabilize is

$$T_w = Ce^{-KT_r} \tag{3.36}$$

The probability of hitting within $\pm T_w$, out of a total cycle time of $1/F$, is

$$\text{Prob(failure)} = 2T_wF = 2FCe^{-KT_r} \tag{3.37}$$

The Actel 1989 *ACT-1 Family Gate Arrays Product Guide* lists constants C and K. Here we have adjusted both constants for our units system of hertz and seconds:

$$C = 0.5 \times 10^{-9} \text{ (sampling switch rise-time constant)}$$
$$K = 4.6052 \times 10^{9} \text{ (amplifier response time constant)}$$

The mean time between failures (MTBF), in hours, may be computed from the probability of failure and R, the input transition rate. Since metastability can happen only when the input changes, the more input transitions there are, the greater the chance of failure:

$$\text{MTBF} = \frac{0.000277}{\text{Prob(failure)} \times R} \tag{3.38}$$

where MTBF = mean time between failure, h
R = input transition rate, Hz
Prob(fail) = probability of failure on any single input transition

Figure 3.31 plots the resulting MTBF-versus-clock frequency. This figure assumes the input transition rate is one-tenth the clock rate. At 35 MHz, the probability of failure is 4×10^{-12}. If the circuit handles 3.5 million inputs per second, it will fail every 19 h (about once a day).

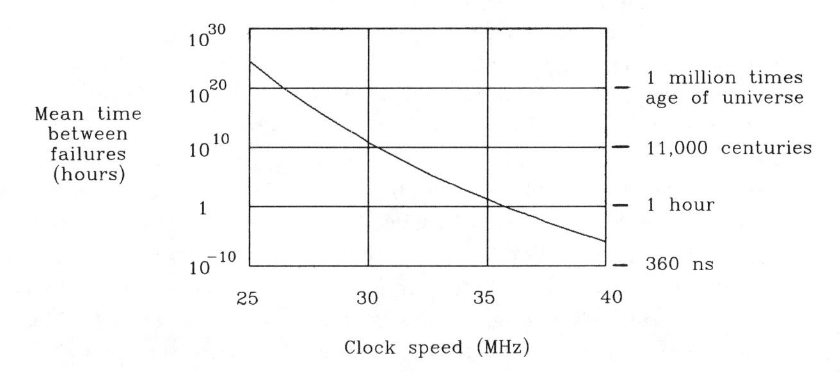

Figure 3.31 Mean time between failures (MTBF) versus clock speed for the circuit in Figure 3.30.

3.11.3 Evidence for Very Long Resolution Times

According to the data presented in Figure 3.28, we must adjust the potentiometer in Figure 3.27 to within 10 ps of the critical switching time to produce a delay, clock-to-output, greater than 20 ns. This is a difficult feat.

Fortunately, there is another way to achieve very accurate delay adjustments. We can construct a feedback network that monitors the output for metastability, controlling the clock-to-D input skew to produce very long resolution times.

The feedback circuit amplifies the number of metastable events so that we can easily see them.

This circuit appears at the top of Figure 3.27. It is composed of a "tee" section *RCR* low-pass filter which monitors the Q output voltage, feeding it back to the DATA buffer U_2.

When the rising edges of the DATA signal arrive too early, the Q output switches ON every cycle. This raises the average voltage of the Q output. The "tee" filter responds by injecting positive current into the input node of U_2, slightly raising the voltage of the delayed CLKA signal at that point. The negative edges of CLKA, now raised more positive than normal, cause U_2 to switch HI slightly later than usual, effectively delaying the rising edges of the DATA signal.

The overall effect is to regulate the position of the DATA edge transitions. The control range is ±100 ps. Once the potentiometer adjustment brings the DATA signal within 100 ps of the critical switching time, the control loop begins to work. The potentiometer adjustment then becomes much less sensitive and easier to adjust.

With the potentiometer adjusted to produce maximum delay, we get Figure 3.32. The first waveform is the DATA input, the second is the CLK input, and the speckled waveform is the Q output. Because the scope samples its input waveforms instead of continuously displaying them, the scope catches only one point on each waveform, causing the speckled dots.

Sometimes the Q output waits 24 ns and goes HI. Sometimes it goes LO. Other times it waits much longer before deciding to go HI.

The *least* amount of delay, clock-to-output, is 24 ns. Remember that the nominal switching delay, for good inputs, is 13 ns (Figure 3.28). This long delay indicates that the DATA timing remains held by the feedback loop to within a few picoseconds of the critical switching point. Within this limit, the actual DATA transition moment slurs back and forth randomly. This random behavior is due to thermal noise inside the flip-flop, and random external noise injected into the circuit. Very close to the critical switching point, the DATA sample is hitting all possible time instants with equal probability.

The digital sampling scope runs in point accumulation mode, keeping every sample point on the screen forever. The scope accumulates points until it has 20 points for each of the 512 possible horizontal positions on the screen. The DATA and CLK signals were prerecorded separately and digitally superimposed on the final picture. The accumulation time for this picture is 3 seconds.

The rightmost points indicate occasional delays, clock-to-Q, of at least 30 ns. How likely are these events?

Reasoning with Equation 3.34, the DATA window width exceeding a given resolution time decreases exponentially with resolution time. If the DATA arrival time distrib-

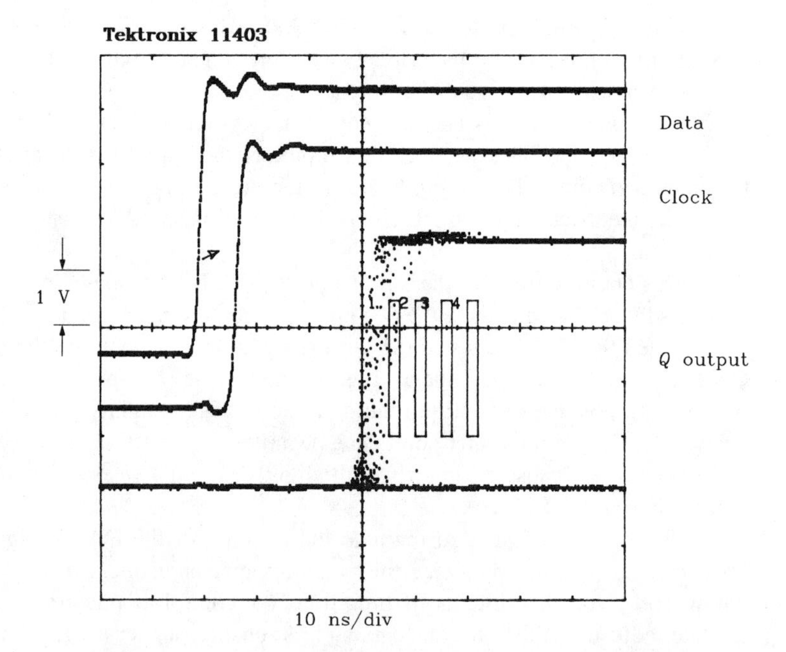

Figure 3.32 Metastability in 74HC174 (CMOS) 3-s point accumulation.

utes evenly near the critical switching point, we expect to see an exponential decay in the probability of long resolution times. That is, for every fixed increase in resolution time, we expect to get a fixed percentage decrease in the number of events exceeding that resolution time.

We can directly test this hypothesis with the Tektronix mask-counting feature. The four boxes in Figure 3.32 define four mask-counting regions. The scope counts every dot inside or touching each mask. The masks are equally spaced at 5-ns intervals (35, 40, 45, and 50 ns after the clock).

In this example, masks 1 and 2 received 13 and 1 hits, respectively. No hits occurred in masks 3 or 4. We expect the number of hits per mask to decay exponentially, but we do not yet have enough hits to verify our calculations.

Figure 3.33 uses the same setup as Figure 3.32 but leaves the point accumulation on for 30 minutes. The mask counts are

Mask 1	30 ns	4685
Mask 2	35 ns	445
Mask 3	40 ns	42
Mask 4	45 ns	4

The decay constant between boxes is a factor of 10. The last box, which got four hits, sits at a clock-to-output delay of 45 ns. If we waited 50 h (100 times longer than in Figure 3.33), we could probably get four hits in a box positioned at 55 ns.

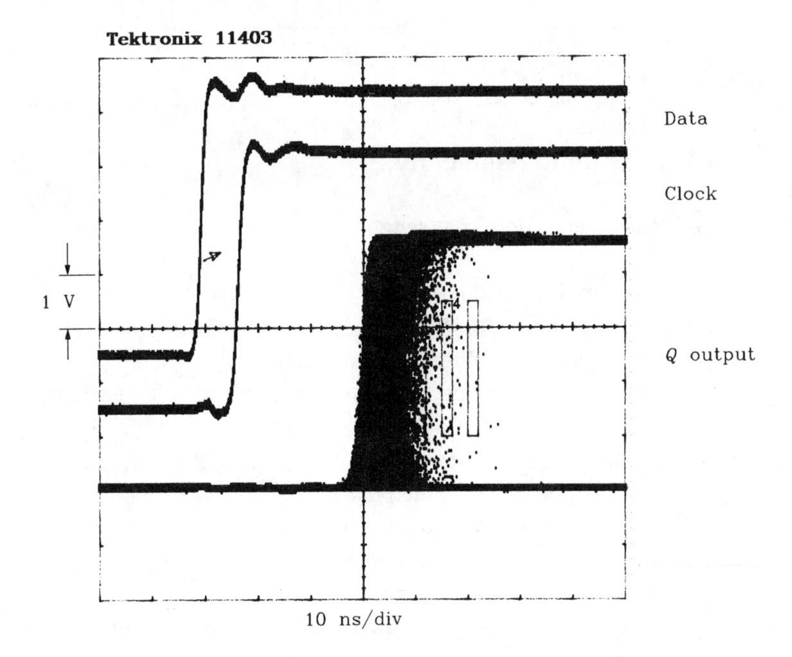

Figure 3.33 Metastability in 74HC174 (CMOS) 30-min. point accumulation.

Figure 3.34 uses a 74F174 flip-flop for the same experiment. Its delay times are much shorter than the 74HC174, but the effect is the same. Note that the 74F174 output, having less buffering than the 74HC174 output, tends to rise to half-mast and then make its decision one way or the other. This output glitch can easily trigger edge-sensitive circuitry following the *Q* output.

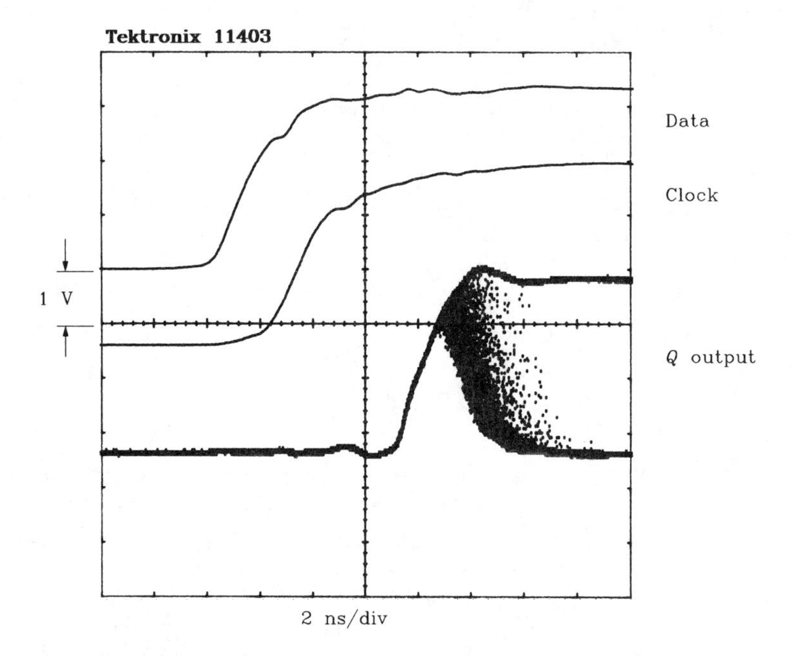

Figure 3.34 Metastability in 74F174 (TTL) 10-s point accumulation.

3.11.4 Cures for Metastability

The following suggestions may help if you are experiencing metastable problems:

(1) Use a faster flip-flop; it will probably have a narrower metastable window.

(2) Put two (or more) flip-flops in series, all clocked by the same clock. A chain of N flip-flops has a probability of error equal to P^N, where P is the chance of metastable failure for one flip-flop. Standard practice calls for at least two, and sometimes three, flip-flops in series for every asynchronous input.

(3) Use a metastable-hardened flip-flop. These devices combine a really fast low-power flip-flop inside, which has a very large K value, with a regular-speed output driver. They have very attractive metastable characteristics.

(4) Sample less often (if possible). This wider clock period reduces the probability of hitting the metastable window and gives the flip-flop more time to resolve its output. The failure rate goes down more than exponentially with decreasing clock speed.

(5) Some flip-flops have worse metastable problems with slow-moving input signals. Use a fast input edge rate.

POINTS TO REMEMBER:

All flip-flops exhibit metastability.

The probability that a flip-flop output will delay more than *T* seconds goes down exponentially with increasing time *T*.

4

Transmission Lines

At high frequencies, transmission lines are superior in three ways to ordinary point-to-point wiring:

- Less distortion
- Less radiation (EMI)
- Less crosstalk.

In exchange for these good properties, transmission lines require more drive power than ordinary point-to-point wiring. For high-speed circuitry, the improved signal performance is well worth the additional power.

Before studying the benefits of transmission lines, let's consider an example highlighting the disadvantages of ordinary point-to-point wiring (wire-wrap technology).

4.1 SHORTCOMINGS OF ORDINARY POINT-TO-POINT WIRING

A now-famous Silicon Valley company, which we will call NEWCO, once built a large prototype of their first high-speed processor. They decided to use point-to-point wiring to avoid the expense and delay of making a printed circuit board. The prototype was constructed using wire-wrap technology on a circuit card measuring 16 in. × 20 in. This prototype included over 600 gates, with 2000 distinct signal nets. Here are some statistics about the signal nets:

Number of nets	2000
Average net length	4 in. (no terminations)
Average wire height above ground	0.2 in.
Wire size (AWG 30)	0.01 in. diameter
Signal rise time	2.0 ns
Knee frequency (Equation 1.1)	250 MHz (= 0.5/2.0 ns)

The next three sections examine how NEWCO's point-to-point wiring performed.

4.1.1 Signal Distortion in Point-to-Point Wiring

NEWCO's planned rise time of 2.0 ns has an electrical length of (Equation 1.3):

$$l = \frac{\text{Rise time (ps)}}{\text{Speed (ps/in.)}} = \frac{2000 \text{ ps}}{85 \text{ ps/in.}} = 23.5 \text{ in.} \qquad [4.1]$$

The critical dimension separating lumped from distributed systems is

$$l/6 = 3.9 \text{ in.} \qquad [4.2]$$

NEWCO believed that, because their average wiring lengths fell near this boundary, their circuits would experience little ringing. They were wrong.

NEWCO realized that circuits larger than $l/6$ behave in a distributed fashion. They knew distributed circuits always ring[1] unless terminated. That subject is treated in Section 4.3. Because their circuits were (mostly) lumped, they incorrectly thought they would not ring.

Lumped circuits may or may not ring, depending on the Q of the circuit. The measure Q of a circuit indicates how quickly signals on the circuit die out. Signals on a low-Q circuit damp quickly, while those on a high-Q circuit bounce around, slowly dissipating over several ringing cycles.

The value of Q is technically defined as the ratio of energy stored to energy lost per radian of oscillation. From this arcane definition follows an approximation[2] showing the maximum overshoot voltage of a particular circuit as a function of Q:

$$\frac{V_{\text{overshoot}}}{V_{\text{step}}} = e^{-\left[\frac{\pi}{(4Q^2-1)^{1/2}}\right]} \qquad [4.3]$$

where $V_{\text{overshoot}}$ = amount the output rises above the steady-state output level, V
V_{step} = expected steady-state level, V
Q = resonance parameter (here we assume $Q > 0.5$)

The ideal second-order circuit in Figure 4.1 decays with time constant $2L/R$, exactly obeying Equation 4.3.

As a rule of thumb, a digital circuit with a Q of 1 displays, in response to a *perfect step input*, a 16% overshoot. A digital circuit with a Q of 2 displays a 44% overshoot. Any Q below $\frac{1}{2}$ has no overshoot or ringing. The ringing experienced by a circuit also is

[1]Ringing on a signal line includes both overshoot (going past the mark) and undershoot (attaining the mark and then falling back before stabilizing).

[2]This approximation is derived from the solution to a second-order linear differential equation describing an *RLC* low-pass filter. First find the point at which the derivative of the solution passes through zero (a maximum point) and then evaluate the solution at that point. The boundary conditions are $F(0) = 0$, $F(\infty) = 1.00$, $F'(0) = 0$.

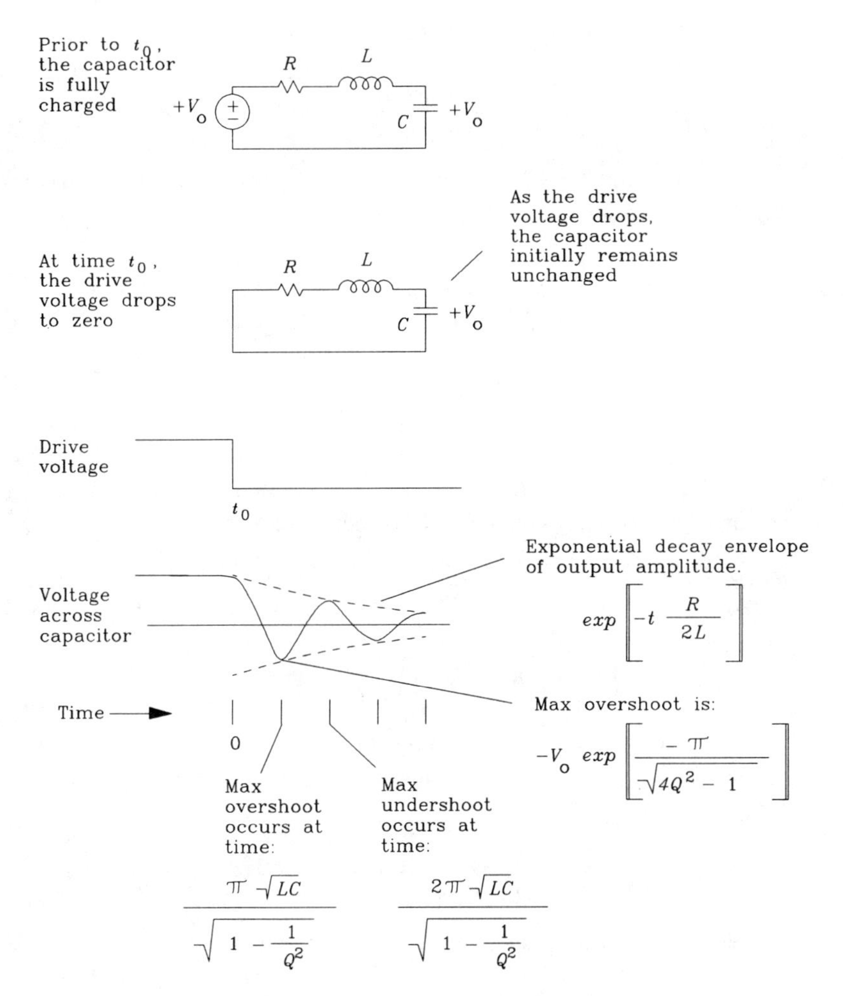

Figure 4.1 Overshoot and ringing calculated by the Q method.

a function of the relation between the natural ringing frequency of the circuit and the rise time of the driver. We will investigate that, too.

Calculating the Q of a digital circuit is easy once we know the circuit inductance. That leads us to the basic problem with point-to-point wiring: high inductance.

A high wiring inductance, when working into a heavy capacitive load, makes a high-Q circuit.

We may use the formula listed in Appendix C for the inductance of round wire suspended above a ground plane to calculate the inductance L of a typical net in the NEWCO system:

$$L = X\left(5.08\times10^{-9}\right)\left(\ln\left(\frac{4H}{D}\right)\right) = 89 \text{ nH} \qquad [4.4]$$

where L = inductance of loop, H
D = diameter of wire-wrap wire, 0 .01 in.
H = height of wire above ground, 0.2 in.
X = length of wire, 4 in.

Using Equation 3.12, we can calculate the Q of the *RLC* circuit formed by the driver source resistance, the series inductance of the wire, and the load capacitance at the receiver:

$R = 30\ \Omega$ (output resistance of a TTL driver)
$L = 89$ nH (average wiring inductance)
$C = 15$ pF (typical load)

$$Q \approx \frac{(L/C)^{1/2}}{R_S} = \frac{(89\ \text{nH}/15\ \text{pF})^{1/2}}{30\ \Omega} = 2.6 \quad [4.5]$$

A value for Q of 2.6 implies that, with a perfect step input, we will get quite a bit of ringing. The expected worst-case overshoot voltage (from equation 4.3):

$V_{\text{step}} = 3.7\ V$ (TTL step output)
$Q = 2.6$ (from Equation 4.5)

$$\text{Overshoot} = V_{\text{step}} \exp\left(\frac{-\pi}{(4Q^2 - 1)^{1/2}}\right) = 3.7e^{-0.616} = 2.0\ \text{V} \quad [4.6]$$

This worst-case overshoot happens only if the NEWCO logic drivers can transmit significant energy at frequencies above the ringing frequency. We can find the ringing frequency using Equation 4.7:

$$F_{\text{ring}} = \frac{1}{2\pi(LC)^{1/2}} = \frac{1}{2\pi[(89\ \text{nH})(15\ \text{pF})]^{1/2}} = 138\ \text{MHz} \quad [4.7]$$

Our standard measure of the spectral content is the knee frequency, defined by Equation 1.1. The knee frequency of NEWCO's logic gates (250 MHz) is well above the ringing frequency (138 MHz), and so there is plenty of electric energy to excite fully the ringing behavior. A knee frequency of exactly 138 MHz would attenuate the ringing by about half. Logic gates with lower knee frequencies induce even less ringing.

Thinking entirely in the time domain, we conclude that when the rise time equals one-half the ringing period, the worst-case ringing is reduced by half. Longer rise times excite less ringing, while rise times much shorter than one-half the ringing period excite worst-case ringing.

There is one more fact we can squeeze from our Q analysis. We know an average NEWCO circuit rings at a frequency of 138 MHz, with a maximum overshoot of 2.0 V. Linear circuit theory tells us the worst overshoot always occurs one-half ringing period after the step edge. We can therefore predict that the maximum overshoot will occur 3.6 ns after each logic transition.

Ringing in the NEWCO prototype is a big problem.

4.1.2 EMI in Point-to-Point Wiring

EMI stands for *electromagnetic interference.* Open-current loops like those found in wire-wrapped products should immediately raise big red EMI flags. Large current loops carrying quickly changing currents generate transient fields. The magnetic fields from these loops radiate directly into the antennas of FCC technicians who will not certify your digital product. See the reference list in the back of this book for more information about FCC regulatory testing.

Transmission lines dramatically reduce EMI. They accomplish this by constraining the flow of return signal currents. With ordinary wiring, current from a logic driver flows out one signal wire and returns somehow along the power wiring. The separation between these two paths, or the total loop area between them, may be several square inches. As measured by FCC test procedures, the resulting magnetic fields are directly proportional to the total area of your product's signal current loops.

Transmission lines are constructed to keep returning signal currents close to the outgoing signal path. The resulting effective current loop area is very small. Magnetic fields from the outgoing and returning current paths cancel, dramatically reducing problems with EMI. The proper engineering of ground paths is discussed in Chapter 5.

In reference to Figure 4.2, a printed circuit board having its signal traces 0.005 in. above a solid ground plane has a current loop area 40 times smaller than open wiring situated 0.2 in. above a ground plane. For the same signal rise time, such a printed circuit board radiates 32 dB less electromagnetic field energy per wire than the NEWCO prototype.

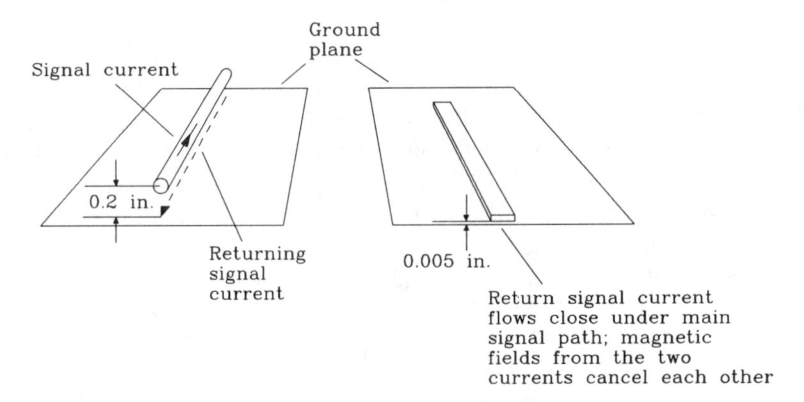

Figure 4.2 EMI is proportional to wire height above ground.

4.1.3 Crosstalk in Point-to-Point Wiring

As diagramed in Figure 4.3, crosstalk arises through the action of changing magnetic fields. Some of the magnetic field lines from currents flowing in loop A penetrate loop B. Changes in the current in loop A therefore change the magnetic flux encircled by loop B. A changing flux in loop B induces noise voltages in loop B, called crosstalk. The constant

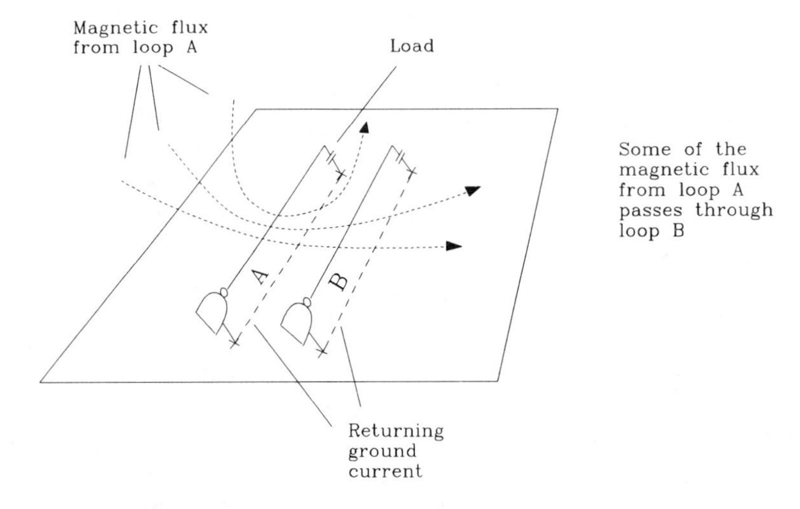

Figure 4.3 Crosstalk in wire-wrapped system.

of proportionality linking the change of current in loop *A* to the voltage in loop *B* is called the mutual inductance between loops *A* and *B*, denoted by the symbol L_M.

Crosstalk in high-speed wire-wrapped systems is a major problem. Let's calculate how much crosstalk NEWCO had. We assume here there are two adjacent loops, each 4 in. × 0.2 in. height, running parallel at a separation of 0.1 in.

Use the formula in Appendix C for the mutual inductance of two parallel wires to compute the mutual inductance. We can use the inductance figure from Equation 4.4 above for the inductance of the transmitting wire.

$$L_M = L\left[\frac{1}{1+(s/h)^2}\right] = 71 \text{ nH} \qquad [4.8]$$

$h = 0.2$ (height above ground plane)
$s = 0.1$ (separation between wires)
$L = 89$ nH (inductance of one wire)

where L_M = mutual inductance between wires

This value is comparable to the self-inductance of an individual net. That says the two nets will be highly coupled; we expect a lot of crosstalk.

The next step in crosstalk calculation is finding the maximum *dI/dt* of the driving loop and multiplying that figure by the mutual inductance to obtain a crosstalk voltage.

At this point, our best guess for the actual rise time of the received signal across the load capacitance is 3.6 ns (that's the time needed to maximum overshoot). Plugging that value into Equation 2.42:

$$\Delta V = 3.7 \text{ V}$$
$$T_{10-90} = 3.6 \text{ ns (estimated)}$$
$$C = 15 \text{ pF (load capacitance)}$$

$$\frac{dI}{dt}(\max) = \frac{1.52 \times \Delta V}{{T_{10-90}}^2} C = \frac{(1.52)(3.7)}{(3.6 \times 10^{-9})^2} 15 \times 10^{-12} = 6.5 \times 10^6 \text{ A/s} \quad [4.9]$$

The crosstalk works out to 12% (0.46 V):

$$\text{Crosstalk} = \frac{dI}{dt}(\max)\, L_M = (6.5 \times 10^6)\,(71 \times 10^{-9}) = 0.46 \text{ V} \quad [4.10]$$

Are you surprised? Only 4 in. of nearby wire generates 460 mV of crosstalk. Within a radius of $\frac{1}{10}$ in., a good technician can easily bundle 10 or 20 wires. The crosstalk from each wire adds linearly. The crosstalk from only 10 nearby wires would be 50%, more than enough to cause serious errors.

Large bundles of wires running in parallel from place to place are a terrible way to build high-speed buses. Technicians like to gather wires together in the rows between chips, so they can see the pin numbers written on the back side of each chip. This practice makes the crosstalk problem much worse. Straight point-to-point wiring, pressed down as close to the ground plane as possible, is much better than gathered or bundled wiring.

As it turns out, NEWCO abandoned its prototype (which had 128-bit bus structures) without ever getting it to function fully. They lost several weeks of valuable design time and were unable to verify key aspects of their design before committing to a printed circuit layout.

POINTS TO REMEMBER:

Distributed circuits always ring if unterminated. Lumped circuits can also ring if their Q is too high.

Point-to-point wiring has a lot of inductance. This inductance, working into a heavy load capacitance, makes a high-Q circuit.

Large current loops carrying quickly changing currents generate transient fields. Reducing current loop area reduces EMI.

Straight point-to-point wiring, pressed down as close to the ground plane as possible, is much better than gathered or bundled wiring.

Systems with thousands of connections warrant extra attention to crosstalk.

4.2 INFINITE UNIFORM TRANSMISSION LINE

Transmission lines have many exotic properties and are the subject of much scientific study. We will study here only those basic aspects which apply to the practice of distributing high-speed digital signals over copper media. Many of the good reference books listed in the back of this text provide more information about transmission lines.

The transmission line forms we will study include the coaxial, twisted-pair, microstrip, and stripline configurations (see Figure 4.4).

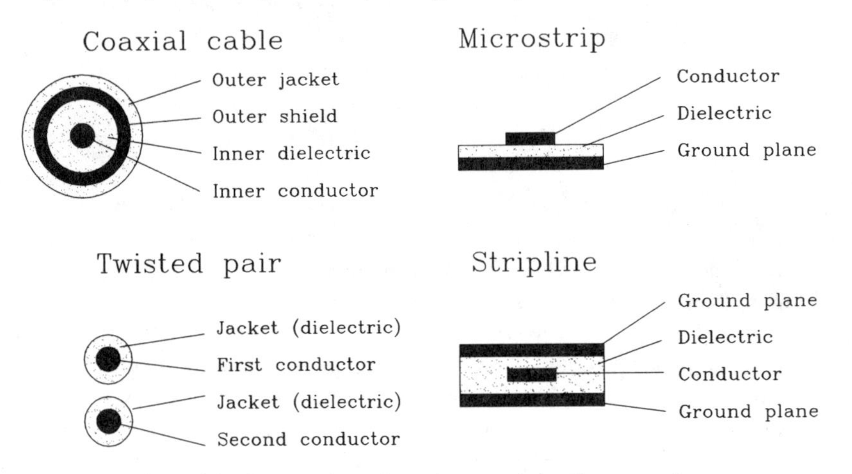

Figure 4.4 Cross sections of popular transmission line geometries.

4.2.1 Ideal Distortionless, Lossless Transmission Line

An ideal transmission line consists of two perfect conductors. These conductors have zero resistance, are uniform in cross section, and extend forever. Figure 4.4 illustrates four popular configurations. These include both balanced (twisted-pair) and single-ended, also called unbalanced, transmission line configurations (coax, microstrip, and stripline). In a balanced transmission line, signal current flows out along one wire and back along the other. In a single-ended transmission line, signal current flows out the signal wire and back along a ground connection. The ground connection in a single-ended transmission line is usually larger than the signal connection and may be shared among many signal wires.

Voltages impressed upon one end of an ideal transmission line propagate forever, at constant velocity, without distortion or attenuation. We will call any transmission line *ideal* which has these three properties:

- It is infinite in extent (it starts here and goes forever in one direction).
- Signals propagating on the line are not distorted as they progress.
- Signals propagating on the line are not attenuated as they progress.

The voltage at any point along an ideal transmission line is a perfect delayed copy of the input waveform. The amount of delay per unit length along a transmission line is called its *propagation delay* and is given in this book in units of picoseconds/inch. *Propagation velocity* and *transmission velocity* both refer to the inverse of propagation delay. Convenient units for velocity are inches/picosecond. Some reference materials rate transmission velocity in percent, where 100% is the velocity of light in a vacuum. The velocity of light in a vacuum is equal to 0.0118 in./ps, or a delay of 84.7 ps/in. A relative velocity of 66% would give a longer delay per inch, equal to

$$\text{Delay (ps/in.)} = \frac{84.7 \text{ ps/in.}}{\text{percent velocity}} = \frac{84.7}{0.66} = 128 \text{ ps/in.} \qquad [4.11]$$

The propagation delay of any transmission line is related to its series inductance per unit length and its parallel capacitance per unit length. It should not surprise you that a section of transmission line would have a parasitic series inductance (conductors always do). All nearby conductors also share some mutual capacitance. In a transmission line, these factors are both proportional to length, and their fine balance is responsible for the distortionless propagation of signals.

Let's measure the capacitance and inductance of RG-58/U coaxial cable, as shown in Figure 4.5. First cut off a 10 in. section of RG-58/U coaxial cable. Using a good-quality impedance meter, measure its capacitance. The correct value is 26 pF, which works out to 2.6 pF/in.

Next short out one end of the same 10-in. section and measure (from the other end) its inductance. Here the correct value is 64 nH, which works out to 6.4 nH/in.

Using a very sensitive four terminal ohmmeter, you may notice that the center conductor of this cable also has a series resistance of 0.009 Ω, or 0.9 mΩ/in. Ideal transmission lines have zero resistance, but for our purposes right now the 10-in. segment of RG-58/U is close enough to ideal.

Electromagnetic wave theory[3] says that the propagation delay is equal to

$$\text{Delay (ps/in.)} = 10^{+12}[(L/\text{in.})(C/\text{in.})]^{1/2} \qquad [4.12]$$

If the inductance and capacitance are specified in units other than inches, then the square root of their product is equal to the delay, in seconds, per unit of distance. Equa-

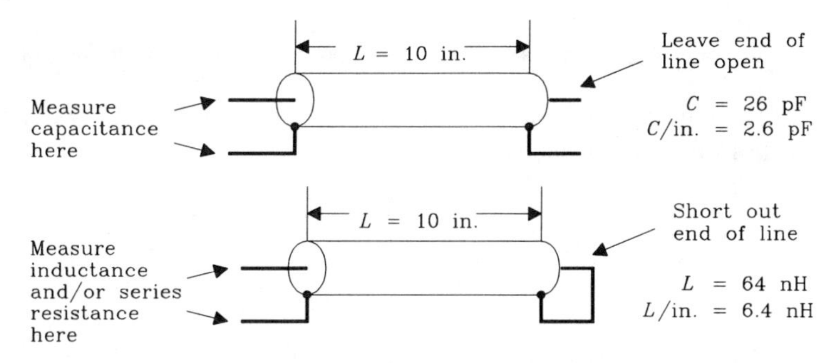

Figure 4.5 Measurement pertaining to inductance and capacitance of transmission lines.

[3]See S. R. Seshadri, *Fundamentals of Transmission Lines and Electromagnetic Fields,* Addison-Wesley, Reading, Mass., 1971.

tion 4.12 computes delay in picoseconds/in., which is convenient for working with printed circuit boards.

Given the capacitance per unit length and the propagation delay, we can determine the input impedance of this transmission line. We will do this by impressing a step voltage at the end of the line and determining how much current must flow into the line to support uniform propagation of the resulting waveform.

Imagine, as shown in Figure 4.6, a step of V volts propagating along the transmission line. Figure 4.6 shows the voltage waveforms as a function of time at the beginning of the cable, and at points X and Y along the cable. At time t_0 the step has passed point X, and T seconds later, at time t_1, the step passes point Y. During interval T the capacitance between points X and Y charges to voltage V.

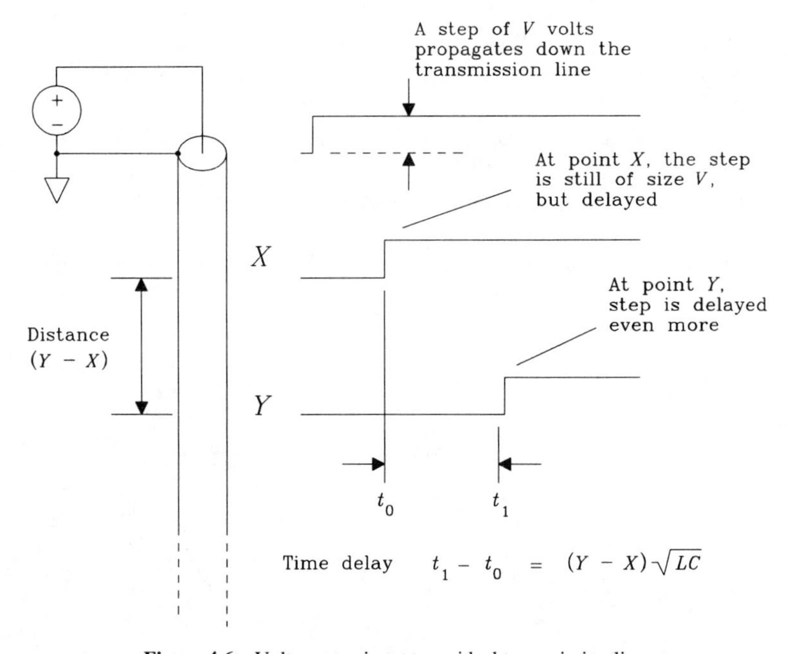

Figure 4.6 Voltage step input to an ideal transmission line.

How much current is required to charge the capacitance between points X and Y to voltage V? First, calculate the capacitance C:

$$C_{XY} = (C/\text{in.})(Y - X) \tag{4.13}$$

The total additional charge (which must come from the driving source) is equal to

$$\text{Charge} = C_{XY}V = (C/\text{in.})(Y - X)V \tag{4.14}$$

The time interval (in seconds) during which C_{XY} must be charged is equal to the separation of the two points times the propagation delay in seconds per unit length:

$$T = (Y - X)[(L/\text{in.})(C/\text{in.})]^{1/2} \tag{4.15}$$

The average current must be equal to the charge supplied per unit time T:

$$I = \frac{\text{charge}}{T} \qquad [4.16]$$

Substituting Equations 4.14 and 4.15 for charge and T, respectively:

$$I = \frac{(C/\text{in.})(Y-X)V}{(Y-X)[(L/\text{in.})(C/\text{in.})]^{1/2}} \qquad [4.17]$$

This gives us the current flow at the input required to sustain a propagating step edge of V volts. Simplifying and solving for the ratio V/I, Equation 4.18 shows us the input impedance of the transmission line, called Z_0, or the *characteristic impedance*:

$$Z_0 = \frac{V}{I} = \left(\frac{L/\text{in.}}{C/\text{in.}}\right)^{1/2} \qquad [4.18]$$

Notice that this ratio, the input impedance, is a constant. It has no imaginary part and is not a function of frequency. This constant ratio is a function of the physical geometry of the transmission line. Characteristic impedance commonly ranges from 10 Ω (between inner and outer shields of triaxial cable) to 300 Ω (in a balanced configuration used for TV antenna connections).

Our piece of RG-58/U has a characteristic impedance of

$$Z_0 = \left(\frac{6.4\text{ nH}}{2.6\text{ pF}}\right)^{1/2} = 50\ \Omega \qquad [4.19]$$

That's the characteristic impedance value for RG-58/U listed in the *Belden Wire and Cable Master Catalog 885*.

Characteristic impedances typically used on printed circuit boards vary from about 50 to 75 Ω. Trace dimensions needed to produce roughly these impedances using an epoxy FR-4 substrate are diagramed in Figure 4.7. Appendix C contains accurate formulas for characteristic impedance. We denote the characteristic impedance of an ideal transmission line with the symbol Z_0.

Let's drive a signal into an ideal transmission line from a drive circuit having a fixed output impedance. The result appears in Figure 4.8. This figure also shows the result of driving the same signal into a resistor and into a capacitor. In all cases the drive signal is a unit step, and the output impedance of the drive circuit is R_S.

The resistive load R_L creates a simple voltage divider. This voltage divider reduces the output voltage at point A to a fixed fraction of the actual drive voltage. If the impedance of the load exceeds that of the driver, most of the drive signal amplitude appears at point A.

The ideal transmission line, having a resistive input impedance, acts exactly like a resistive load. The voltage appearing at point B, which actually propagates down the cable, is a fixed fraction of the unloaded drive voltage V_0. Equation 4.20 is called the *input acceptance equation* for a transmission line:

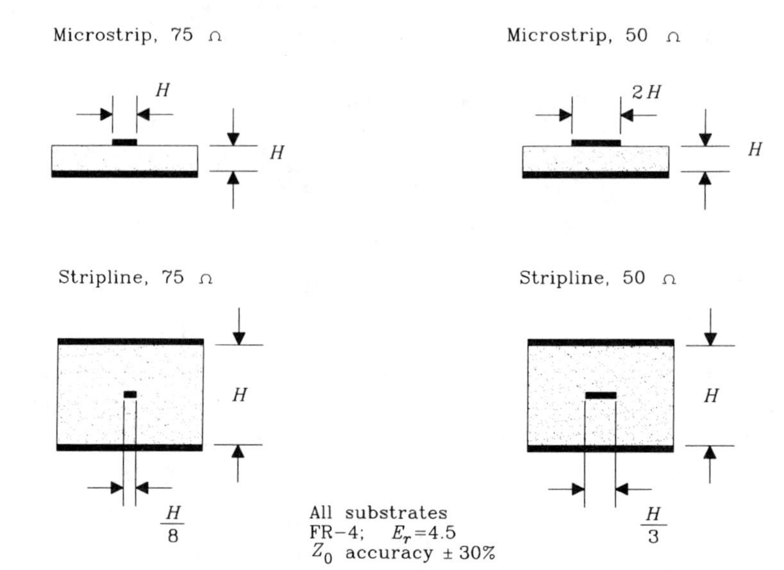

Figure 4.7 Cross sections of approximate trace geometries needed to produce 50- and 75-Ω transmission lines.

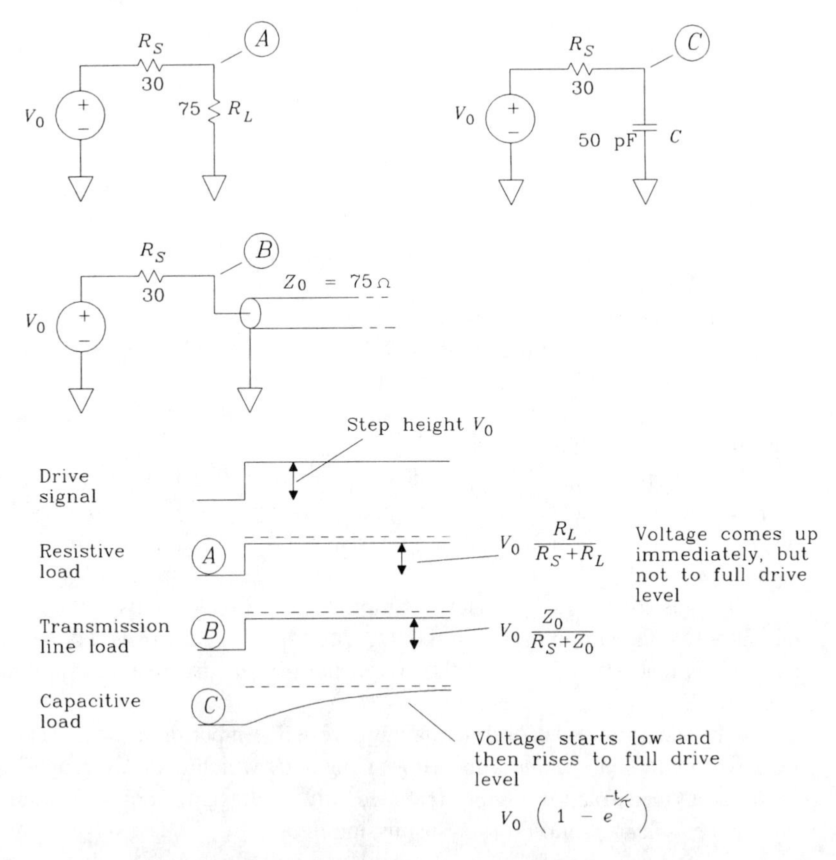

Figure 4.8 How an ideal transmission line differs from a capacitor.

$$V_{\text{accepted}} = V_0 \frac{Z_0}{R_S + Z_0} \quad [4.20]$$

An ideal transmission line does not behave like a capacitor. The short-term impedance of a capacitor is quite low, attenuating the drive signal at first. As time progresses, current flowing through resistor R_S charges up the capacitor, and the output voltage at point C rises, asymptotically approaching its final value. The final value is equal to the drive voltage.

There are circumstances under which a transmission line, driven by too high a source impedance, can load a drive circuit like a capacitor. Section 4.4 explains this effect. For now, keep in mind that the input to an ideal infinite transmission line looks resistive, not capacitive.

4.2.2 Lossy Transmission Lines

An ideal transmission line has zero resistance. A practical transmission line always has some small series resistance. The nonzero resistance of practical transmission lines causes both attenuation (loss) and distortion in propagating signals. This section shows how to estimate the resistance of a transmission line and how to estimate its attenuation.

Series resistance, for long cables, is measured in ohms/1000 feet. When dealing with twisted-pair cables, this resistance includes the series resistance of both the outgoing and return wires. For coaxial cables the resistance includes the center conductor and the outer shield. For computing attenuation accurately, the inner conductor resistance and the shield resistance of the coax must be added, because current flows equally in both.

Here are eight rules of thumb for calculating the resistance of round copper wires:

(1) Twenty-four gauge wire (AWG 24), being 0.02 in. in diameter, has a resistance of 25 Ω/1000 ft (at room temperature).
(2) Twisted-pair 24 AWG cable has a total series resistance of 50 Ω/1000 ft at room temperature (1000 ft out on one wire and 1000 ft back on the other).
(3) RG-58/U coaxial cable uses a stranded core of AWG 20 having a resistance of 10.8 Ω/1000 ft (at room temperature).
(4) The American Wire Gauge (AWG) system is a logarithmic measure of wire diameter. The larger the gauge size, the smaller the cable.
(5) Every three AWG points doubles the wire resistance.
(6) Every three AWG points halves the wire cross-sectional area.
(7) Diameter being proportional to the square root of area, every six AWG points halves the diameter.
(8) The resistance of copper increases 0.39% with every 1°C increase in temperature. Over a 70°C temperature range, that's a variation of 31%.

Here are handy equations for working with AWG sizes and inches:

$$\text{AWG} = (-10) - 20\log_{10} (\text{diameter in inches}) \quad [4.21]$$

$$\text{Diameter in inches} = 10^{-(\text{AWG}+10)/20} \qquad [4.22]$$

$$R \text{ per 1000 ft} = \frac{0.01\ \Omega}{(\text{diameter})^2} \qquad (25°\text{C}) \qquad [4.23]$$

$$R \text{ per 1000 ft} = 10^{(\text{AWG}-10)/10} \qquad (25°\text{C}) \qquad [4.24]$$

Printed circuit traces have resistances which are a function of the copper thickness and the trace width. The trace thickness is rated in plating weight, typically 1- or 2-oz plating, corresponding to a 0.00135- or 0.0027-in. thickness, respectively.[4] The resistance per inch of a printed circuit trace may be calculated from thickness and width:

$$R = \frac{0.65866 \times 10^{-6}}{WT}\ \Omega/\text{in.} \qquad [4.25]$$

where R = series resistance of line, Ω/in.
W = width of line, in.
T = thickness of line, in.

If the copper plating weight is known, we can use it directly:

$$R = \frac{0.000487}{(W)(\text{oz})}\ \Omega/\text{in.} \qquad [4.26]$$

where R = series resistance of line, Ω/in.
W = width of line, in.
oz = thickness of copper plating, oz

Series resistance in cables adds both attenuation (loss) and distortion to our ideal transmission line model. Signal attenuation means that signals shrink progressively as they move along the cable. Signal distortion means signals of different frequencies are attenuated (and phase-shifted) by different amounts as they propagate. The relation among attenuation, phase shift, and frequency at a point X inches from the head of the cable is prescribed by Equation 4.27. This equation applies only to infinite transmission lines. A cut, or terminated, line no longer follows Equation 4.27. Section 4.3 studies the effects of cuts and terminations.

$$H_X(w) = e^{-X[(R+jwL)(G+jwC)]^{1/2}} \qquad [4.27]$$

where R = series resistance of line, Ω/in.
L = series inductance of line, H/in.
C = parallel capacitance of line, F/in.
G = parallel conductance of line, mhos/in.
$H(w)$ = complex function specifying amplitude and phase response of transmission line at frequency $w = 2\pi f$
X = length of cable, in.

[4]Plating weight refers to the number of ounces of copper deposited on a flat surface per square foot.

The term G is practically zero for most digital transmission problems. This term models the effects of current leakage, due to wet or imperfect insulation, between the signal conductors in long cables. It also crops up when dealing with dielectric losses.

For printed circuit boards, ribbon cables, or indoor coaxial installations at frequencies below 1 GH we may safely assume G is zero.

The zero G assumption reduces the form of Equation 4.27:

$$H_X(w) = e^{-X[(R+jwL)(jwC)]^{1/2}} \tag{4.28}$$

Splitting the exponent of Equation 4.28 into real and imaginary components, the real component controls attenuation, whereas the imaginary component controls phase shift:

$$H_X(w) = e^{-X\,\mathrm{Re}[(R+jwL)(jwC)]^{1/2}} e^{-Xj\,\mathrm{Im}[(R+jwL)(jwC)]^{1/2}} \tag{4.29}$$

$$\text{Attenuation at frequency } w = e^{-X\,\mathrm{Re}[(R+jwL)(jwC)]^{1/2}} \tag{4.30}$$

$$\text{Phase shift at frequency } w = e^{-Xj\,\mathrm{Im}[(R+jwL)(jwC)]^{1/2}} \tag{4.31}$$

The term $-\mathrm{Re}[(R + jwL)(jwC)]^{1/2}$, the logarithm of signal amplitude at a unit length, is proportional to the transmission line attenuation in decibels.

The term $-\mathrm{Im}[(R + jwL)(jwC)]^{1/2}$ is the transmission line phase shift per unit length, in radians. Together, the attenuation and phase-shift terms form the *propagation coefficient* of a transmission line.

Series resistance in transmission lines perturbs the characteristic impedance. Equation 4.32 describes the characteristic impedance of a transmission line as a function of frequency:

$$Z_0(w) = \left(\frac{R + jwL}{jwC}\right)^{1/2} \tag{4.32}$$

Characteristic impedance is a strong function of frequency. At low frequencies, where R exceeds wL, the characteristic impedance given in Equation 4.32 is inversely proportional to the square root of frequency. At high frequencies, where wL exceeds R, the characteristic impedance flattens out to a constant value. Practical transmission lines always exhibit both modes of operation. Depending on the frequency regime in which we choose to operate a line, it behaves either as an R-C transmission line (low frequencies) or a low-loss transmission line (high frequencies). These cases are distinguished by the relative magnitudes of the inductive and resistive terms in Equation 4.32:

$$\begin{aligned}&RC \text{ case:}\\ &w \ll R/L \qquad (\text{also } R \gg wL)\end{aligned} \tag{4.33}$$

$$\begin{aligned}&\text{Low-loss case:}\\ &w \gg R/L \qquad (\text{also } R \ll wL)\end{aligned} \tag{4.34}$$

We will study low-loss transmission lines first, as they play a more important role in high-speed digital engineering.

4.2.2.1 Low-loss transmission line

As w rises above R/L, the phase angle of the term $[(R + jwL)(jwC)]^{1/2}$ approaches $+\pi/2$. As this happens, the imaginary part practically equals $w(LC)^{1/2}$, while the real part stabilizes at $\frac{1}{2}[R(C/L)^{1/2}]$.

Figure 4.9 shows the result for RG-58/U coaxial cable. This figure plots the real and imaginary parts of the propagation coefficient versus frequency. The plot is drawn on a log-log axes to highlight the w^0, $w^{1/2}$, and w^1 relationships in the curves. Below frequency R/L, both the real part (log of attenuation) and the imaginary part (phase in radians) are proportional to $w^{1/2}$. Above R/L, the imaginary part (phase) grows linearly with increasing frequency, while the real part (attenuation) stays fixed.

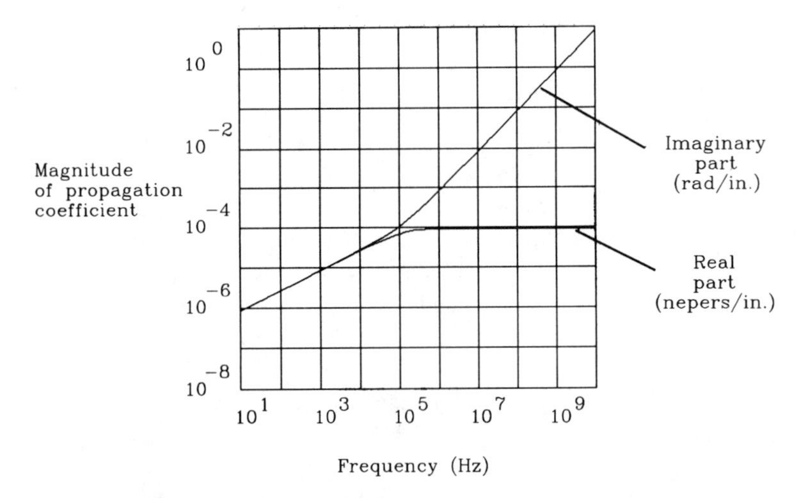

Figure 4.9 Propagation of a cable with fixed series resistance (no skin effect).

The properties of linear phase and fixed attenuation imply that the transmission line, for frequencies above R/L, is just a simple time-delay element. The delay is proportional to distance. Double the distance yields twice the delay.

The gain of this time-delay element is always less than 1 (we have a lossy circuit). The loss, measured in decibels, is proportional to distance. Double the distance yields twice as much decibel loss. One neper equals an 8.69 dB loss.

At frequencies above R/L, the characteristic impedance levels off to a constant value of $(L/C)^{1/2}$. At high frequencies, the characteristic impedance is real-valued, like an ordinary resistor.

A good model for transmission lines operated in their low-loss region is

$$\text{Characteristic impedance, } Z_0 = (L/C)^{1/2} \qquad [4.35]$$

$$\text{Attenuation at } X \text{ inches } = e^{-\left[\frac{RX}{2(L/C)^{1/2}}\right]} \qquad [4.36]$$

$$\text{Loss per inch} = 4.34\left[\frac{R}{(L/C)^{1/2}}\right]\text{dB} \tag{4.37}$$

$$\text{Delay per inch, } T_p = (LC)^{1/2}\ \text{(s/in.)} \tag{4.38}$$

From Equations 4.35 and 4.38 we can deduce the following handy relations:

$$L = Z_0 T_p \tag{4.39}$$

$$C = \frac{T_p}{Z_0} \tag{4.40}$$

where L = inductance, H/in.
C = capacitance, F/in.
T_p = propagation delay, s/in.
Z_0 = characteristic impedance, Ω

Ordinary digital logic gates tolerate very little signal loss. Any slight droop in received signal amplitude takes a big bite out of the received data noise margin. For that reason, on-card digital signal transmission networks are designed for very low loss. Low loss implies low resistance. We'll use Equation 4.42 to determine how much resistance is tolerable.

First set the loss given in Equation 4.37 equal to a nominal low value of 0.2 dB (2% loss):

$$(X)\,4.34\left[\frac{R}{(L/C)^{1/2}}\right] = 0.2 \tag{4.41}$$

where X = transmission line length, in.
R = transmission line resistance, Ω/in.
L = transmission line inductance, H/in.
C = transmission line capacitance, F/in.

Rearranging terms in Equation 4.41, bringing the total wire resistance RX to the left side, shows what the maximum allowable total resistance is. Equation 4.42 concludes that for low loss, the total wiring resistance RX must be a small fraction of the transmission line characteristic impedance.

$$RX = 0.046(L/C)^{1/2} \tag{4.42}$$

where RX = total line resistance, Ω
L = transmission line inductance, H/in.
C = transmission line capacitance, F/in.

Remember that these equations hold only for infinite, unterminated transmission lines. The line must be driven at one end and observed some distance X from the driver. Also note that R, the DC resistance of the line, appears as a constant in these formulas. Section 4.2.3 discusses implications of the skin effect, which causes R to grow larger at high frequencies. Finally, we assumed here a loss of no more than 0.2 dB, so that our signals would be attenuated only 2%.

4.2.2.2 *RC* transmission line

What happens at frequencies below *R*/*L*? Figure 4.9 shows that for frequencies below *R*/*L*, attenuation diminishes (the received signal gets bigger). At the same time, the phase is proportional to the square root of frequency, not linear with frequency as in the low-loss case. This nonlinear phase relationship introduces signal distortion, as different parts of a signal are phase-shifted by varying amounts. Equation 4.32 also shows the characteristic impedance going up dramatically for frequencies below *R*/*L*.

A transmission line operated in this region is termed an *RC transmission line*. The partial differential equations describing the behavior of such a line are called *diffusion equations*, and the line is sometimes called a *diffusion line*.

EXAMPLE 4.1: *RC* Transmission Line

You probably have *RC* transmission lines in your house. The two wires running from the nearest central telephone switching office to your phone are usually AWG 24 wire. These wires are twisted in a configuration yielding the following values for *L*, *C*, and *R*.

$$Z_0(w) = \left(\frac{R + jwL}{jwC} \right)^{1/2} = |648| \angle -45° \quad [4.43]$$

$R = 0.0042\ \Omega/\text{in.}$
$L = 10\ \text{nH/in.}$
$C = 1\ \text{pF/in.}$
$w = 10{,}000\ \text{rad/s}\ (1600\ \text{Hz})$ nominal operating frequency for voice signals

At 1600 Hz, the center of the voice band on telephone lines, we find a characteristic impedance magnitude of 648 Ω, at a phase angle of –45 degrees. Do you think this has anything to do with the telephone company practice of terminating phone lines in 600 Ω?

Very large-scale integrated circuits incorporating long lines (0.2 in.) of polysilicon or other relatively high-resistance materials often exhibit *RC* transmission line behavior at digital frequencies. Also, very long lines operated at low frequencies, such as the original transoceanic telephone cables, operate in this region.

Engineers interested in exploiting the very low loss attainable in the *RC* operating region must restrict their signals to the frequency band below *R*/*L*. In other words, the digital knee frequency (as defined in Equation 1.1) must lie below *R*/*L*.

In typical short-distance digital applications the digital rise times are so short that the digital knee frequency lies well above the *R*/*L* boundary.

These short rise times force us to accommodate the higher, but fixed, attenuation available in the low-loss operating region. Once we have designed wide-bandwidth circuits that work in the low-loss area, they are guaranteed to work at frequencies down in the *RC* zone. For that reason, we will study *RC* transmission lines no further.

4.2.3 Skin Effect

For every electrical parameter, we must consider the frequency range over which that parameter is valid. The series resistance of transmission lines is no exception. It, like everything else, is a function of frequency. Figure 4.10 depicts the equivalent series resistance of RG-58/U plotted as a function of frequency. The plot uses log-log axes. Figure 4.10 also shows, on the same axes, the inductive reactance wL.

At frequencies below $w = R/L$, the resistance exceeds the inductive reactance and the cable behaves as an *RC* transmission line (characteristic impedance changes with frequency, nonlinear phase delay). Above $w = R/L$, the cable is a low-loss transmission line (characteristic impedance is constant, linear phase delay).

Above 10^5 Hz the series resistance begins going up. This causes more attenuation (more loss) but maintains linear phase. This increase in resistance is called the *skin effect.*

The real and imaginary parts (loss in nepers and phase in radians) of the propagation coefficient $[(R + jwL)(jwC)]^{1/2}$ are plotted in Figure 4.11. One neper equals 8.69 dB of loss. This figure shows the *RC* zone, the constant attenuation zone, and the skin-effect zone where attenuation rises but with linear phase. As illustrated, the low-loss zone is very narrow compared to the *RC* and skin-effect regions.

What causes the skin effect, and what does it have to do with skin?

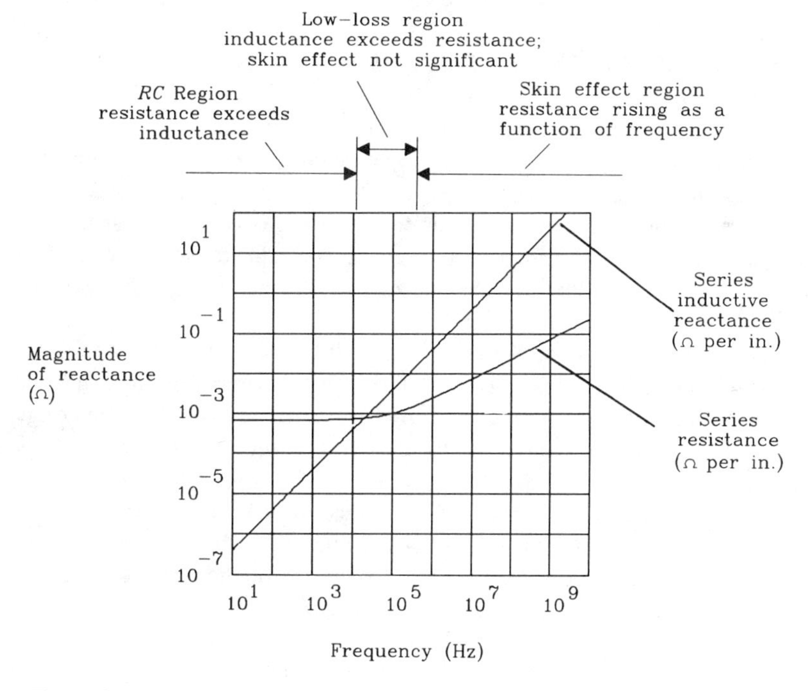

Figure 4.10 Series resistance and series inductive reactance of RG-58/U coax versus frequency.

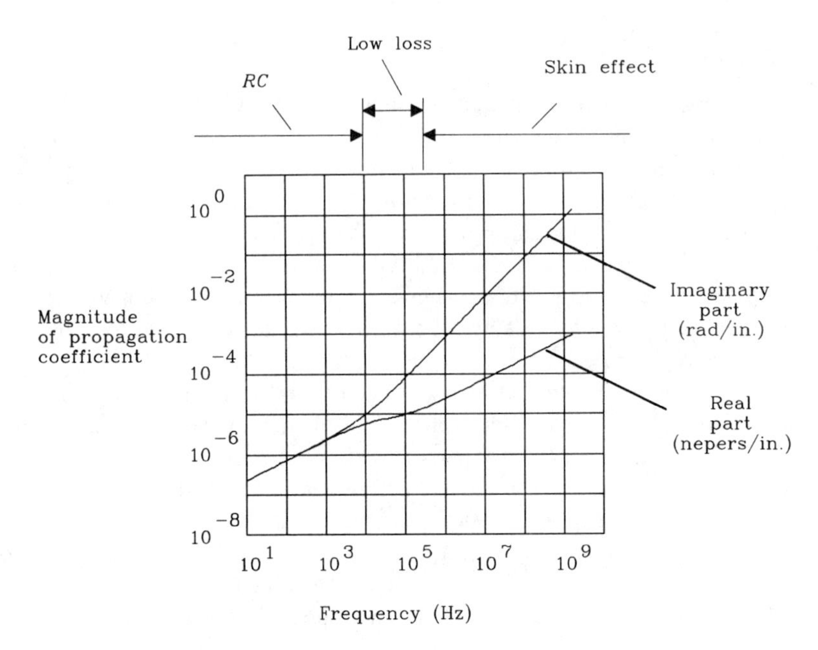

Figure 4.11 Propagation coefficient of RG-58/U includes skin effect.

4.2.3.1 Mechanics of skin effect

At low frequencies, the distribution of current density inside a conductor is uniform. Looking at a cross section of a conducting wire, there is as much current flowing in the center as near the edge.

At high frequencies the current density grows larger near the surface of the wire, and almost no current flows in the center. This adjustment in current distribution is drawn in Figure 4.12. At low frequencies, current fills the wire evenly. At high frequencies, current flows only near the surface.

To mentally justify the current distribution at high frequencies, first imagine the wire is sliced lengthwise into concentric tubes, like growth rings on a tree trunk.

Natural symmetry prevents any current from passing between rings, so cutting them apart makes no difference. All the current proceeds absolutely parallel to the wire's central axis.

Now that the wire is sliced into rings, we may separately consider the inductance of each ring. The inner rings, like long, skinny pipes, have more inductance than the outer rings, which are fatter. We know that at high frequencies current follows the path of least inductance. Therefore, at high frequencies we expect more current in the outer tree rings than in the inner. That is exactly what happens. At high frequencies, most of the current bunches up near the outside surface of a conductor.

The skin-effect mechanism is even more pronounced than one might predict based merely on the individual inductances of the tree ring slices. Mutual inductance

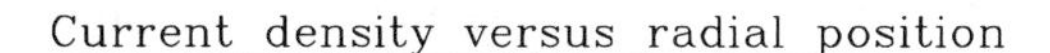

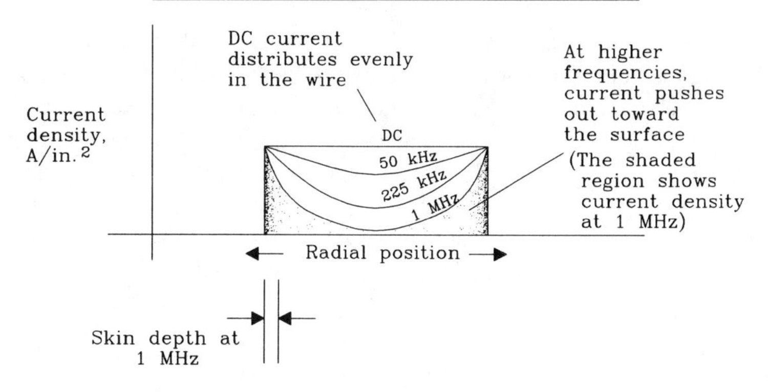

Cross section of wire

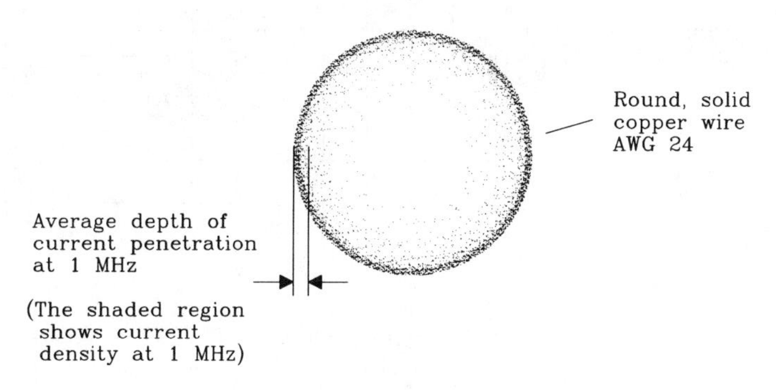

Figure 4.12 Distribution of current in a round wire.

between the rings actually forces current tightly against the outer surface (the skin) of a conductor.

At high frequencies, the average depth of current penetration, called the *skin depth*, is quite shallow. The distribution of current within a conductor experiencing the skin effect falls off exponentially as we approach the interior. The average current depth (skin depth) is a function of frequency w (in rad/s), the magnetic permeability of the conductor μ, and the volume resistivity p.

$$\text{Skin depth} = \left(\frac{2p}{w\mu}\right)^{1/2} \qquad [4.44]$$

With most of the current flowing in a shallow ring near the surface of a conductor, one may imagine that the apparent resistance of that conductor would increase a lot. How much is a function of skin depth. The conductor's apparent resistance is inversely proportional to the depth of current flow (the skin depth). Equation 4.44 says that the skin depth is inversely proportional to the square root of frequency. Combining these facts, the AC resistance of conductors grows proportionally to the square root of frequency.

Skin depth is a material property that varies with the bulk conductivity of the conducting medium. It is not a function of conductor shape.[5] Figure 4.13 plots skin depth for copper versus frequency. The second plot in Figure 4.13 gives the resistance of AWG 24 round copper wire versus frequency. At frequencies low enough that the skin depth is comparable to, or greater than, the wire radius, we measure only the total DC resistance of the wire (the current distribution fills the wire evenly). When the skin depth is less than the wire radius, the resistance per inch goes up proportional to the square root of frequency. In the skin depth-limited region, the resistance is given in Equation 4.45.

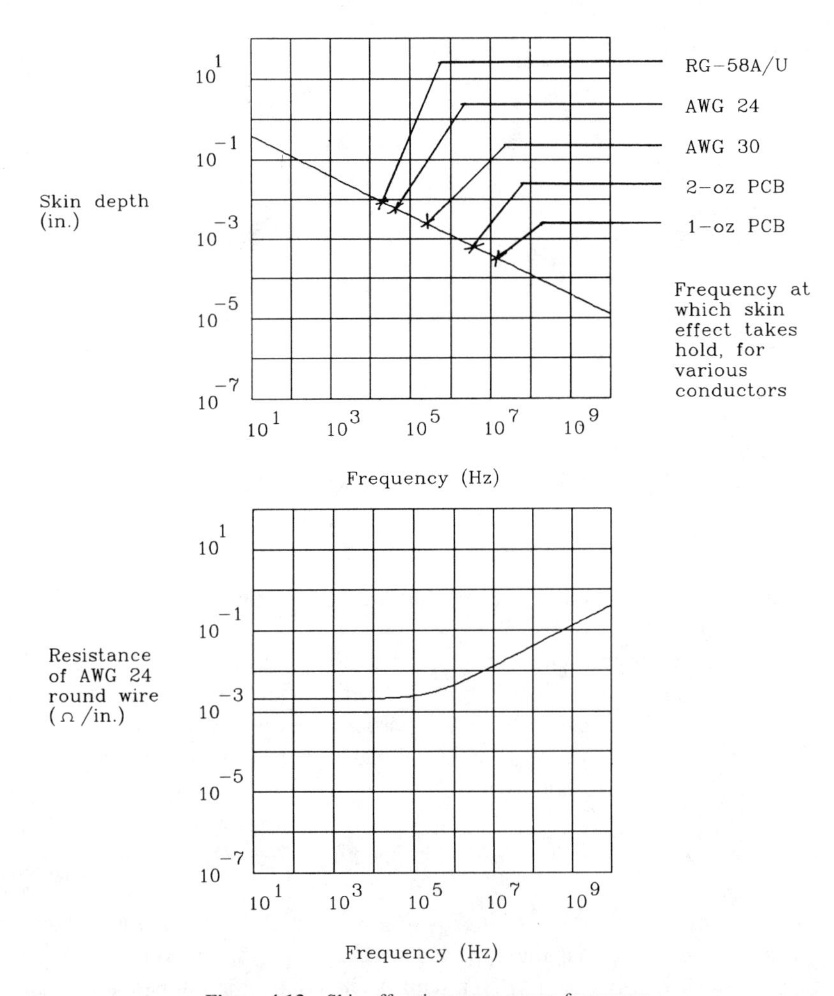

Figure 4.13 Skin effect in copper versus frequency.

[5]Surface finish does play a role at frequencies above 10 GHz.

$$R_{AC}(f) = \frac{(2.61 \times 10^{-7})(fp_r)^{1/2}}{\pi D} \qquad [4.45]$$

where D = wire diameter, in.
R_{AC} = AC resistance, Ω/in.
p_r = relative resistivity, compared to copper = 1.00
f = frequency, Hz

The problem with applying Equation 4.45 in practical situations is that at low frequencies it predicts zero resistance. We know that at DC the wire has a finite resistance. Equation 4.46 attempts to combine both AC and DC resistance models into one formula. There is no closed-form solution for the combined model; Equation 4.46 is just a useful approximation.

$$R(f) = \left\{(R_{DC})^2 + \left[R_{AC}(f)\right]^2\right\}^{1/2} \qquad [4.46]$$

This equation better models the physical reality: At low frequencies resistance stays constant, while at high frequencies resistance grows proportional to the square root of frequency. The frequency at which resistance begins to grow is the frequency where the skin depth becomes smaller than the thickness of the conductor. For a round conductor, the critical depth equals the conductor radius. For a flat, rectangular conductor such as a printed circuit trace, the critical depth is half the conductor thickness.

Model square conductors using Equations 4.45 and 4.46, by substituting the square conductor perimeter, measured in inches, for the quantity πD.

Table 4.1 lists, for various conductors, the frequency at which the skin effect begins to take hold.

TABLE 4.1 SKIN-EFFECT FREQUENCIES FOR CONDUCTORS

Round conductors	Radius	Skin-effect frequency (KHz)
RG-58/U	0.017	21
AWG 24	0.010	65
AWG 30	0.005	260
Printed circuit trace	**Copper weight (oz)**	**Skin-effect frequency (MHz)**
0.010 width	2	3.5
0.005 width	2	3.5
0.010 width	1	14.0
0.005 width	1	14.0

If skin effect is a surface phenomenon, then increasing the surface area should help. Litz wire does just that. A section of Litz wire is composed of multiple strands of wire, all insulated from each other and woven in a specific twist pattern. Their weave ensures that each strand is exposed to similar magnetic forces, causing equal currents to flow in each strand. The large surface area represented by the multiple strands reduces the skin-effect resistance. Litz wire is useful in large superconducting magnet windings and motor armatures at frequencies up to 1 MHz. Beyond that, it becomes almost impossible to balance the currents in each strand.

4.2.3.2 Frequency response in the skin-effect region

Substituting the Equation 4.46 for R in Equation 4.28, we can predict both attenuation and phase shift on transmission lines operating in the skin-effect region.

Transmission loss, in decibels, is proportional to resistance (Equation 4.37). Resistance is proportional to the square root of frequency. Therefore, attenuation in decibels must be proportional to the square root of frequency. This effect shows clearly in the attenuation plot for RG-174/U (Figure 4.14).

Introductory texts on transmission line theory often focus on the central region in Figure 4.14, between the *RC* zone and the skin effect. In this central area, the cable's attenuation is flat with frequency, there is no phase distortion, and the characteristic impedance is also flat. In this zone the cable (except for a fixed attenuation) looks ideal. In the real world, the ideal operating zone is very narrow if it exists at all.

In the skin-effect region, halving a cable's length improves its frequency response by a factor of 4. This happens because attenuation is proportional to the product of resistance (square root of frequency) and length. When we halve the length, we halve the attenuation. When we quadruple the frequency, the attenuation doubles back up to normal.

For ordinary digital transmission, the total resistance limitation of Equation 4.42 still applies, but with resistance taking on a frequency-dependent behavior. Plugging in the skin-effect resistance, as measured at the digital knee frequency,[6] into Equation 4.42

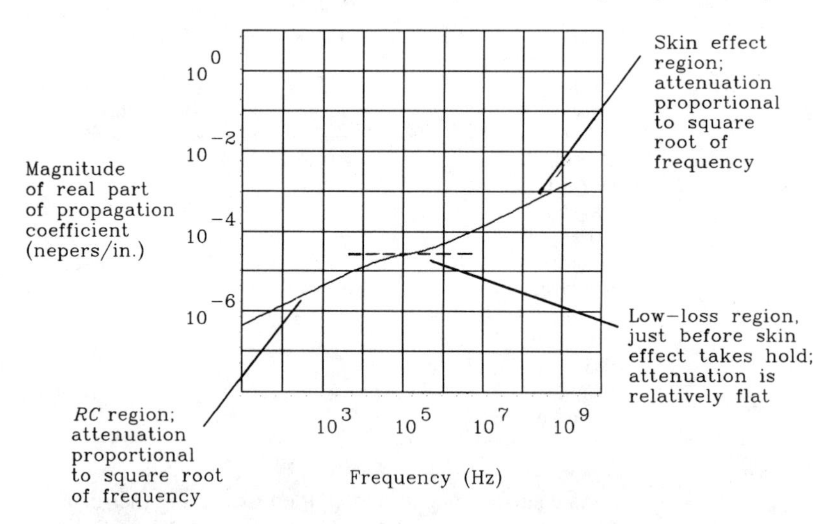

Figure 4.14 Attenuation coefficient of RG-174/U showing skin effect.

[6]See Equation 1.1 defining digital knee frequency.

we get a nicely conservative result. Adhering to this guideline, our transmission circuits will always work well, with rising edges that come through practically undistorted.

Long-distance data transmission systems, which use data receivers with larger voltage margins than ordinary TTL, can tolerate more than 0.2 dB loss. A greater loss budget implies the circuit operates at longer distances.

Use loss Equation 4.30 to directly figure the expected loss at your digital knee frequency.[7] In Equation 4.30, plug in the skin-effect resistance Equation 4.46 for the R term.

Limiting the loss to 0.5 dB at the digital knee frequency allows each rising edge to come through at 95% amplitude. If you can tolerate degradation in the rise time, then figure what rise time you want to come through to the receiver and use its knee frequency when figuring the loss limitation of 0.5 dB.

Another trick for long-distance communication involves coding the data to have an equal number of ones and zeros and then passing it through an AC coupling network. The AC coupling network removes any DC bias impressed upon the signal by the data drivers. The resulting waveform has equal numbers of excursions high and low. A data receiver for this signal should have an accurate switching threshold of precisely zero. This approach can tolerate large amounts of attenuation (3 dB or more at the maximum alternation rate, equal to half the clock speed).

A coding which restricts the maximum run length of consecutive ones or zeros can tolerate even a little more attenuation. Figure 4.15 illustrates the worst-case data pattern for a run length-limited coding system. At point A, the data transmitter starts a long run of ones. At point B, the limited frequency response of our long cable has ramped up to a maximum value. At point C, the small data pulse comes through. The effective frequency of the small data pulse is $F_{CLK}/2$, while the effective frequency of the long data pattern is $F_{CLK}/4N$. If the cable frequency response at $F_{CLK}/2$ is half the amplitude of the response at $F_{CLK}/4N$, then the pulse at C never crosses the zero threshold and the receiver never sees it.

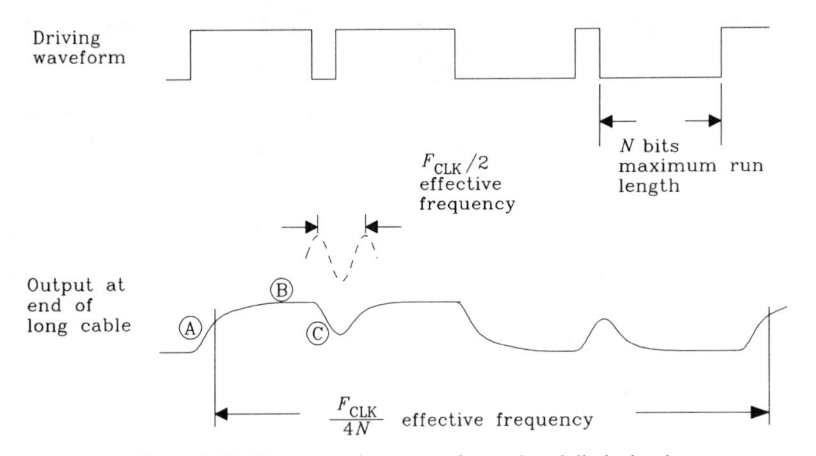

Figure 4.15 Worst-case data pattern for run-length limited code.

[7]See Equation 1.1 defining digital knee frequency.

It is a good practice to keep cabling short enough that the response ratio for a run length-limited system is greater than seven-tenths:

$$\frac{\left|H(2\pi_{F_{CLK}}/2)\right|}{\left|H(2\pi_{F_{CLK}}/4N)\right|} > 0.7 \qquad [4.47]$$

To exceed this distance limitation you will need some form of analog signal equalization.

4.2.3.3 Transmission line impedance in the skin-effect region

Once past the critical frequency R/L, the wL term grows linearly with w, while the $R(w)$ term grows, due to the skin effect, proportional to $w^{1/2}$. The $R(w)$ term stays small compared to wL, and so the impedance given by Equation 4.32 remains fixed at $(L/C)^{1/2}$. The input impedance of a transmission line is not much affected by the skin effect.

4.2.4 Proximity Effect

The *proximity effect* is a physical phenomenon that causes opposing currents in adjacent wires to draw toward one another (Figure 4.16). The proximity effect is caused by changing magnetic fields and so only disturbs the flow of high-frequency currents. Steady currents, which have static magnetic field patterns, do not respond to the proximity effect.

The proximity effect is distinct from Ampere's discovery that adjacent wires carrying opposing currents repel. While Ampere's forces push the atomic lattice structure of the two wires apart, the proximity effect merely increases the current density at the inside surfaces of the two wires. The proximity effect exerts no net mechanical force on the wires.

Like the skin effect, the proximity effect redistributes current density, resulting in a larger effective resistance at high frequencies. Unlike the skin effect, the proximity effect does not continue to get worse with increasing frequency. The proximity effect reaches equilibrium at a rather low frequency.

The proximity effect must be multiplied by a wire's AC skin-effect resistance (Equation 4.45).

The magnitude of the proximity effect at equilibrium is determined by the ratio of wire separation, between centers, to wire diameter. Figure 4.17 plots the magnitude of the

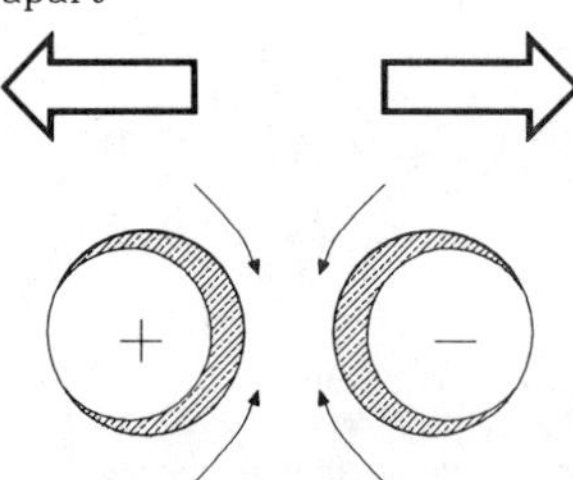

Figure 4.16 Proximity effect on two round wires carrying opposite high-frequency currents.

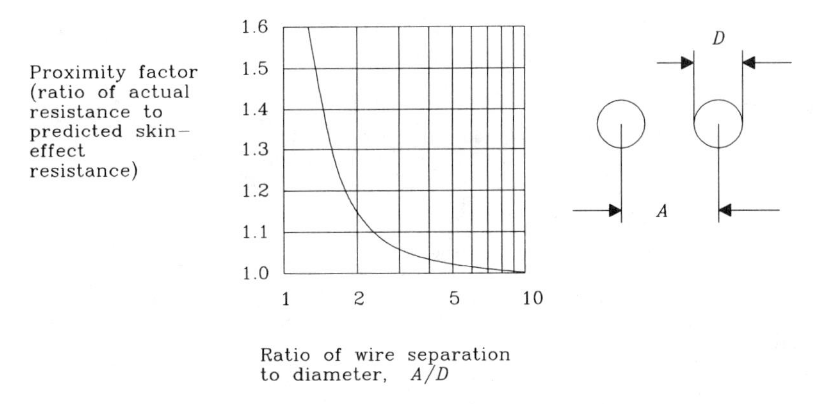

Figure 4.17 Proximity factor for parallel round wires. (Reproduced from Frederick Terman, *Radio Engineer's Handbook*, McGraw-Hill, New York, 1943, p. 36.)

proximity effect R_{AC}/R_{DC} versus the separation-to-diameter ratio. The proximity effect is most noticeable when the two wires are almost touching.

The same forces responsible for the proximity effect also cause return current flowing in a ground plane to follow closely underneath a signal conductor. In general, current flows in whatever formation minimizes the total loop inductance. This is the same as saying that, given an alternative, nature picks a current density distribution which minimizes the total energy stored in magnetic fields surrounding a system of conductors.

4.2.5 Dielectric Loss

Put a piece of glass epoxy printed circuit board material (with no copper on either side) into a microwave oven and bake it on full power for 1 min. It will be noticeably warmed by the microwaves. For that matter, try a Pyrex baking dish. It heats up, too.

In fact, just about any dielectric material heats up in a microwave oven. The heat absorbed by a dielectric material in the presence of changing electric fields is proportional to the *dielectric loss factor* for that material.

When used as the insulating dielectric of a transmission line, dielectric losses translate into signal attenuation. The higher the dielectric loss, the more attenuation.

Dielectric loss is a function of frequency. The glass epoxy material commonly used for printed circuit boards (FR-4) has negligible loss for digital applications below 1 GHz. At higher frequencies, the dielectric loss of FR-4 grows larger. For the highest-frequency circuitry, designers choose ceramic substrates, like alumina, having much better dielectric loss factors in the gigahertz regime.

Analog designers worry more about dielectric loss in FR-4 materials at low frequencies. The loss problem is particularly acute when constructing high-Q circuits intended to ring without signal amplitude loss for many cycles. Digital applications typically avoid high-Q circuit topologies and so are not very sensitive to dielectric loss.

For digital circuit board applications below 1 GHz, ignore dielectric losses.

In long cables, the dielectric properties are more significant. Typical PVC insulated telephone wire has a measurable dielectric loss at 10 MHz. This dielectric loss, which increases with frequency, is often lumped with skin effect losses into an overall dB loss model proportional to f^y where y is slightly bigger than 1/2.

POINTS TO REMEMBER:

The input to an *infinite* transmission line looks resistive, not capacitive.

Handy relations for the inductance and capacitance of a transmission line:

$$L = Z_0 T_p \quad [4.48]$$

$$C = \frac{T_p}{Z_0} \quad [4.49]$$

Total wiring resistance is usually a small fraction of transmission line impedance for ordinary digital applications.

The skin effect seriously limits the frequency response of long transmission lines.

For short-haul digital applications, transmission line attenuation in decibels is proportional to the square root of frequency (skin effect).

The proximity effect has only a minor effect on transmission line attenuation.

For digital applications below 1 GHz, ignore dielectric losses.

4.3 EFFECTS OF SOURCE AND LOAD IMPEDANCE

Now, for the bad news. Equation 4.29, describing the loss and phase shift of an infinite transmission line, is a theoretical best case. Any combination of practical source and load impedances connected to a real (finite-length) transmission line will degrade its performance. This degradation may be slight or it may be devastating, depending on the particular source and load impedances used with the transmission line.

In signal transmission problems you must first confirm the capability of a transmission cable to carry your signals. For digital signals, sufficient capability is established by checking that the signal transmission loss $H_X(w)$ at the digital knee frequency[8] is less than a few tenths of a decibel. Then consider the effects of source and load impedance.

This section shows how to estimate the effect of a particular source and load impedance combination. We will also discuss how to select good practical values for source and load components.

4.3.1 Reflections on a Transmission Line

Referring to Figure 4.18, when a signal is impressed on the end of transmission line, a fraction of the full source voltage propagates down the line. This fraction is a function of

[8]See Equation 1.1 defining digital knee frequency.

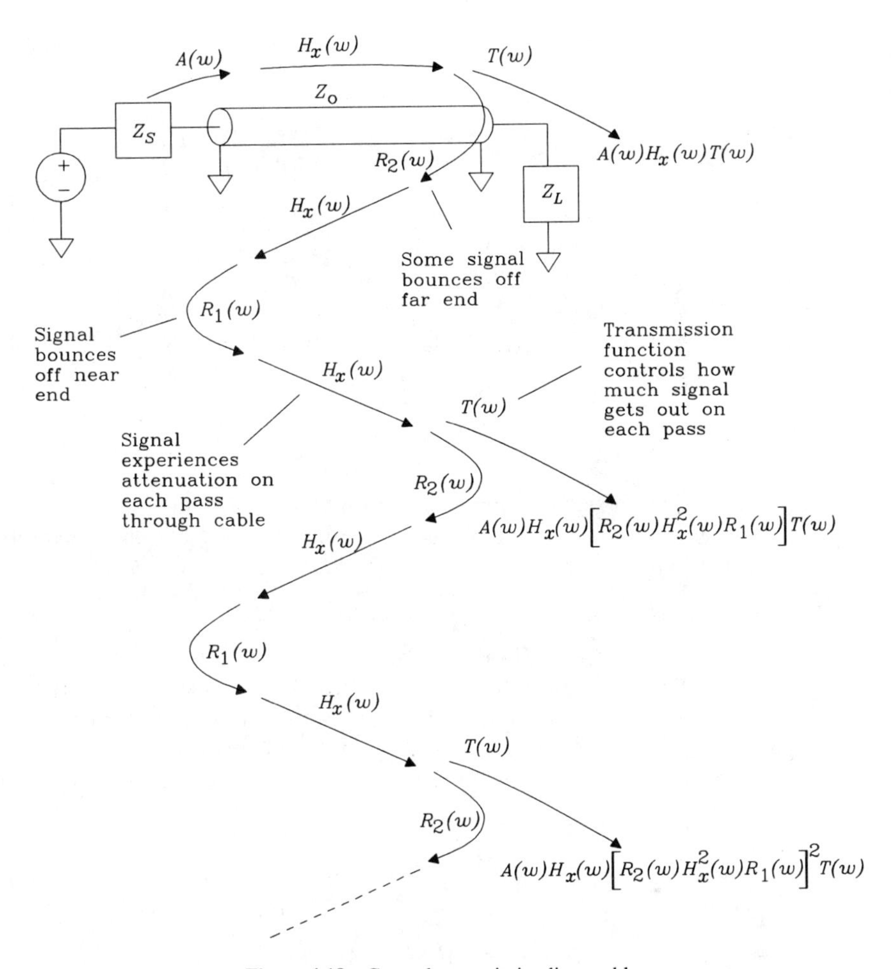

Figure 4.18 General transmission line problem.

frequency and is called $A(w)$, the input acceptance function. The value of $A(w)$ is determined by the source impedance Z_S, the transmission line impedance given by Equation 4.32, and the input acceptance Equation 4.50.

$$A(w) = \frac{Z_0(w)}{Z_S(w) + Z_0(w)} \qquad [4.50]$$

As the signal propagates, it is attenuated by the propagation function $H_X(w)$. This equation resembles Equation 4.30 except that, due to the skin effect, the $R(w)$ term is now a function of frequency.

$$H_X(w) = e^{-X[(R(w)+jwL)(jwC)]^{1/2}} \qquad [4.51]$$

At the far end of the cable a fraction of the attenuated signal amplitude emerges. This fraction is a function of frequency and is called $T(w)$, the output transmission function. The value of $T(w)$ is determined by the load impedance Z_L, the transmission line impedance given by Equation 4.32, and the output transmission Equation 4.52. This fraction $T(w)$ ranges from 0 to 2.

$$T(w) = \frac{2Z_L(w)}{Z_L(w) + Z_0(w)} \tag{4.52}$$

When fraction $T(w)$ of the propagating signal emerges from the far end, a reflected signal also travels back along the cable toward the source. As it reflects, this signal crosses over the tail of the incoming signal. Both signals propagate simultaneously in opposite directions, neither interfering with the other.

The fraction of the propagating signal that reflects back toward the source is called $R_2(w)$, the far-end reflection function.

$$R_2(w) = \frac{Z_L(w) - Z_0(w)}{Z_L(w) + Z_0(w)} \tag{4.53}$$

The reflected signal is attenuated again by $H_X(w)$ as it travels back to the head end, where it reflects a second time off the source impedance. The source end reflection coefficient is $R_1(w)$.

$$R_1(w) = \frac{Z_S(w) - Z_0(w)}{Z_S(w) + Z_0(w)} \tag{4.54}$$

After the head-end reflection, the signal is attenuated a third time by $H_X(s)$, and then part of it again emerges through the transmission function $T(w)$. Part of this second signal also reflects back toward the source, in an endless cycle.

The first signal to emerge from the cable is attenuated by $A(w)$, $H_X(w)$, and $T(w)$.

$$S_0(w) = A(w)H_X(w)T(w) \tag{4.55}$$

The second signal to emerge, after having reflected off both load and source ends, is attenuated by

$$S_1(w) = A(w)H_X(w)\left[R_2(w)H_X{}^2(s)R_1(w)\right]T(w) \tag{4.56}$$

Successive emerging signals are characterized by

$$S_N(w) = A(w)H_X(w)\left[R_2(w)H_X{}^2(s)R_1(w)\right]^N T(w) \tag{4.57}$$

Eventually, all signals $N = [0, 1, ..., \infty]$ emerge. The sum of all these emerging signals is

$$S_\infty(w) = \sum_{n=0}^{\infty} S_n(w) \tag{4.58}$$

Fortunately, there is a closed-form equivalent for this infinite sum:

$$S_\infty(w) = \frac{A(w)H_X(w)T(w)}{1 - R_2(w)H_X{}^2(w)R_1(w)} \tag{4.59}$$

Equation 4.59 is the frequency response, from source to load, of the transmission system depicted in Figure 4.18.

Figure 4.19 diagrams a hypothetical case. In this case the total wiring resistance amounts to only 1.2 Ω. Compared to the high-frequency characteristic impedance $(L/C)^{1/2}$

of 50 Ω, we can ignore the DC resistance for rough calculations, setting $Z_0(w) = 50$. This simplification makes all the reflection functions real numbers, which are easy to manipulate in hand calculations. An exact analysis is usually carried out by computer, in which case the exact formulas should always be used.

The four reflection coefficients for Figure 4.19 are

$$A(w) = 0.847 \text{ (input acceptance function)}$$
$$R_2(w) = 0.200 \text{ (far-end reflection coefficient)}$$
$$R_1(w) = -0.695 \text{ (near-end reflection coefficient)}$$
$$T(w) = 1.2 \text{ (far-end transmission function)}$$

The magnitude of the propagation constant, at a length of 15 in., is +0.940.

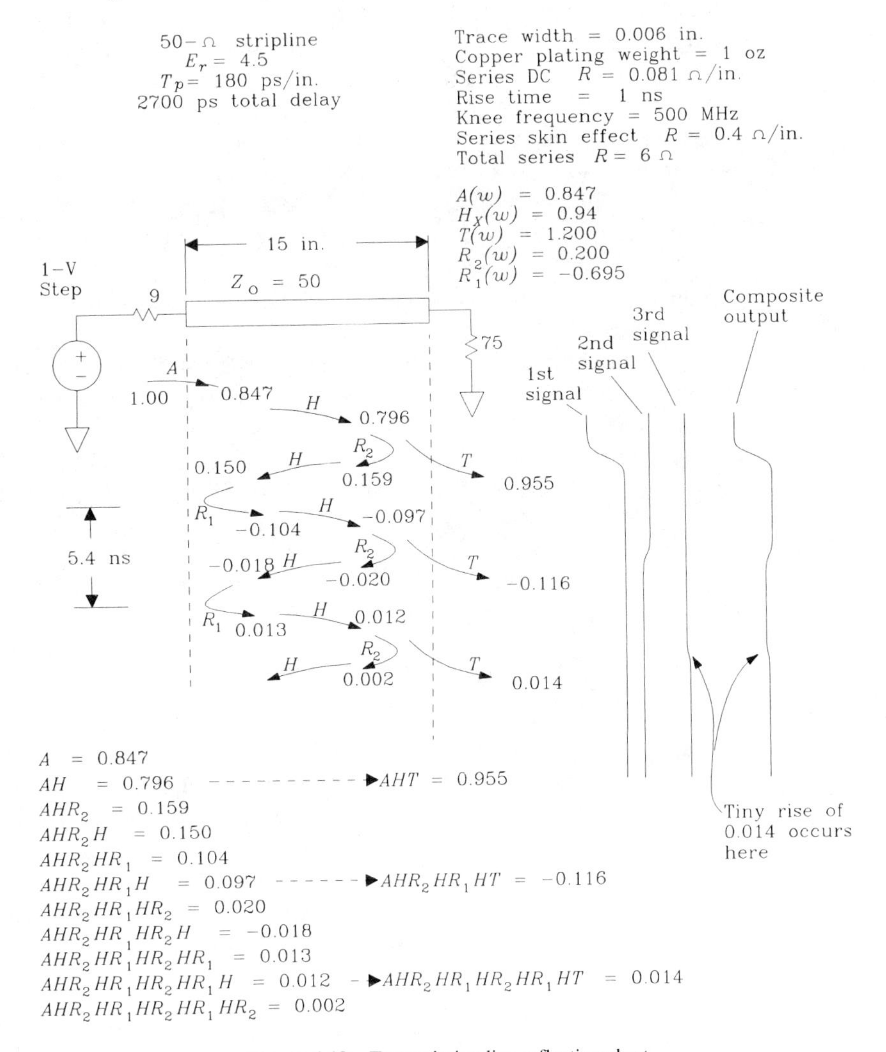

Figure 4.19 Transmission line reflection chart.

The phase delay associated with $H_X(w)$ equals 2700 ps.

Notice we ignore both the *RC* operating region and the skin effect, assuming $H_X(w)$ to be a constant. This omission is justified only by the necessity of producing a readable example. A good computer model includes both effects. In our example, $H_X(w)$ is so near unity that variations in its value would make little difference in the outcome.

From left to right, the chart shows signals progressing along the cable. The time axis extends vertically, starting with time zero at the top and running toward the bottom of the page. Assume the voltage generator makes a 1.00-V step at time zero.

The leading edge of an incoming signal proceeds down the cable from left to right. This signal is subject to the input acceptance function and so begins with amplitude 0.847 V. It arrives at the far end 2700 ps later a little smaller. The amplitude at the far end is $A(w)H(w) = 0.796$ V.

After being multiplied by $T(w)$, this first signal emerges from the cable with an amplitude of 0.955 V. At each stage, as the signal reflects back and forth, Figure 4.19 scales the bouncing signal by appropriate reflection coefficients. As each signal emerges, it is scaled by the transmission coefficient.

On the right of Figure 4.19 are the various emerging waveforms, drawn to scale. Each waveform aligns with its time of arrival. At the far right is the composite sum of all the pieces.

Eventually, the waveform decays to its steady-state value of 0.893 V. This is equal to the DC response of the system $S_\infty(0)$.

If the incoming signal rise time were sufficiently slow, all the composite pieces would mush together. The result would not show any ringing. It is only when the rise time is comparable to (or shorter than) the round-trip delay that the opportunity for overshoot and ringing arises.

Let's now consider the various means of controlling reflections on transmission lines. First, combine Equations 4.52 and 4.53 to relate the T and R_2 coefficients:

$$T(w) = R_2(w) + 1 \qquad [4.60]$$

This simplification reduces Equation 4.59:

$$S_\infty(w) = \frac{H_X(w)A(w)(R_2(w)+1)}{1 - R_2(w)R_1(w)H_X{}^2(w)} \qquad [4.61]$$

Assuming the cable parameter $H_X(w)$ is fixed, we have two parameters under our control, the source and load impedances. The source impedance controls terms $A(w)$ and $R_1(w)$ in Equation 4.61. The load impedance controls only term $R_2(w)$. For good digital transmission, we usually want a frequency response that is flat up to at least the digital knee frequency.[9]

Engineers have long standardized on three accepted methods for ensuring a flat frequency response from Equation 4.61, *end termination*, *series termination*, and *short line*.

[9]See Equation 1.1 defining digital knee frequency.

4.3.2 End Termination

This method sets $R_2(w)$ to zero. The resulting simplification in Equation 4.61 is:

$$S_{\text{end term}} = H_X(w)A(w) \tag{4.62}$$

Physically, this eliminates the first reflection. Signal energy in Figure 4.18 enters the cable, propagates to the far end, and exits the cable with no reflections. Since no delayed versions of the original signal exist, there is little possibility of fouling up the frequency response.

The method for achieving $R_2(w) = 0$ is easy. Just set the load resistance Z_L equal to the cable characteristic impedance Z_0. Under these conditions the reflection coefficient R_2 (Equation 4.53) goes to zero.

For very long cables operating in the *RC* mode, finding a terminating network that matches the characteristic impedance over a wide frequency range is a challenge.

4.3.3 Source Termination

This method sets $R_1(w)$ to zero. The resulting simplification in Equation 4.61 is:

$$S_{\text{source term}} = H_X(w)A(w)\left[R_2(w)+1\right] \tag{4.63}$$

Physically, this eliminates the second reflection but not the first. Signal energy in Figure 4.18 enters the cable, propagates to the far end, and exits the cable. The reflected energy travels back up toward the source but does not reflect at that point ($R_1 = 0$). No remaining energy reflects back to the load a second time.

The method for achieving $R_1(w) = 0$ is easy. Just set the source resistance Z_S equal to the cable characteristic impedance Z_0. Under these conditions the reflection coefficient R_1 (Equation 4.54) goes to zero.

With Z_S set equal to Z_0, the acceptance function becomes 1/2. This is usually compensated for, at the far end of the line, by leaving the line unterminated ($Z_L = \infty$). This procedure sets $T(w) = 2$ (and also $R(w) = 1$). This doubling of the line voltage at its end compensates for the halving which happened at its input. The disadvantage to this approach is that, because $R_2(w) = 1$, a large signal reflects back toward the input.

Circuits connected intermediate to the line between the source and the open end see a mixed-up signal due to this massive reflection. They see a half-sized signal propagating down toward the far end, followed later (after one round trip to the far end and back) by another half-sized step bringing the voltage up to full par value.

4.3.4 Very Short Line

This method uses such a short line that $H_X(w)$ is practically unity. There is then no significant attenuation or phase delay. The resulting simplification in Equation 4.61 is

$$S_{\text{short line}}(w) = \frac{A(w)\left[R_2(w)+1\right]}{1 - R_2(w)R_1(w)} \tag{4.64}$$

Substituting Equations 4.50, 4.53, and 4.54 for $A(w)$, $R_2(w)$, and $R_1(w)$, respectively:

$$S_{\text{short line}}(w) = \frac{\dfrac{Z_0}{Z_S + Z_0} \cdot \dfrac{2Z_L}{Z_L + Z_0}}{1 - \dfrac{Z_L - Z_0}{Z_L + Z_0} \cdot \dfrac{Z_S - Z_0}{Z_S + Z_0}} \qquad [4.65]$$

Multiplying both numerator and denominator by $(Z_L + Z_0)$ and collecting terms in the denominator:

$$S_{\text{short line}}(w) = \frac{Z_0 2Z_L}{2Z_L Z_0 + 2Z_0 Z_S} \qquad [4.66]$$

Dividing both top and bottom by $2Z_0$:

$$S_{\text{short line}}(w) = \frac{Z_L}{Z_L + Z_S} \qquad [4.67]$$

The line does just what we would expect: nothing. We get in this case a simple impedance divider network formed by the load and source impedances Z_L and Z_S.

For these assumptions to work, the line must act as a lumped circuit element. It must be much shorter than one-sixth the electrical length of a rising edge.

CONDITIONS FOR SHORT TRANSMISSION LINE

$$\text{Length} << \frac{1}{6} \frac{T_{\text{rise}}}{(LC)^{1/2}} \qquad [4.68]$$

where T_{rise} = rise time, s
L = line inductance, H/in.
C = capacitance, F/in.
length = maximum line length, in.

4.3.5 Settling Time of a Poorly Terminated Transmission Line

In reference to Figure 4.18, portions of the incoming signal rattle back and forth inside the transmission line many times on their way to the load. At each round trip, the amplitude of the rattling signal is reduced by the factor R_1R_2. As time progresses, the signal size decreases exponentially.

If the product R_1R_2 is sufficiently low, the second and subsequent reflections are so small they have no effect. The cable reaches its steady-state condition as soon as the first pulse emerges from the far end of the cable.

If the product R_1R_2 is larger, the cable may require several round trips to stabilize.

The time required for one complete round trip is equal to the cable length times its propagation delay.

$$T = (\text{length})(LC)^{1/2} \qquad [4.69]$$

During that period, the signal amplitude shrinks by R_1R_2. A reasonable high-level model for this behavior is

$$\text{Signal size}(t) = \left|R_1(w)R_2(w)\right|^{(t/T)} \quad [4.70]$$

The magnitude of the product R_1R_2 is always less than unity, and so the above equation always decays with increasing time. We can use Equation 4.70 to find how long it takes (roughly) for a cable to settle to within some tolerance ε of its final value.

$$\text{Time to settle within } \varepsilon = T\frac{\ln(\varepsilon)}{-\ln\left(\left|R_1(w)R_2(w)\right|\right)}$$
$$= (\text{length})(LC)^{1/2}\frac{\ln(\varepsilon)}{-\ln\left(\left|R_1(w)R_2(w)\right|\right)} \quad [4.71]$$

Equation 4.70 is good for thinking about underdamped backplanes or lengthy unterminated lines. In those cases a clocking system might need to wait for the lines to stop ringing before clocking in data. Usually, reflections get worse at high frequencies, so always use $w = 2\pi F_{\text{knee}}$ in Equation 4.70.[10]

Equation 4.70 should impress upon us all the importance of reducing R_1, reducing R_2, or making sure the line is short.

POINTS TO REMEMBER:

Any combination of practical source and load impedances connected to a transmission line degrade its performance.

The frequency response of a transmission line system is

$$S_\infty(w) = \frac{A(w)H_X(w)T(w)}{1 - R_2(w)H_X^2(w)R_1(w)} \quad [4.72]$$

Only when the round-trip delay exceeds your signal rise time will overshoot and ringing arise.

Eliminate reflections by reducing R_2 (end termination), reducing R_1 (series termination), or making sure the line is short (sets $H_x = 1$).

4.4 SPECIAL TRANSMISSION LINE CASES

4.4.1 Unterminated Line

Lines are called *unterminated* when neither source nor load impedance matches the characteristic impedance of the transmission line. Usually, the load impedance of an untermi-

[10]See Equation 1.1 defining digital knee frequency.

nated line is higher than the transmission line characteristic impedance. The source impedance may be higher or lower. Unterminated lines behave differently depending on whether the source impedance is much greater, or much less, than the line impedance. We will consider the two cases separately.

In both cases, the load impedance is very high, so $R_2(w) \approx 1$ (see Equation 4.53) and $T(w) \approx 2$ (see Equation 4.52). The difference in the two cases is in the sign of $R_1(w)$, and the magnitude of $A(w)$.

4.4.1.1 Low source impedance with an unterminated line

This case arises when a resistive, low-impedance output (like ECL, or a high-powered TTL bus driver) drives a transmission line.

We can sketch the unit step response of this sort of line without resorting to a detailed reflection chart. The input acceptance function $A(w)$ in this case is near unity (see Equation 4.50), and the transmission function $T(w)$ is near +2 (see Equation 4.52). Their product, the initial step output, will be near +2.0 V.

Because the reflection coefficient $R_1(w)$ (see Equation 4.54) is near –1, the product R_2R_1 will be almost –1. There is always some loss in a line, so the magnitude of the R_1R_2 product is slightly less than unity.

The negative sign on R_1R_2 means that successive reflected signals emerging from the line will have opposite signs. As the response to an input step damps out, it oscillates back and forth around the final value. Two complete round trips (four line traversals) occur between the emergence of successive signals having the same polarity. The oscillation period is therefore equal to four times the line time delay. The decay time follows from Equation 4.71.

We now know the step response starts out with a near-100% overshoot, oscillates with a period equal to four line time delays, and decays with a known time constant. The final value is equal to the input step size since there is no DC load impedance. Figure 4.20 sketches this step response.

If your signal rise times are shorter than the round-trip delay, the overshoot will show clearly in the output signal. This causes excessive current flow in the input protection diodes of most TTL and CMOS logic inputs. The excessive current returns through the chip's ground pin, causing ground bounce between the internal ground reference and the external ground plane. In extreme cases, overshoot from low-impedance lines can damage the input protection circuitry.

4.4.1.2 High source impedance with an unterminated line

This case arises when a very resistive high-impedance output (like an unbuffered CMOS output) drives a transmission line.

We can sketch the unit step response of this sort of line without resorting to a detailed reflection chart. The input acceptance function $A(w)$, in this case is very low (see Equation 4.50), while the transmission function $T(w)$ is near +2 (see Equation 4.52). Their product, the initial step output, will be small.

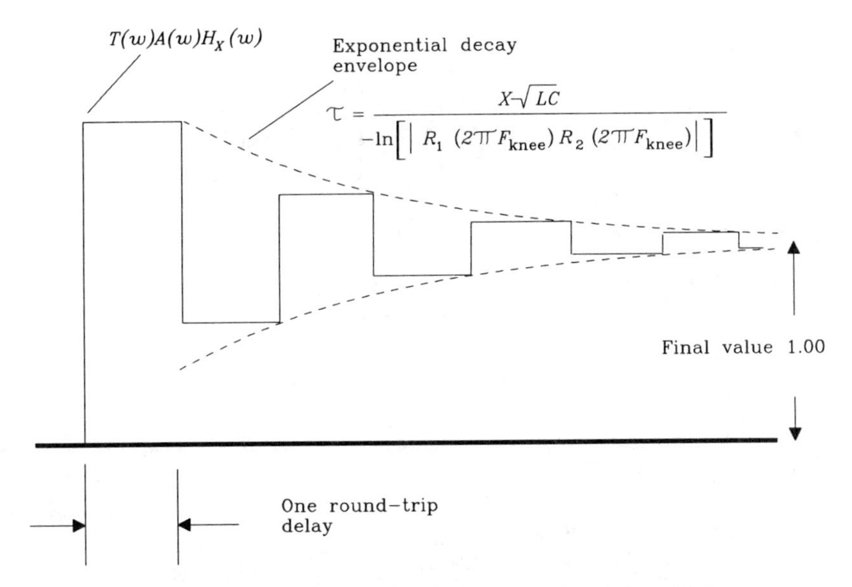

Figure 4.20 Estimating the step response of an unterminated line with low source impedance.

This time the reflection coefficient $R_1(w)$ (see Equation 4.54) is near +1, so the product R_2R_1 will be almost +1. There is always some loss in a line, so the magnitude of R_1R_2 is slightly less than unity.

The positive sign on R_1R_2 means that successive reflected signals emerging from the line will have the same sign. The output waveform must then build up monotonically to its final value. The decay time of the successive emerging signals (equal to the buildup time of the output signal) is given by Equation 4.71.

We now know the step response begins small, building with a known time constant. The final value is equal to the input step size, since there is no DC load impedance. Figure 4.21 sketches this step response. It looks like the response of an *RC* filter.

The time constant of this step response is close to the product of the source impedance and the total line capacitance. Such an estimate treats the line as a lumped element and is appropriate for short lines.

The correspondence between unterminated lines driven by a high impedance and the response of an *RC* filter is responsible for the popular myth that the input to a transmission line looks like a capacitive load.

4.4.2 Capacitive Loads Connected in the Middle of a Line

Figure 4.22 illustrates a long line having a single capacitor hooked in the center. A signal entering from the left impinges on the capacitor and is split in two. One portion of the signal reflects backward, and a second portion propagates through.

The tricky aspect of this problem is that the reflection coefficient is a function of frequency. We will deal separately with estimating the size of the reflected signal and estimating the effect on the propagated signal.

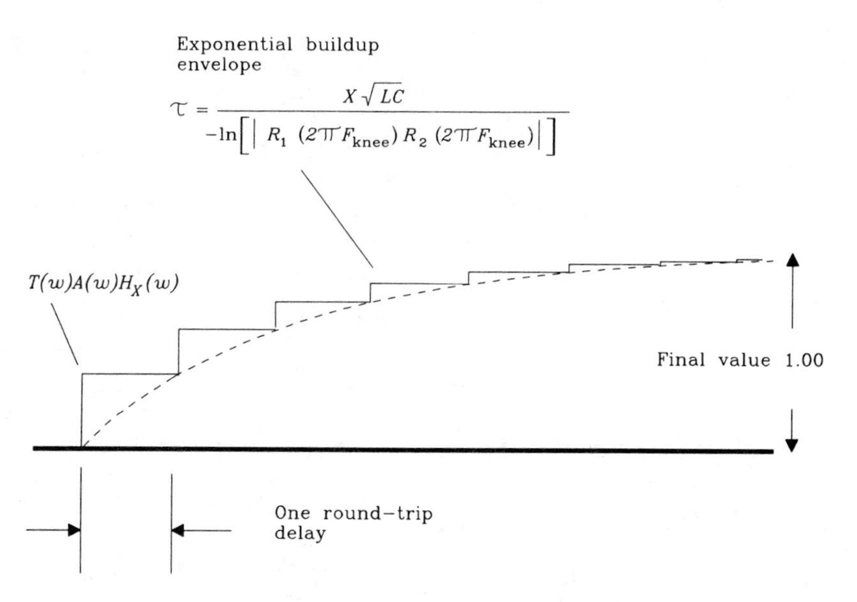

Figure 4.21 Estimating the step response of an unterminated line with a high source impedance.

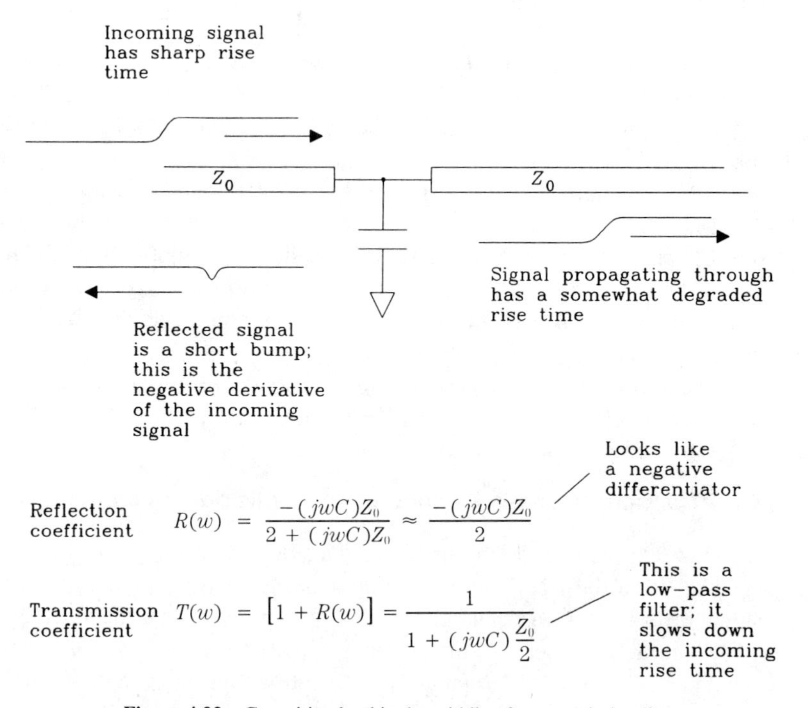

Figure 4.22 Capacitive load in the middle of a transmission line.

4.4.2.1 Signal reflected from a capacitive load

As with any other reflection problem, let's try using the reflection Equation 4.53. This equation requires that we specify the line and terminating impedances. For now, we will call the transmission line impedance Z_0, and work on the termination impedance.

The left section of the transmission line in Figure 4.22 terminates at the capacitor. The total terminating load at that point equals the reactance of the capacitor in parallel with the input impedance of the remainder of the line. Without knowing the termination conditions of the right-hand line, we can make few assumptions about its input impedance. How, then, can we calculate the total terminating load?

To work our way out of this dilemma, first assume that we are dealing with a low-loss line (not an *RC* case). Further assume that the right-hand line is end-terminated. Its input impedance is therefore equal to $Z_0 = (L/C)^{1/2}$, independent of frequency. Equivalently, we might assume the right-hand line is very long, and so signals reflecting off its far end arrive too late to influence the immediate reflection from capacitor *C*. Either way, we will assume the input impedance of the right-hand end equals Z_0.

We may now substitute the parallel combination of capacitor *C* and Z_0 for the term Z_L in Equation 4.53. Simplifying and rearranging terms, we arrive at this conclusion for the reflection coefficient at a capacitive load:

$$R_C(w) = \frac{-jwCZ_0}{2 + jwCZ_0} \qquad [4.73]$$

For frequencies above $f_{max}=(CZ_0\pi)^{-1}$ reflection is almost total. Don't use the transmission line above that frequency. For frequencies below f_{max} the reflection coefficient differentiates. It effectively returns a pulse equal to the derivative of the input step. The constant of differentiation is equal to $-C(Z_0/2)$.

If your digital knee frequency[11] is below f_{max}, then you may estimate the peak amplitude of the reflected pulse:

$$P = C\frac{Z_0}{2}\frac{-(\Delta V)}{T_{rise}} \qquad [4.74]$$

where ΔV = incoming voltage step size
P = reflected pulse amplitude, V
T_{rise} = 10–90% rise time of incoming signal, s
C = load capacitance, F
Z_0 = high-frequency line impedance, $(L/C)^{1/2}$

4.4.2.2 Signal transmitted through a capacitive load

Assume, as above, that both lines are long, and so their effective impedance as seen by the capacitor, for short-time durations, is equal to $Z_0 = (L/C)^{1/2}$. That assumption made, we can quickly calculate the transmission coefficient:

[11]See Equation 1.1 defining digital knee frequency.

$$T_C(w) = 1 + R_C(w) = \frac{1}{1 + jwC(Z_0 / 2)} \quad [4.75]$$

This is the equation for a low-pass filter having a time constant equal to $C(Z_0/2)$. The 10–90% rise time of this step response will be 2.2 times the time constant, or

$$T_{10-90}(\text{step response}) = 2.2C\frac{Z_0}{2} \quad [4.76]$$

The capacitive load deteriorates the rise time of signals propagating past it. To find the rise time of the propagating signal, use Equation 3.1. It mixes the incoming rise time and the capacitor rise time to find the outgoing rise time.

The approximations of this section (and Section 4.4.2.1) hold if

(1) The transmission line is terminated in both directions, or

(2) The transmission line is longer (in both directions) than a rising edge.

With a low-impedance driver connected too near the load capacitance, the effective drive impedance, as seen by the capacitor, goes down. The net result is less reflection and less rise-time distortion.

4.4.3 Equally Spaced Capacitive Loads

The situation in Figure 4.23 arises often in large bus formations, especially on memory cards containing large arrays of single in-line memory modules (SIMMs). The capacitive loads are of equal value and spaced evenly.

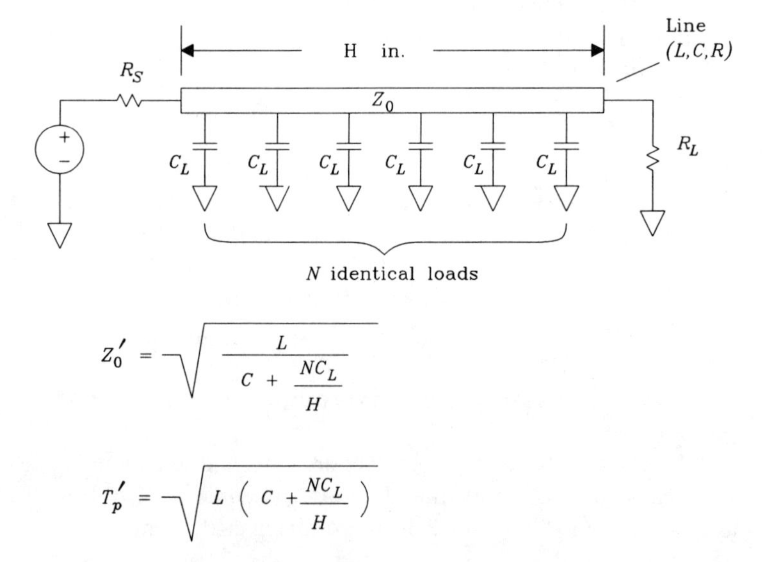

Figure 4.23 Equally spaced capacitive loads.

If the length of a rising edge exceeds the spacing between loads, we can derive a simplified approximation to the circuit behavior. This approximation will tell us two things:

(1) The effective impedance of the line is reduced.
(2) The propagation delay of the line is increased.

Both items critically affect the performance of high-speed signal buses.

4.4.3.1 Effective impedance of a uniformly loaded bus

For rising edges comparable (or shorter) in length to the load spacing, the signal bounces back and forth according to Equation 4.73. For sufficiently light loads (small capacitors), just sum the reflections from each load independently to compute a total reflected pulse height. Summing the reflections is a worst-case operation because the reflected pulses do not all arrive at any one point at the same time.

Secondary and tertiary reflections are greatly reduced and not usually worth computing.

For rising edges longer in length than the load spacing, the effects of individual capacitors smear evenly along the rising edge. The resulting effect would be no different if we used twice as many capacitors of half the value or if we *distributed the capacitance uniformly at an equivalent rate of picofarads per inch.*

Distributing the capacitance uniformly is the key to understanding this circuit.

Form a new transmission line model having the same inductance and resistance per inch as the original but with a new capacitance. Divide the total load capacitance by the length of the bus in inches to compute the load capacitance per inch. Then add this capacitance to the existing capacitance per inch of the transmission line to find the new model capacitance.

$$C' = C_{\text{line}} + \frac{NC_{\text{load}}}{\text{length}} \qquad [4.77]$$

where C_{load} = load capacitance, pF
N = number of loads
length = length of bus, in.
C_{line} = transmission line capacitance, pF/in.
C' = new model of effective pF/in.

Now using this model, we can recompute the effective transmission line impedance Z':

$$Z_0' \approx \left(\frac{L}{C'}\right)^{1/2} \qquad [4.78]$$

4.4.3.2 Propagation delay of a uniformly loaded bus

$$\text{Effective delay} = (LC')^{1/2} \text{ ps/in.} \qquad [4.79]$$

where C' = new model effective pF/in.
L = existing inductance, pH/in.

The effective characteristic impedance of a uniformly loaded bus can be ridiculously low. This makes it difficult for the drive circuit to impress a full-sized signal on the bus. Even changing to a lower impedance drive circuit still leaves the problem of time delay. This problem is due to the distributed inductance of the transmission line structure and cannot be avoided.

EXAMPLE 4.2: Uniformly Loaded Bus

Sam is building a large memory board using single in-line modules (SIMMs). He plans to combine 16 SIMMs into a giant memory array, as shown in Figure 4.24. The address lines for all 16 SIMMs are driven in parallel from one end, marked gate *A*.

Here are the critical parameters for each line:

$$C_{\text{load}} = 50 \text{ pF}$$
$$N = 16$$
$$\text{Length} = 8 \text{ in.}$$
$$C_{\text{line}} = 2.9 \text{ pF/in.}$$
$$L = 7250 \text{ pH/in.}$$

First compute the effective capacitance along the line:

$$C' = C_{\text{line}} + \frac{NC_{\text{load}}}{\text{length}} = 102.9 \text{ pF/in.} \qquad [4.80]$$

Use the new capacitance value to refigure Z_0 and the propagation delay:

$$Z_0 = \left(\frac{L}{C'}\right)^{1/2} = 8.4\ \Omega \qquad [4.81]$$

$$\text{Delay/in.} = (LC')^{1/2} = 864 \text{ ps/in.} \qquad [4.82]$$

The total line delay is

$$\text{Delay} = (\text{length})(\text{delay/in.}) = 6900 \text{ ps} \qquad [4.83]$$

The last SIMM will receive its address information 6.9 ns after the first SIMM. This skew degrades the memory timing margin. Not only that, the termination value and the drive impedance will both have to be impossibly low.

Possible solutions all involve breaking the SIMM address bus into multiple buses with fewer loads on each.

As a check, Sam should use a circuit similar to that in Figure 1.6 to measure the total line capacitance ($C' \times$ length). Sam may need to use smaller resistors than those shown in Figure 1.6 to source the current needed to drive the SIMM inputs through their transition region.

4.4.4 Right-Angle Bends

At the right-angle bend in Figure 4.25, the effective transmission line width increases. An increase in width contributes extra unwanted parasitic capacitance. The right-angle bend looks like a capacitive load attached to the transmission line.

We could round the outside corner of the bend, leaving a constant width. This reduces the amount of reflection and signal rise-time degradation for signals rounding the

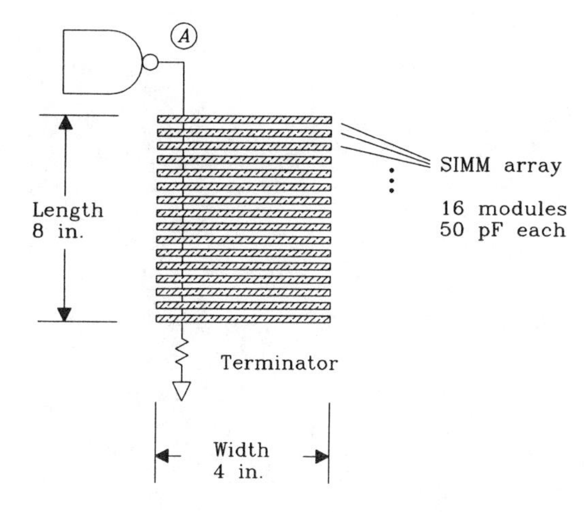

Figure 4.24 SIMM loading example.

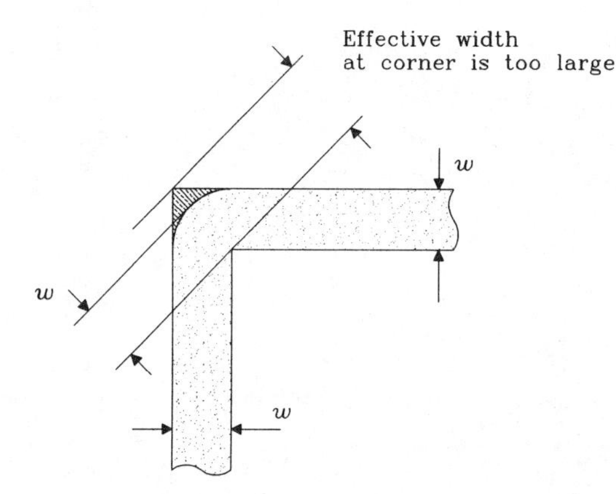

Figure 4.25 Right-angle bend in a transmission line.

corner. A simpler approach which holds promise for speeds up to 10 GHz chamfers the corner according to Figure 4.26.[12] Chamfering the corners may be easier to accomplish than rounding, depending on your layout software.

The amount of load capacitance represented by the shaded area in Figure 4.25 roughly equals

$$C \approx \frac{61we_r^{1/2}}{Z_0} \quad [4.84]$$

where w = width of line, in.
e_r = relative electric permeability, compared to air
Z_0 = characteristic impedance at high frequency, Ω
C = load capacitance of corner, pF

[12]T. C. Edwards, *Foundations for Microstrip Circuit Design,* John Wiley and sons, New York, New York, 1983.

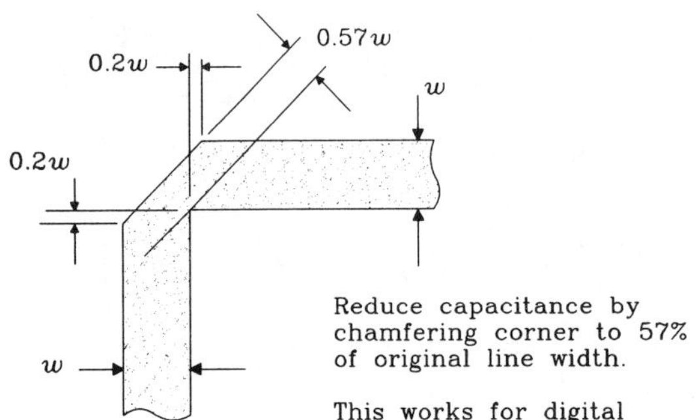

Figure 4.26 Chamfering the corner of a PCB trace to reduce capacitive loading.

We can figure the 10–90% rise-time constant associated with this lumped load using Equation 4.76:

$$T_{10\text{-}90} = 2.2\left[\frac{61we_r^{1/2}}{Z_0}\right]\frac{Z_0}{2} = 67w(e_r)^{1/2}\text{ ps} \qquad [4.85]$$

This is a very tiny rise time indeed. For circuits having rise times less than 100 ps and very wide lines (as is common in microwave engineering) it may be a serious issue.

Don't worry about 45-degree turns; they are fine as is. Chapter 7 discusses the effect of vias.

4.4.5 Delay Lines

When configured in a serpentine shape, a transmission line can serve as an effective delay line. This helps overcome problems associated with hold-time requirements on very fast flip-flops and other digital timing issues. A printed circuit delay line is very cheap, compared to external delay elements.

Figure 4.27 shows the input and output waveforms associated with a 4.9 ns digital delay, diagramed in Figure 4.28. The input rise time is 638 ps, while the output rise time

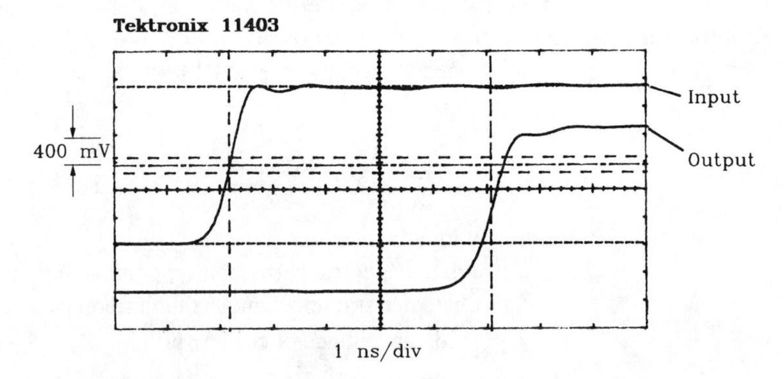

Figure 4.27 Delay line implemented with a printed circuit board trace.

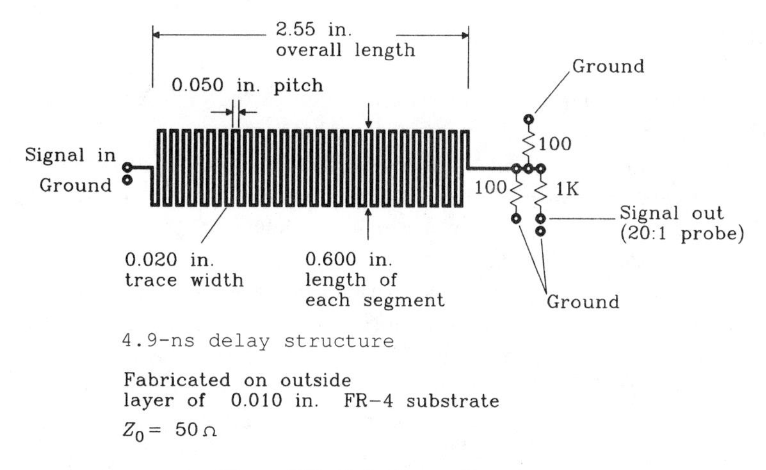

Figure 4.28 Delay line configuration.

is 888 ps. Delay lines usually disperse the incoming rising edges somewhat. In this configuration, reducing the trace separation between serpentine sections increases their cross-coupling, leading to more rising-edge dispersion. This particular design yields 560 ps of dispersion (that's the 10–90% rise time of the step response when driven with a perfect input signal).

When shrinking this design to fit on thinner FR-4 substrates, shrink the width in proportion to the substrate thickness. This keeps the impedance constant. You may also shrink the intertrace separations in proportion to the substrate thickness. This provides the same cross-coupling between traces and preserves the rise time. Shrinking both width and thickness while keeping the intertrace separation constant will yield less cross-coupling and a better rise time.

The dielectric constant of epoxy glass FR-4 circuit board material varies with temperature. The total variation is about 20% over a temperature range of 0–70°C. This variation in dielectric constant leads to a variation in the delay of an FR-4 circuit board trace of about 10% over temperature. Traces in FR-4 are slower at higher temperatures.

POINTS TO REMEMBER

Capacitive loads degrade the rise time of passing signals and reflect pulses back upstream.

Uniformly distributed capacitive loads reduce a transmission line's effective impedance and slow it down.

A printed circuit trace makes an effective small delay line.

4.5 LINE IMPEDANCE AND PROPAGATION DELAY

Transmission line impedance is a function of the geometry of the conductors and the electric permittivity of the material separating them.

For printed circuit board traces, the most critical dimension is the ratio of trace width to height above ground. For coaxial cables the most critical dimension is the ratio of center conductor diameter to shield diameter. For twisted-pair lines it is the ratio of wire diameter to wire separation.

In all cases impedance is inversely proportional to the square root of electric permittivity. Propagation delay is a function only of the electric permittivity.

Figures 4.29–4.35 illustrate the use of the formulas listed in Appendix C for computing transmission line parameters. These formulas are grouped according to transmission line type: coax, twisted pair, microstrip, and stripline.

The transmission line formulas listed in Appendix C for microstrip and stripline transmission structures are the most reliable formulas the authors could find. They come from the microwave literature and are supplied with the original references should you wish to explore them further. The overall accuracy of each formula and the range of parameters over which it maintains that accuracy are listed along with the microstrip and stripline formulas. This is in contrast to the popular formula set published, among other places, in the *Motorola MECL System Design Handbook*. Motorola popularized this formula set during the 1970s in conjunction with its ECL logic family. Here we refer to this formula set as the *simple formula set*.

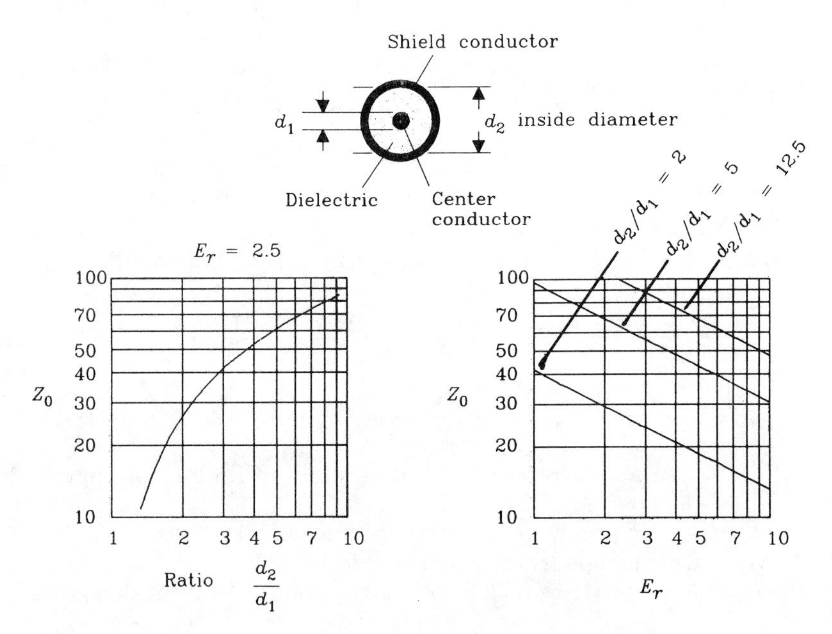

Figure 4.29 Characteristic impedance of coaxial cable versus geometry and permittivity.

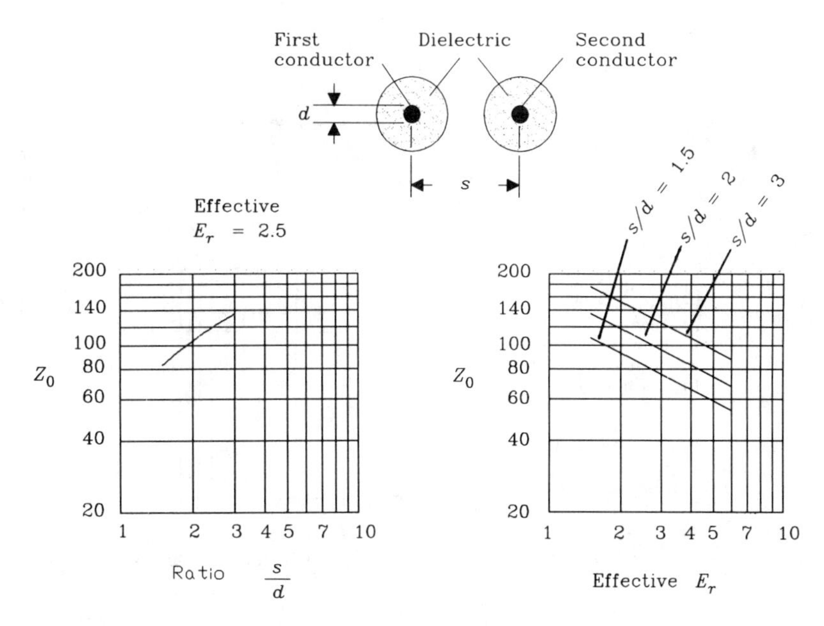

Figure 4.30 Characteristic impedance of a twisted-pair cable versus geometry and permittivity.

The advantage of the simple formula set is that it may be easily applied using a hand calculator. It gives reasonable answers for line impedances in excess of 75 Ω, if the trace height is greater than 0.020 in. At the time these formulas first appeared, a trace height of 0.020 in. was the norm.

Today's circuits are often fabricated at heights above ground of 0.005 in. or less. At these low heights, the effect of trace thickness becomes very pronounced. The formula sets in Appendix C accurately predict the impact of trace thickness. Working with these formulas, we can predict how a change from 1- to 2-oz copper weight will affect the final impedance value.

The most obvious failing of the simple formula set occurs at low trace impedances. When the trace's width exceeds its height by a factor of 7, the simple formula blows up, delivering *negative* answers. That effect appears in Figure 4.32. If you need to implement low-impedance clock distribution traces (perhaps 20 Ω), the simple formula set will not be useful.

4.5.1 Control of Transmission Line Parameters

Clearly, accurate control over impedance requires accurate control of both physical geometry and electric permittivity.

4.5.1.1 How tightly must we control impedance

In reference to Equation 4.53, a 10% error in transmission line impedance makes a 5% reflection. That helps a lot. Given a percentage budget for mismatch reflections, we

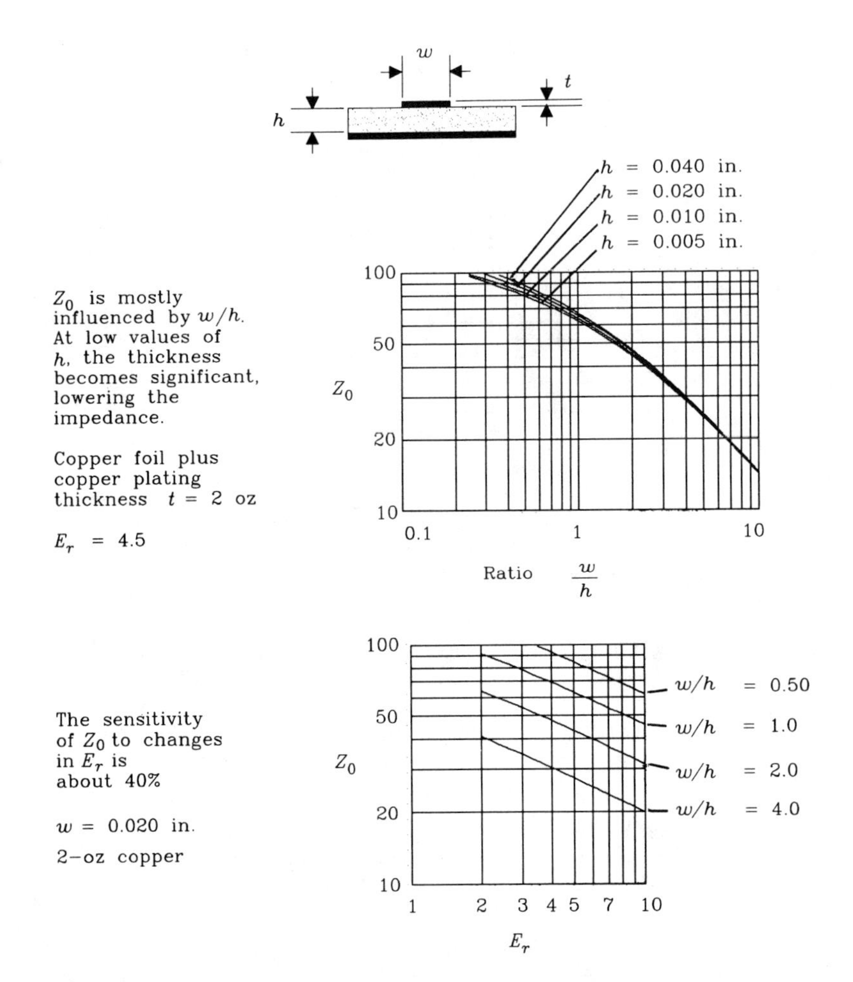

Figure 4.31 Characteristic impedance of a microstrip transmission line versus geometry and permittivity. (See formulas in Appendix C.)

can double it to find the allowed mismatch between the characteristic impedance and the terminating resistors. For example, a 10% reflection budget leaves room for a 10% tolerance in the characteristic impedance plus a separate 10% tolerance for the terminating resistors. Commonly, more accurate termination values are specified (perhaps 2%), leaving more tolerance for impedance variations.

For coaxial and twisted-pair situations, the designer works with whatever impedance tolerance is available. The situation on printed circuit boards is another matter. By specifying various board parameters on the fabrication artwork, the designer can attain almost arbitrary control over impedance variations.

Don't specify overly-tight tolerances. It costs extra to meet tight specifications, because of the extra testing required, lower yields, and other production hassles. (See Section 4.5.1.4.)

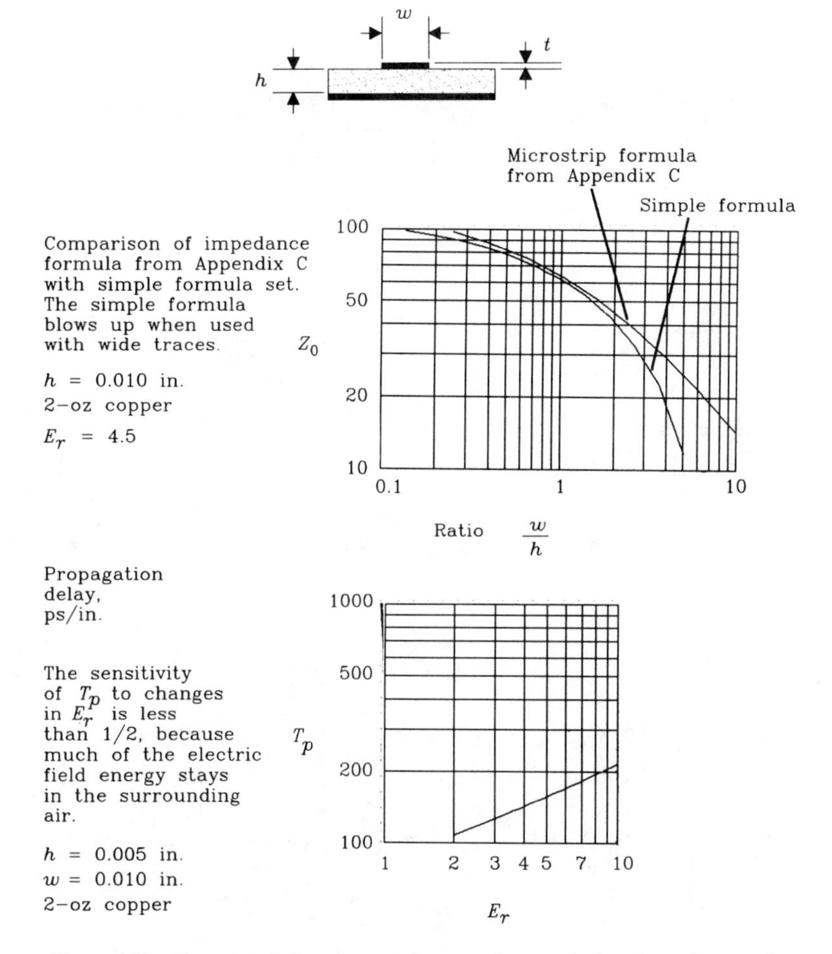

Figure 4.32 Characteristic impedance and propagation speed of a microstrip transmission line. (See formulas in Appendix C.)

4.5.1.2 How physical dimensions affect impedance

In most transmission line impedance formulas, the physical dimensions appear as arguments to a natural logarithm. The logarithm function is particularly slow-moving, meaning that large variations in physical dimensions make a small impact on the resulting impedance. This factor works in your favor.

The sensitivity of impedance to changes in physical size is low. *Sensitivity* is defined as the percent change in impedance per percent change in line width. A log-log plot shows sensitivity directly. The slope of any function plotted on log-log paper is equal to the sensitivity of the function to changes in its argument.

A slope of 1 means the function is directly proportional to the input. A 1% change in the input results in a 1% change in output. A slope of 1/2 means the function is propor-

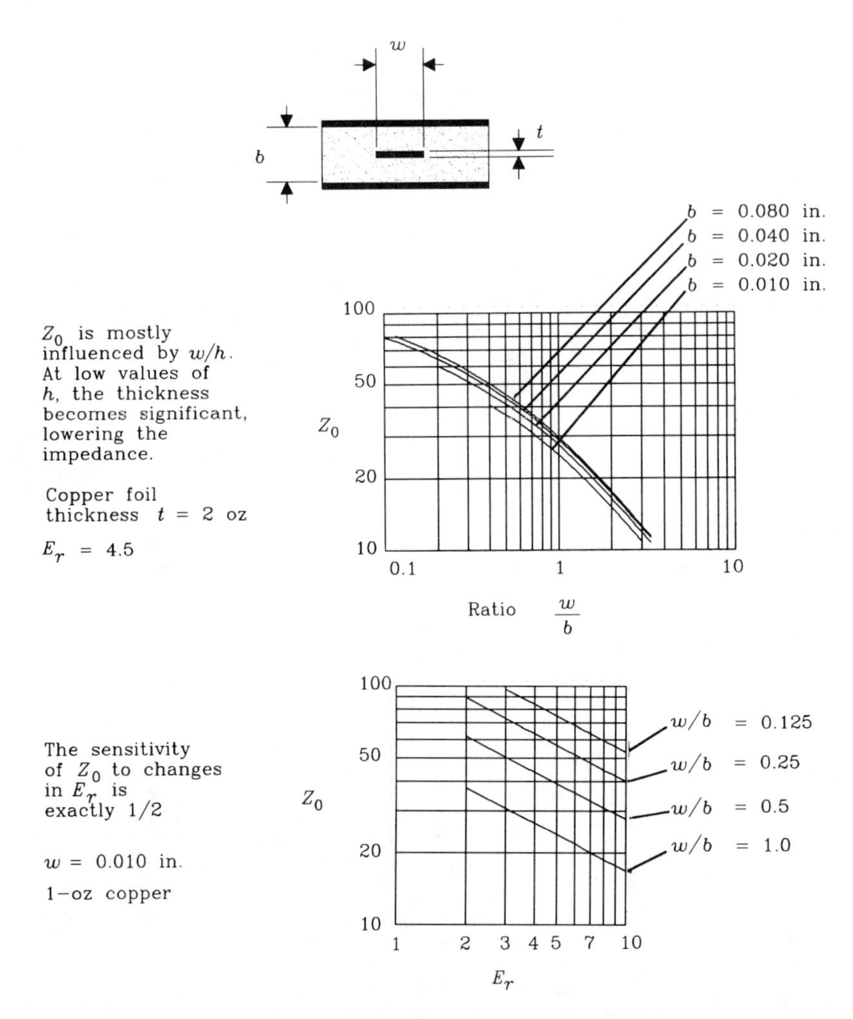

Figure 4.33 Characteristic impedance of a stripline transmission line versus geometry and permittivity. (See formulas in Appendix C.)

tional to the square root of the argument. A 1% change in the input results in a 0.5% change in output.

The plots in Figures 4.29–4.35 are formatted on log-log axes to help with the determination of sensitivity.

For critical applications, plan a two-pass fabrication cycle. This gives you an opportunity to try out a controlled-impedance design and then make adjustments for the inevitable parasitic effects during the second pass. Have the first-pass board(s) professionally microsectioned and dimensioned to determine whether the fabrication process accurately reproduced your design. This data, combined with a high-frequency dielectric test and an impedance measurement of test traces on the board will determine whether the design needs adjustment.

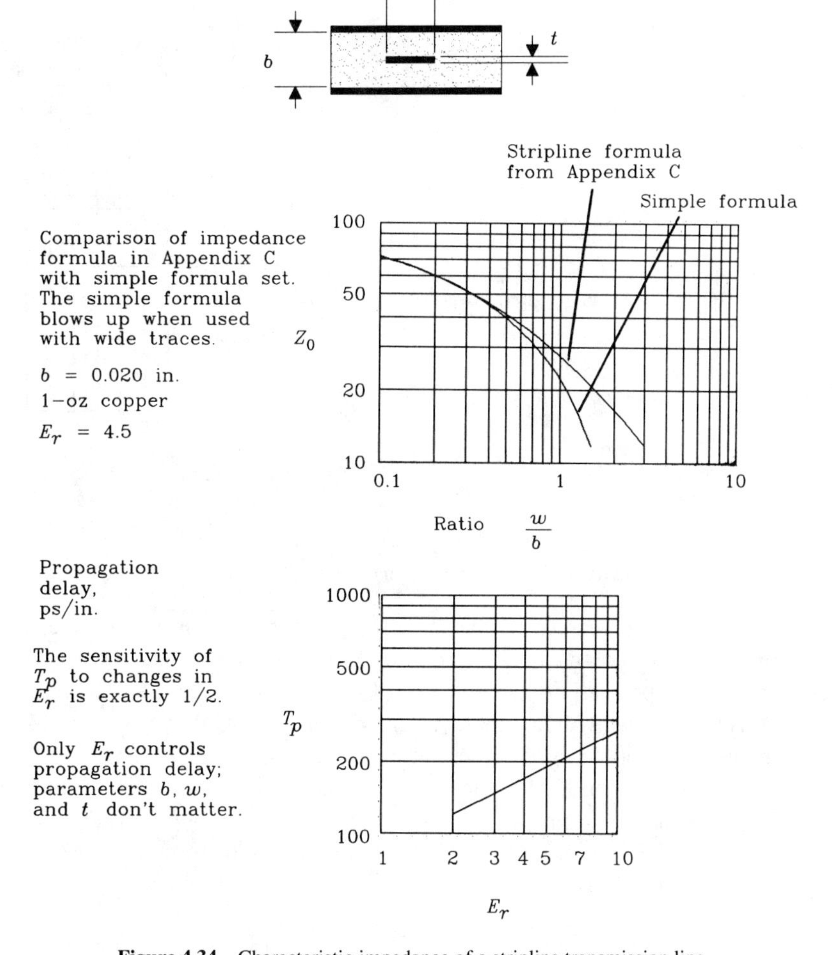

Figure 4.34 Characteristic impedance of a stripline transmission line.

4.5.1.3 Effective electric permittivity

All formulas for transmission velocity are inversely proportional to the *effective* square root of electric permittivity. The effective permittivity is sometimes difficult to determine.

For example, in a coaxial cable all the electric fields stay within the cable, in the region between shield and center conductor. The effective permittivity is just the permittivity of the dielectric insulating material.

In a loosely bound twisted-pair cable, or one with a high ratio of wire separation to diameter, the field arches between conductors in sweeping curves, passing primarily through air. The effective permittivity is an average of the relative permittivity of air (1.00) and the relative permittivity of the insulating material.

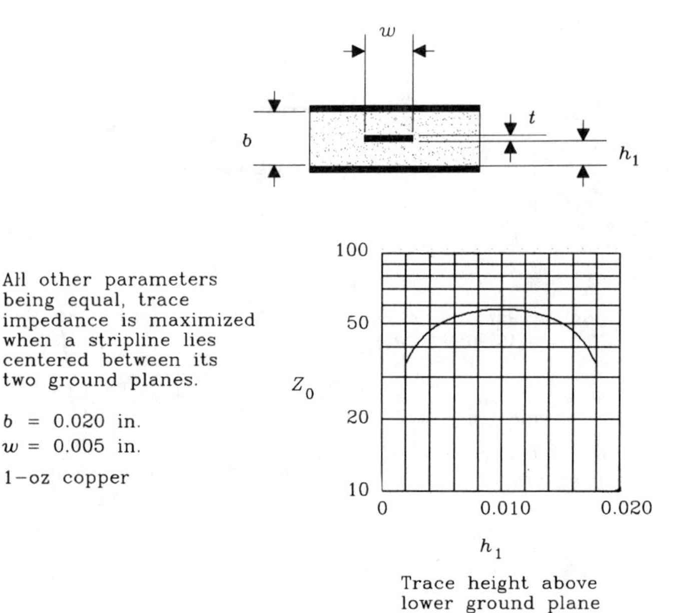

Figure 4.35 Impedance of an offset stripline. (See formulas in Appendix C.)

Flat ribbon cables are particularly susceptible to this effect. Thick cables with insulation material completely surrounding the conducting members have, *for adjacent wires*, an effective permittivity close to that of the insulating material. The effective permittivity for calculations involving nonadjacent wires is practically unity (most of the electric field arches up out of the insulator and through the air).

Some ribbon cable manufacturers use a flat, hard, thin insulating material that supports the wires but does not surround them. The wires make visible bumps in the surface of the thin insulating ribbon. Since most of the field is in the air, this cable has a lower effective permittivity, and thus higher propagation velocity, than thickly insulated cable.

Dielectric permittivity changes with temperature. The permittivity of FR-4 epoxy glass circuit board material varies as much as 20% over a temperature range of 0–70°C. Dielectrics designed for use in coaxial cables exhibit less variation.

4.5.1.4 Reasonable manufacturing tolerances

For printed circuit boards, the manufacturing tolerances achievable depend on both the substrate material and the etching or plating processes used to make up the board.

The popular FR-4 substrate can be manufactured with a variety of epoxy/glass ratios, all of which lead to a relative electric permittivity in the range of 4.00–5.25. Printed circuit board manufacturers control these parameters by buying high-quality, low-tolerance base materials. A requirement for 4.5 ± 0.1 is not unreasonable.

Relative electric permittivity varies with frequency. At low frequencies, 50% resin FR-4 has a relative permittivity of 4.7, decreasing to 4.5 at 1 MHz and 4.35 at 1 GHz. A typical test frequency for permittivity is 1 MHz. Always specify the test frequency for dielectric testing. For impedance calculations, use the permittivity value at the digital knee frequency[13] of your circuits.

The relative permittivity of FR-4 varies dramatically with temperature. If this factor is important to you, consider using a ceramic or Teflon substrate having more stable dielectric properties.

Military standard MIL-STD-275, "Printed Wiring for Electronic Equipment," and the related commercial standard[14] IPC-ML-950, "Performance Specifications for Rigid Multi-layer Printed Boards," both establish guidelines for mechanical and electrical tolerances. The military standard calls out three classes of board according to their difficulty of manufacture: preferred, standard, and reduced producibility (a term invented by military experts). The commercial standard calls out three classes of board according to their application: consumer, general, and high reliability.

Trace width tolerances allowed in the military standard depend on the fabrication technique. Simple etching processes used for inner circuit layers have the best tolerances. The additional plating processes required for outer layers adds uncertainty to their finished trace dimensions. For any layer, thinner copper gives better dimensional control but reduces the current-carrying capacity of wiring. The tolerance for 2-oz copper weight on an outer layer listed in Table 4.2, is the worst case. One-ounce copper on an external layer is a little better.

The military guidelines are a good starting point for understanding typical tolerances. Work with your printed circuit board manufacturer to understand their capabilities and always ask, How much does it cost?

TABLE 4.2 PRINTED CIRCUIT BOARD TOLERANCES MIL-STD-275

	Preferred	Standard	Reduced producibility
Minimum layer thickness (figure at least 10% or 0.001 tolerance on thickness)	0.008	0.006	0.004
Minimum conductor width			
Inner	0.015	0.010	0.008
Outer	0.020	0.015	0.008
Width tolerance			
Inner, 1-oz	+0.002	+0.001	+0.001
	−0.003	−0.002	−0.001
Inner, 2-oz	+0.004	+0.002	+0.001
	−0.006	−0.005	−0.003
Outer, 2-oz	+0.008	+0.004	+0.002
	−0.006	−0.004	−0.002

[13]See Equation 1.1 defining digital knee frequency.

[14]Institute for Interconnecting and Packaging Electronic Circuits (IPC).

4.5.1.5 Software for transmission line calculations

Most digital engineers use the simple formula set listed below, fabricate a batch of circuit cards, and then adjust the line widths and trace spacings as needed.

For better accuracy, use the more complex formula set listed in Appendix C. For your convenience, all the formulas in Appendix C have been implemented in MathCad. They are available from the authors in magnetic form, which will save you the time of retyping them. See the order card in the back of the book.

If you require better prediction of characteristic impedance or crosstalk before fabricating a printed circuit board, you may need a more sophisticated computer model. At the time of publication, the following companies provide elaborate software packages for calculating characteristic impedance and crosstalk:

B. V. Engineering, Chicago, Illinois
  Micro-3
Quad Design, Camarillo, California
  Crosstalk Tool Kit
Quantic Laboratories, Winnepeg, Manitoba, Canada
  Greenfield
  TR line

4.5.2 Formulas Involving Coaxial Cable (Figure 4.29)

Diameter of inner conductor, d_1

Diameter of inside surface of shield, d_2 ($d_2 > d_1$)

Effective relative permittivity, ε_r

(For solid core cable, this equals the electric permittivity of the dielectric material. For foam core, spiral wrap, or other cores containing lots of air, ε_r is less)

Impedance (Ω):

$$\frac{60}{\sqrt{\varepsilon_r}} \ln\left(\frac{d_2}{d_1}\right) \quad [4.86]$$

Propagation delay (ns/in.):

$$85\sqrt{\varepsilon_r} \quad [4.87]$$

4.5.3 Formulas Involving Twisted-Pair Cable (Figure 4.30)

Diameter of conductor, d

Separation between wire centers, s ($s > d$)

Effective relative permittivity, ε_r

(For widely separated wires, use $\varepsilon_r = 1$, and for wire pairs with the insulation touching use the electric permittivity of the insulating material)

Impedance (Ω):

$$\frac{120}{\sqrt{\varepsilon_r}}\ln\left(\frac{2s}{d}\right) \quad [4.88]$$

Propagation delay (ps/in.):

$$85\sqrt{\varepsilon_r} \quad [4.89]$$

4.5.4 Simple Formulas Set for Microstrips (Figures 4.31–4.32)

The values plotted in Figures 4.31–4.32 were calculated using the accurate formulas listed in Appendix C. The simple formulas listed below deliver reasonable approximations to those values. Figure 4.32 includes a plot comparing the simple impedance formula with that in Appendix C.

Height above ground (in.), h

Trace width (in.), w

Line thickness[15] (in.), t

Relative permittivity of substrate, ε_r

(The simple formulas take into account how the electric field splits between the substrate and air, reducing the effective permittivity below the relative permittivity of the substrate. Enter here the relative permittivity of the substrate)

For narrow microstrips only:

Use these formulas when $0.1 < w/h < 2.0$ and also when $1 < \varepsilon_r < 15$.

Impedance (Ω):

$$\frac{87}{\sqrt{\varepsilon_r + 1.41}}\ln\left(\frac{5.98h}{0.8w + t}\right) \quad [4.90]$$

Propagation delay (ps/in.):

$$85\sqrt{0.475\varepsilon_r + 0.67} \quad [4.91]$$

4.5.5 Simple Formula Set for Striplines (Figures 4.33-4.35)

The values plotted in Figures 4.33-4.35 were calculated using the accurate formulas listed in Appendix C. The simple formulas listed below deliver reasonable approximations to those values. Figure 4.34 includes a plot comparing the simple impedance formula with the formula for striplines given in Appendix C.

Separation between grounds (in.), b

Trace width (in.), w

[15]Sometimes reported as plating weight. Each ounce of plating weight is 0.00135 in. thick.

Line thickness[16] (in.), t

Effective relative permittivity, ε_r

(Equals the relative permittivity of the surrounding medium)

For narrow striplines:

Use these formulas when $w/b < 0.35$ and also when $t/b < 0.25$.

Impedance (Ω):

$$\frac{60}{\sqrt{\varepsilon_r}} \ln\left(\frac{1.9b}{0.8w + t}\right) \quad [4.92]$$

Propagation delay (ps/in.):

$$85\sqrt{\varepsilon_r} \quad [4.93]$$

POINTS TO REMEMBER:

For printed circuit board traces, the most critical dimension is the ratio of trace width to height above ground.

Double the reflection budget to find the allowed mismatch between the characteristic impedance and the terminating resistors.

Large variations in physical dimensions make a small impact on the resulting impedance.

The slope of any function plotted on log-log paper is equal to the sensitivity of the function to changes in its argument.

All formulas for transmission velocity are inversely proportional to the *effective* square root of electric permittivity.

[16] Sometimes reported as plating weight. Each ounce of plating weight is 0.00135 in. thick.

5

Ground Planes and Layer Stacking

Ground and power planes in high-speed digital systems perform three critical functions:

- Provide stable reference voltages for exchanging digital signals
- Distribute power to all logic devices
- Control crosstalk between signals.

This chapter focuses on signal crosstalk. Sections 5.1–5.6 assume relatively short traces, for which a lumped analysis of mutual inductance is appropriate. Section 5.7 treats the case of long lines, where we separate coupling into its forward and reverse parts.

Section 5.8 summarizes the rules for designing good printed circuit board layer stacks that control crosstalk.

The formulas in this chapter are accurate only to within a factor of 2. For better accuracy use the formulas as a guide, build a prototype, and then measure its actual performance. Many transmission line structures are easily built out of copper tape and blank circuit board material.

The formulas in this chapter are excellent for showing how electromagnetic effects adjust in response to physical changes. For example, if crosstalk is 30% too high, the formulas can show just how much further to separate your traces. They are less useful for predicting the absolute value of any particular effect.

5.1 HIGH-SPEED CURRENT FOLLOWS THE PATH OF LEAST INDUCTANCE

At low speeds, current follows the path of least resistance. In reference to Figure 5.1, low-speed current transmitted from A to B returns to the driver along the ground plane. This return current flows in wide arcs on its way back to the driver. The current density along each arc corresponds to the conductance of that path.

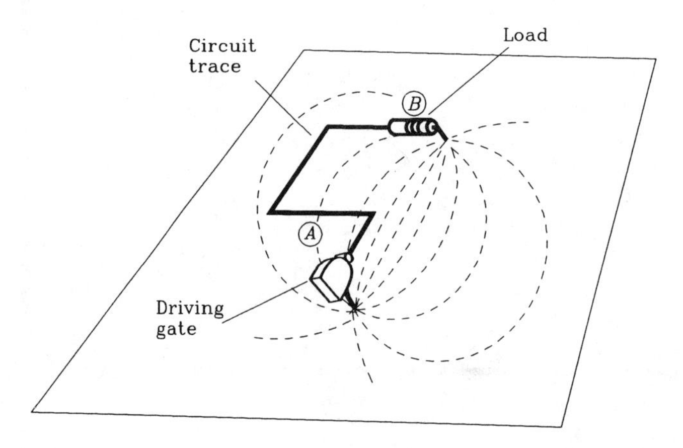

Figure 5.1 At low frequencies current follows the path of least resistance.

At high speeds, the inductance of a given return-current path is far more significant than its resistance. High-speed return currents follow the path of least inductance, not the path of least resistance.

The lowest inductance return path lies directly under a signal conductor, minimizing the total loop area between the outgoing and returning current paths.

Returning signal currents tend follow this direct path, close underneath a signal conductor. Figure 5.2 shows a typical high-frequency return-current path.

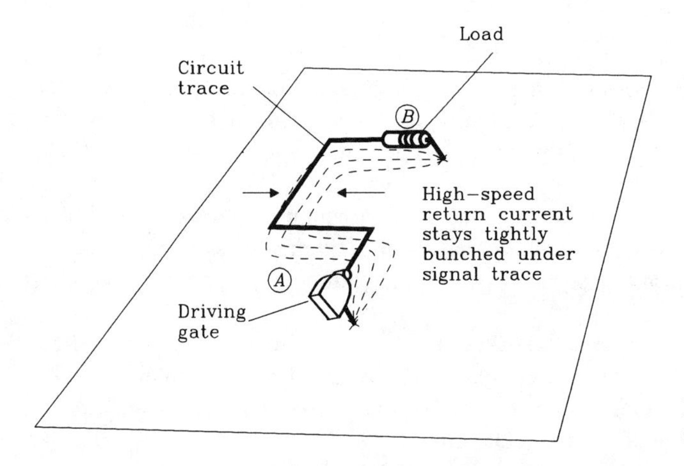

Figure 5.2 At high frequencies current follows the path of least inductance.

Figure 5.3 presents the cross section of a typical printed circuit board trace along with its return-current distribution. The peak current density lies directly under the trace. The current density then falls off sharply to each side.

An approximate relation for the return-current density at a point D inches away from a signal trace is

$$i(D) = \frac{I_0}{\pi H} \cdot \frac{1}{1+(D/H)^2} \qquad [5.1]$$

where I_0 = total signal current, A
H = height of trace above circuit board, in.
D = perpendicular distance from signal trace, in.
$i(D)$ = signal current density, A/in.

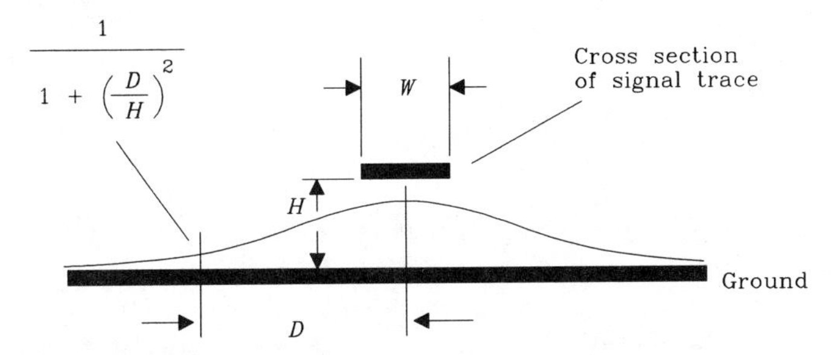

Figure 5.3 Distribution of high-frequency current density underneath a signal trace.

The current distribution (Equation 5.1) balances two opposing forces. Were the current more tightly drawn together, it would have higher inductance (a skinny wire has more inductance than a broad, flat one). Were the current spread farther apart from the signal trace, the total loop area between the outgoing and returning signal paths would increase, raising the inductance (inductance is proportional to loop area). Equation 5.1 specifies the optimum distribution minimizing the total loop inductance of both outgoing and returning current paths.

The current distribution (Equation 5.1) also minimizes the total energy stored in the magnetic field surrounding the signal wire.

POINTS TO REMEMBER:

High-speed current follows the path of least inductance.

Returning signal currents tend to stay near their signal conductors, falling off in intensity with the square of increasing distance.

5.2 CROSSTALK IN SOLID GROUND PLANES

Crosstalk between two conductors depends on their mutual inductance and their mutual capacitance. Usually, in digital problems the inductive crosstalk is as big or larger than capacitive crosstalk, so we will henceforth discuss mainly the inductive coupling mechanism.

The theory behind mutual inductive coupling for lumped circuits appears in Section 1.10. That theory postulates that returning signal currents will generate magnetic fields. Those magnetic fields in turn induce voltages in other circuit traces.

The induced voltages are proportional to the derivative of the driving signal. They become markedly worse as rise times get shorter.

Because the returning current density and its associated local magnetic field strength drop off according to Equation 5.1, we suspect that mutual inductive crosstalk also drops off as we move the two traces in Figure 5.4 away from each other:

$$\text{Crosstalk} \approx \frac{K}{1+(D/H)^2} \qquad [5.2]$$

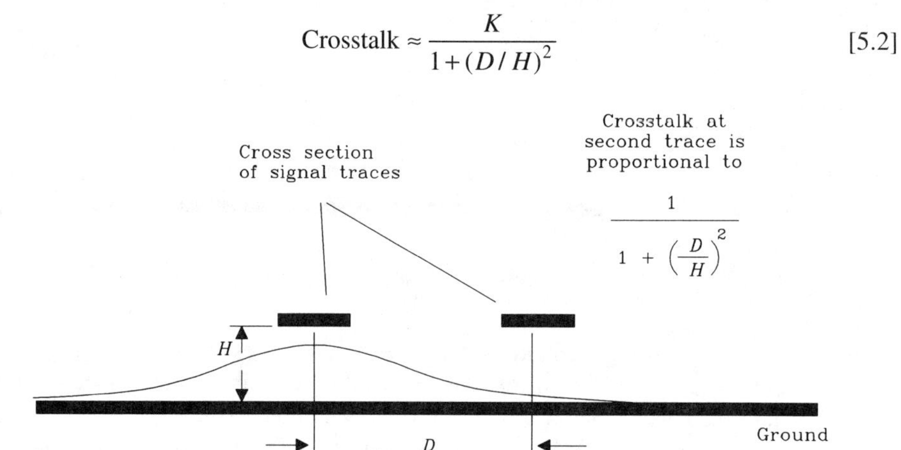

Figure 5.4 Cross section of two traces showing crosstalk.

Here we are expressing crosstalk as a ratio of measured noise voltage to the driving step size. The constant *K* depends on the circuit rise time and the length of the interfering traces. It is always less than 1.

We can check this hypothesis with a simple experiment. The traces configured in Figure 5.5 are 26 in. long and separated by 0.080 in. center to center. They lie on a single-layer main circuit board. The ground plane for this experiment is a solid copper sheet fastened below the main circuit board. Together, the main board and the ground layer sandwich a pile of dielectric spacers of known thickness. By this arrangement we can simultaneously vary the effective height above ground of the driven and receiving traces.

As in problems involving characteristic impedance, the ratio of sizes is more important than the absolute dimensions. In this case it is the ratio *D/H* that determines the crosstalk. By varying the trace height above ground, we can control the ratio *D/H*.

Figure 5.6 plots the resulting step responses, as measured at point *D*, when driving the input with a 3.5-V step. Figure 5.6 uses ground separations of 0.010, 0.020, 0.030, 0.040, and 0.050 in. The last trace (biggest noise pulse) was taken with no ground plane at all.

Figure 5.7 compiles this measured data into a chart showing the mutual inductance between traces as a function of the ratio *D/H*. Area is used as a measure of mutual coupling, as explained in Section 1.8. By measuring the area we can factor out the tendency

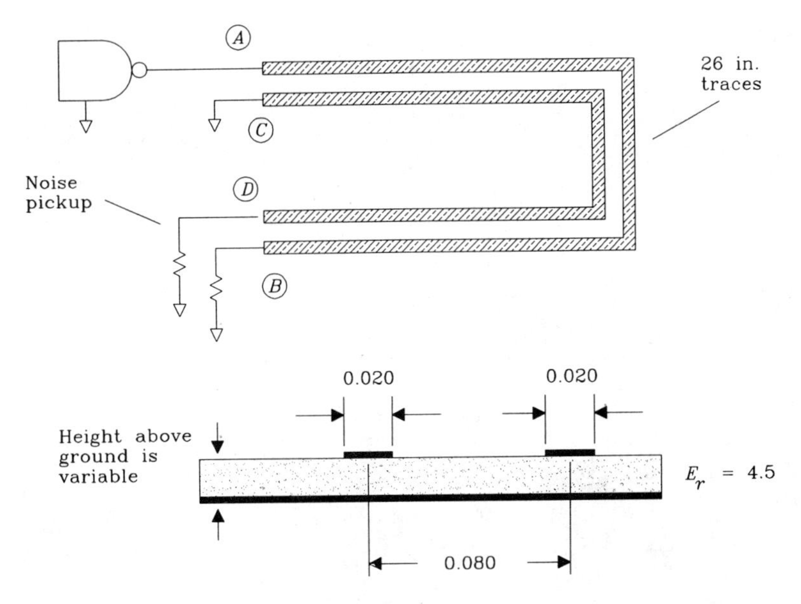

Figure 5.5 Mutual coupling experiment.

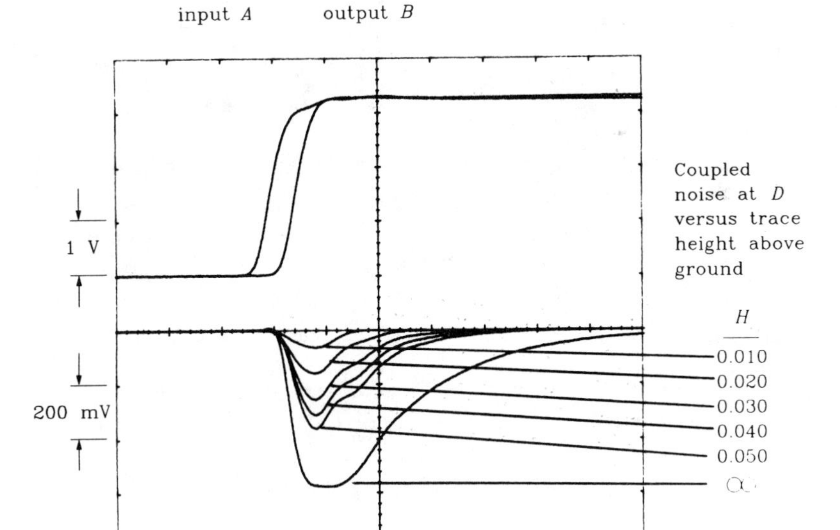

Figure 5.6 Step response of a mutual coupling experiment.

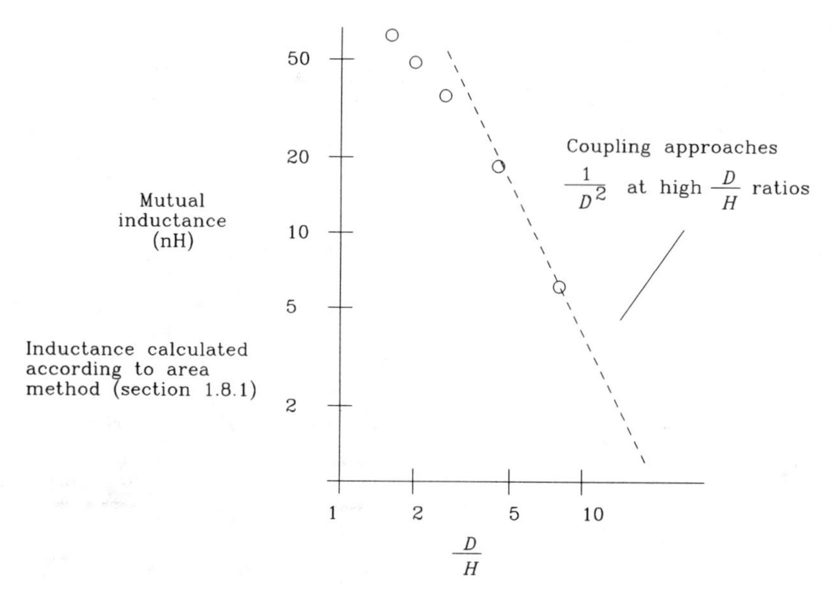

Figure 5.7 Measured data on mutual coupling.

of the driving waveform to slow down when confronted with a high loop inductance. This effect shows in the resulting noise plots as a lengthening of the noise pulse at high coupling factors.

> **POINTS TO REMEMBER:**
>
> Returning signal currents generate magnetic fields, which in turn induce voltages in other circuit traces.
>
> The induced noise coupled into adjacent traces falls off with the square of increasing distance.

5.3 CROSSTALK IN SLOTTED GROUND PLANES

The crosstalk situation depicted in Figure 5.8 is a classic layout mistake called a *ground slot*.

Ground slots happen when a layout engineer runs out of room on the regular routing layers and decides to cram in a trace on the ground plane layer. This is done by cutting a long slot in the ground plane and laying the trace in the slot. Ground slots add inductance to traces passing perpendicularly over the slot, and increase crosstalk. Do not tolerate this practice.

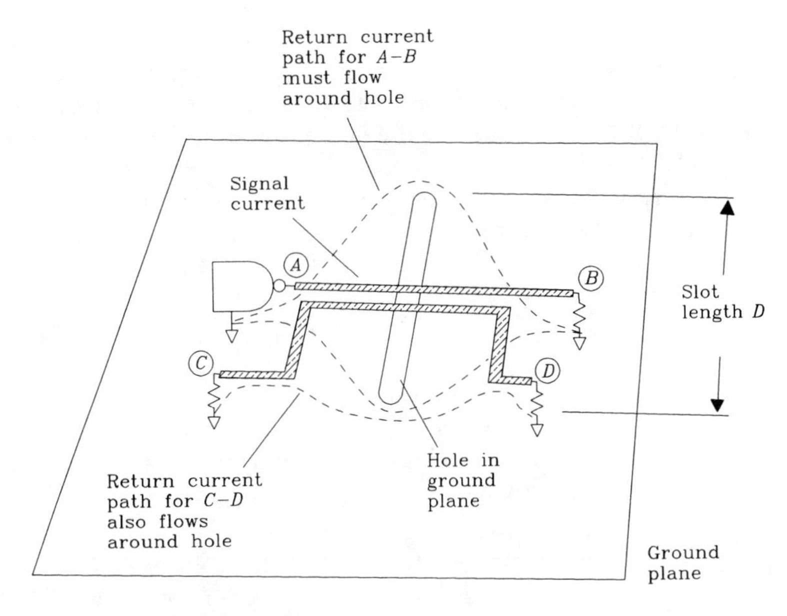

Figure 5.8 Crosstalk in a slotted ground plane.

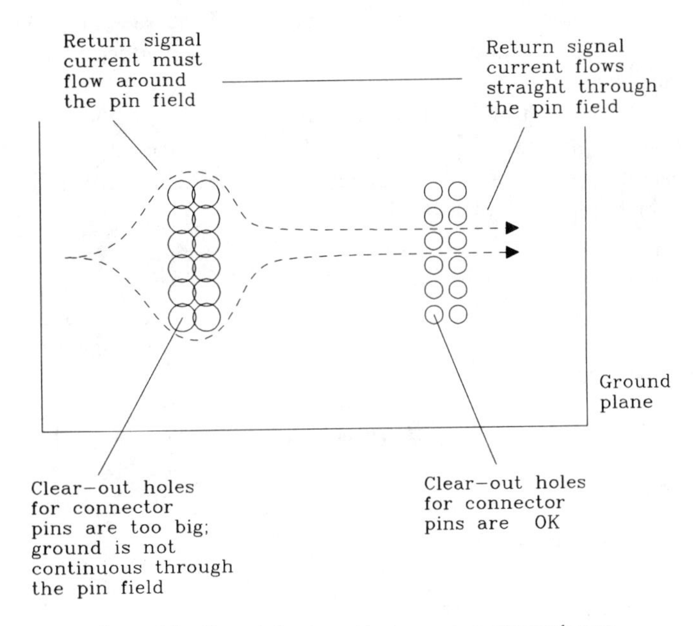

Figure 5.9 Ground slot caused by improper connector layout.

Ground slots also happen on dense backplanes which pass through fields of connector pins. Always make sure the ground clear-outs around each pin have ground continuity between all pins (Figure 5.9).

As shown in Figure 5.8, the return current from driver at *A* cannot flow directly under trace *A-B*. Instead, it diverts around the ends of the ground slot.

The diverted current makes a large loop, dramatically increasing the inductance of signal path *A-B*, which slows the rise time of signals received at *B*.

The diverted current also overlaps heavily with the current loop formed by trace *C-D* and its return-current path. This overlap leads to a large mutual inductance between the signal trace *A-B* and trace *C-D*.

The effective inductance in series with trace *A-B* is

$$L \approx 5D \ln\left(\frac{D}{W}\right) \qquad [5.3]$$

where L = inductance, nH
D = slot length (perpendicular extent of current diversion away from signal trace), in.
W = trace width, in.

The slot width (length of exposed signal trace passing across the slot) has almost no bearing on the signal trace inductance. A slot of any width, no matter how thin, causes current diversion around the ends of the slot. Any slot width from zero up to as big as the slot length will have the same effect.

If the trace lies offset toward one end of the slot, the inductance is less. Slots as small or smaller than the trace width have almost no effect. Slots near but not overlapping a trace have little effect.

The amount of rise-time degradation caused by the slot inductance varies, depending on the termination conditions used. The worst case is with a long line, for which the apparent source resistance on either side of the inductance is Z_0. The resulting 10–90% rise time of the *L*/*R* filter thus formed is

$$T_{10-90\;L/R} = 2.2\frac{L}{2Z_0} \qquad [5.4]$$

Combine this rise time with the natural signal rise time using the square root of sum of squares rule:

$$T_{\text{composite}} = \left[\left(T_{10-90\;L/R}\right)^2 + \left(T_{10-90\;\text{signal}}\right)^2\right]^{1/2} \qquad [5.5]$$

For a short line driving a heavy capacitive load *C*, the 10–90% rise time (assuming critical damping) is

$$T_{10-90} = 3.4(LC)^{1/2} \qquad [5.6]$$

Such a circuit might ring. The *Q* of this circuit depends on R_S, the source resistance of the driver:

$$Q = \frac{(L/C)^{1/2}}{R_S} \qquad [5.7]$$

When Q is greater than 1, the circuit rings. When Q is near 1, the circuit rise time follows Equation 5.6. For Q less than 1, the circuit rise time is slower than Equation 5.6.

If a second trace close to the first trace also intersects the same slot, the two traces couple tightly together. The mutual inductance L_M between it and the first trace is the same as L in Equation 5.3.

If the second trace lies offset toward one end of the slot, its mutual inductance with the first trace decreases linearly with the distance between it and the slot's end.

The cross-coupling voltage between the two traces can be figured from knowledge of their mutual inductance and the time rate of change in current in the driving circuit:

$$V_{\text{crosstalk}} = \frac{\Delta I}{T_{10-90}} L_M \qquad [5.8]$$

For a long line, the ΔI is just equal to the drive voltage divided by the characteristic impedance:

$$V_{\text{crosstalk}} = \frac{\Delta V}{T_{10-90} Z_0} L_M \qquad [5.9]$$

For a short line driving a heavy capacitive load C, the time rate of change in current is a second derivative of voltage:

$$V_{\text{crosstalk}} = \frac{1.52\,\Delta VC}{\left(T_{10-90}\right)^2} L_M \qquad [5.10]$$

Equations 5.4–5.10 apply equally well for inductance caused by any interruption in a ground plane.

POINTS TO REMEMBER:

Slots in a ground plane create unwanted inductance.

Slot inductance slows down rising edges.

Slot inductance creates mutual inductive crosstalk.

5.4 CROSSTALK IN CROSS-HATCHED GROUND PLANES

The *power and ground grid* diagrammed in Figure 5.10 saves board area, but at the expense of increased mutual inductance. This technique does not require separate power and ground plane layers. You may connect ordinary signals on the same layers as the power and ground connections. It is appropriate for small low-speed CMOS and ordinary TTL designs but provides inadequate grounding for high-speed logic.

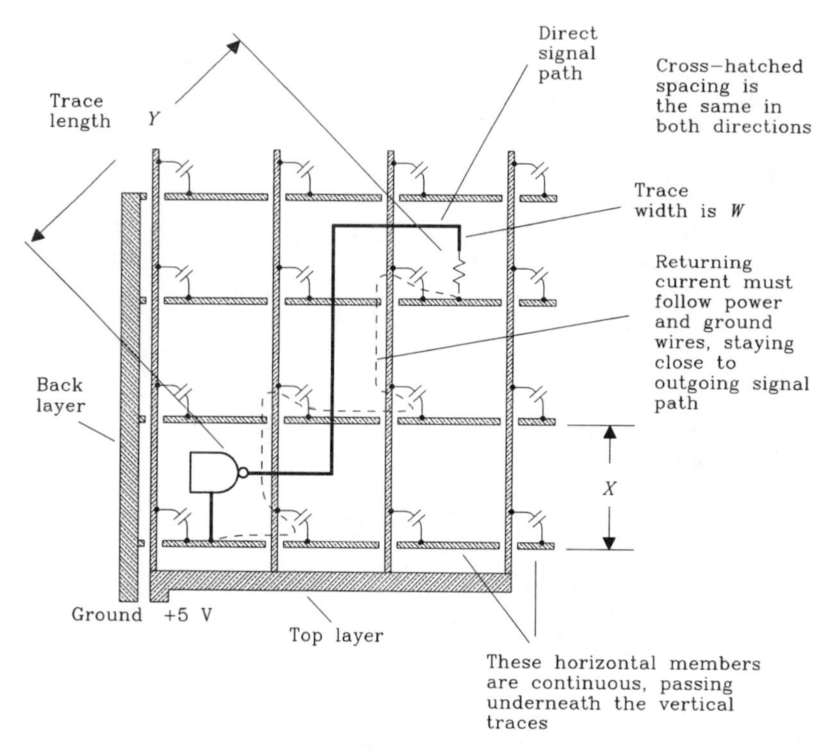

Figure 5.10 Power and ground grid on two layers.

In the ground plane grid scheme, ground lines lie in a horizontal pattern on the bottom of a board, while power traces lie in a vertical pattern on the top. Connecting the two sets of lines at every intersection with a bypass capacitor makes a cross-hatched pattern. Current returns equally well to its source along either the ground or power wiring.

The bypass capacitors used in this system must be particularly good, as some returning current traverses several bypass capacitors on its way back to a driving gate.

The open pattern of power and ground traces leaves plenty of room for routing other signals on the power and ground layers. After completing power and ground connections, horizontal routing channels remain on the ground side and vertical routing channels remain on the power side of the board. If you must use a two layer board, this is a good way to do it.

A related layout pattern is called *cross-hatched ground plane*. This pattern, entirely routed on one layer, consists of vertical and horizontal traces covering the board. The cross-hatched ground plane connects only to ground. No other signals may be routed on this layer.

The cross-hatched ground helps implement high-impedance transmission structures on a thin board. Sometimes on a thin dielectric the narrow width required to implement a satisfactory impedance is too small to fabricate reliably. In this case, a cross-hatched ground plane pattern etched into the ground layer adds series inductance and reduces shunt capacitance, thus raising the line's characteristic impedance. Don't try to implement controlled impedance lines on a cross-hatched ground plane unless the lines run at 45 degrees to the hatch direction. The hatches must be much smaller than the length of a rising edge for this approach to work.

Both the power and ground grid and cross-hatched ground plane layouts induce a lot more mutual inductance between traces than solid ground planes. The question is, will your circuit work with that much mutual inductance?

First let's estimate the self-inductance of a single trace running across a cross-hatched ground plane. This estimation applies equally well to a power and ground grid layout:

$$L \approx 5Y \ln\left(\frac{X}{W}\right) \qquad [5.11]$$

where L = inductance, nH
X = hatch width, in.
W = trace width, in.
Y = trace length, in.

If the trace lies offset near one cross-hatched line, the inductance is a little less. Cross-hatched patterns that are as small or smaller than the width of a trace have almost no effect.

If a second trace, close to the first trace, runs between the same two cross-hatched members, the two traces couple tightly together. The mutual inductance L_M between it and the first trace is the same as L in Equation 5.11.

If the second trace is offset by a good distance D, its mutual inductance with the first trace decreases with a denominator similar to Equation 5.2, but with the cross-hatched dimension X replacing the term H:

$$L_M \approx \frac{5Y \ln(X/W)}{1+(D/X)^2} \qquad [5.12]$$

Use the formulas from Section 5.3 for computing rise-time degradation and crosstalk voltage from this self-inductance and mutual inductance.

POINT TO REMEMBER:

If you must work with only two layers, use the power and ground grid system.

5.5 CROSSTALK WITH POWER AND GROUND FINGERS

The *power and ground fingers layout*, diagrammed in Figure 5.11, like the power and ground grid, allows some mutual inductive coupling but saves even more board area.

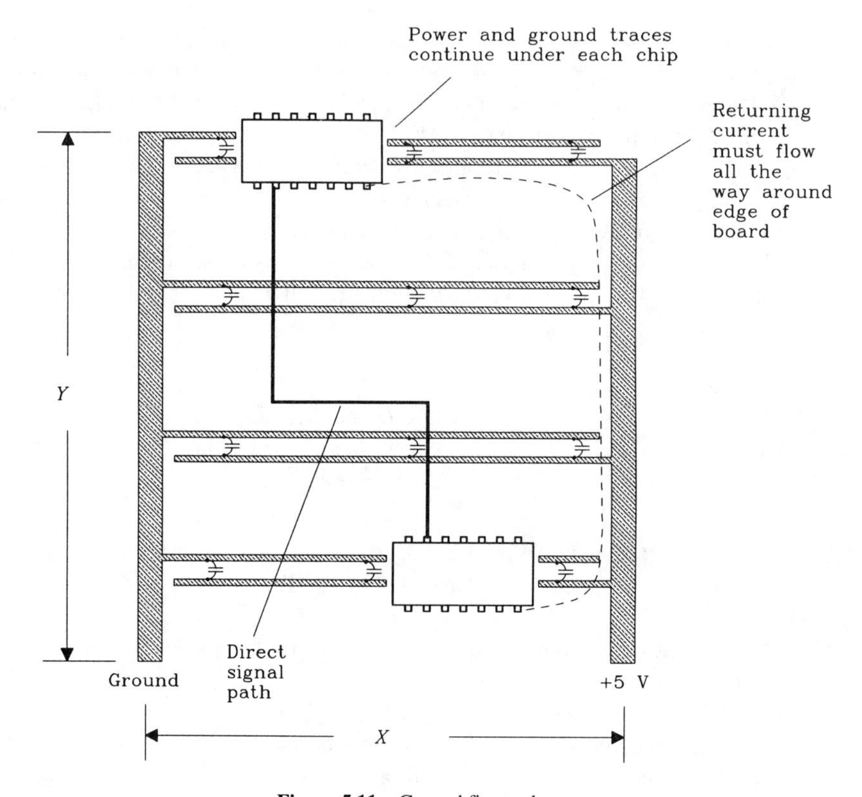

Figure 5.11 Ground fingers layout.

This old layout appears in old computer equipment (like the PDP-8) manufactured before the era of FCC-mandated radiation guidelines. The power and ground fingers layout is also used on cheap wire wrap frames. Don't use it.

The power and ground fingers technique works only with very-low-speed logic implemented on small circuit cards. It's main benefit is that the power and ground wiring can be implemented on a single layer. Signal traces require a second layer.

In the power and ground fingers scheme, one ground trace lies on the right side of the board and one power trace lies on the left. These traces extend from left to right when needed, like long fingers or rungs of a ladder.

Integrated-circuit packages straddle the rungs, attached with short connections to ground or power traces. Bypass capacitors between adjacent power and ground wires complete the picture.

The problem with this layout is that most returning signal currents must go all the way around the edge of the board to get back to their driver. This diversion introduces massive amounts of self-inductance and mutual inductance.

If you must work with only two layers, use the power and ground grid approach in Section 5.4. If some unknown force compels you to use the ground fingers layout, first build up a sample board and measure the mutual inductance between traces. Then calculate whether your circuit has a chance of working. It might work with low-speed CMOS logic or with the old LS-TTL series, but not with any fast logic families. In addition to the danger of the product simply not functioning, electromagnetic radiation from the open current loops will almost certainly fail FCC radiation tests.

Here is an approximation for loop inductance on a power and ground fingers layout:

$$L \approx 5Y \ln\left(\frac{X}{W}\right) \quad [5.13]$$

where L = inductance, nH
X = board width, in.
W = trace width, in.
Y = trace length, in.

Notice that doubling the trace width has almost no effect on the overall inductance. Fat ground traces do not help; what is needed is a web of smaller ground traces covering the surface of the board.

If a trace lies offset to one side, the inductance is a little less.

Because returning currents divert around the outside edge of the board, magnetic fields are everywhere. Any second trace intersects these fields, coupling the traces tightly together. The mutual inductance L_M between any two traces is practically the same as L in Equation 5.13. There is not much of a decrease in mutual coupling with distance.

Use the formulas from Section 5.3 for computing rise-time degradation and crosstalk voltage from this self-inductance and mutual inductance.

POINT TO REMEMBER:

For high-speed-logic, avoid the ground fingers layout.

5.6 GUARD TRACES

Guard traces appear extensively in analog design. At audio frequencies, on a two layer board having no solid ground plane, a pair of grounded traces running parallel to a sensitive input circuit can reduce crosstalk by an order of magnitude.

In the digital world, a solid ground plane provides most of the benefits of grounded guard traces. Beyond the solid ground plane, guard traces provide little additional benefit.

As a rule of thumb, the coupling between microstrips is halved by inserting a third line, grounded at both ends, between them. Their coupling is halved yet again if the third line connects through vias to the local ground plane at frequent intervals.[1] If you have

[1] J. A. Coekin, *High-Speed Pulse Techniques,* Pergamon Press, Oxford, 1975, pp. 203–205.

more than one ground plane layer, then ground the guard trace at each end, but not in the middle.

In digital problems, if two traces lie separated far enough to permit introduction of a guard trace, the coupling is usually already low enough that guarding is unnecessary. See Example 5.1.

EXAMPLE 5.1: Guard Trace Calculations

Two traces, shown in Figure 5.12, lie separated by three trace widths. That is just enough room to fit in a guard trace.

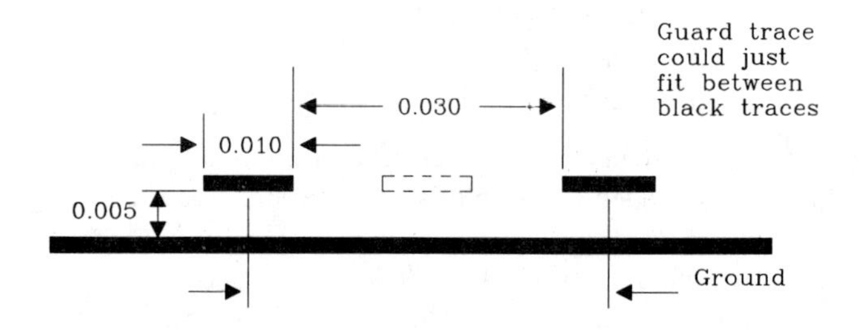

Figure 5.12 Guard trace positioning.

What is the estimated crosstalk?

Using Equation 5.1, the crosstalk fraction can't be any worse than

$$\text{Crosstalk} < \frac{1}{1+(D/H)^2} \qquad [5.14]$$

The centerline separation is 0.040, and the trace height is 0.005, so the ratio D/H is 8.

$$\text{Crosstalk} < \frac{1}{1+(8)^2} = 0.015 \qquad [5.15]$$

This is not enough crosstalk to worry about in a digital system.

How much crosstalk is too much? In analog systems, high-powered signals which cross over to low-level inputs require very high crosstalk immunity. Heterogeneous digital systems that mix logic families are sensitive to crosstalk when higher voltage signals (like TTL) get near lower-voltage parts (like ECL).

For ordinary homogeneous digital systems, a crosstalk level of 1–3% between adjacent wires is fine. This assumes there exists a solid ground plane such that each wire interacts only with its nearest neighbors. Cross-coupling from other more remote wires is negligible. Using a hatched or fingers ground system, where many wire pairs interact, we must sum all the crosstalk contributions before arriving at the composite crosstalk level on a given signal.

Figure 5.13 illustrates a typical guard trace application. The driver sends a known voltage step down trace *A*. Crosstalk from this signal can be received on trace *B* or trace *C*. The traces are 26 in. long, with a characteristic impedance of 50 Ω.

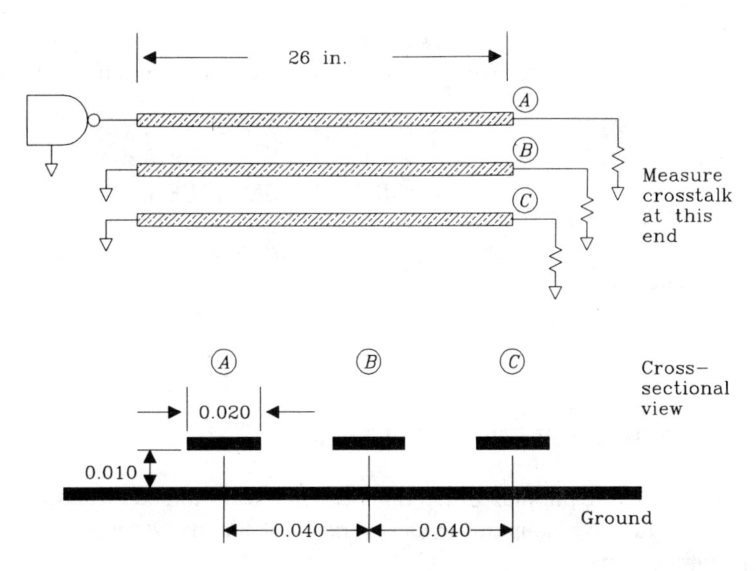

Figure 5.13 Guard trace demonstration.

The various step responses for the system of microstrips appear in Figure 5.14. The large impulse is the crosstalk between wires A and B, with C left disconnected. The middle impulse is crosstalk from A to C, with B left disconnected. It is four times smaller than the A-to-B interference, as predicted by Equation 5.2. With trace B shorted to ground

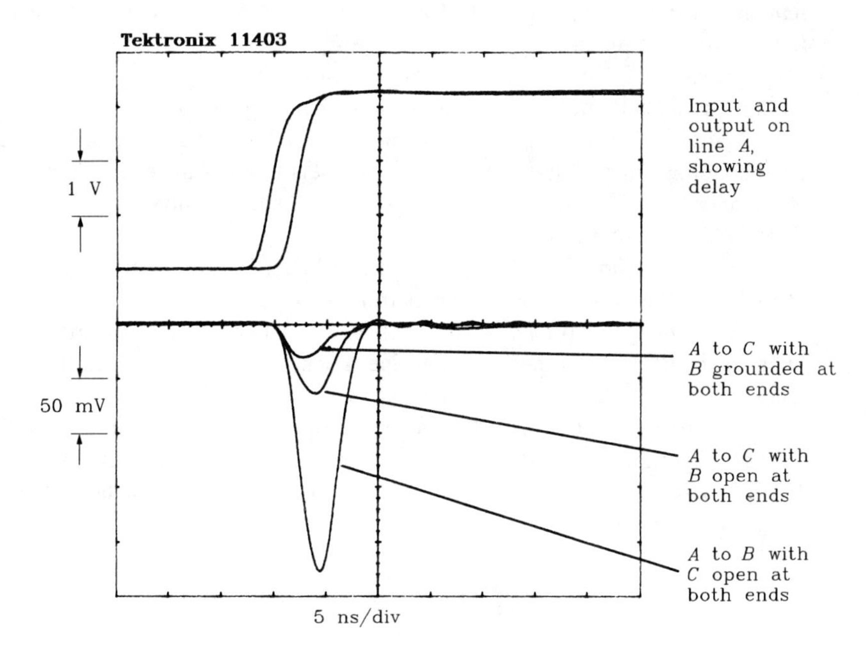

Figure 5.14 Example showing guard trace effect on coupling.

at both ends, we get the smallest coupling from A to C. This coupling is about one-half as big as the middle trace. This halving of coupling is the guard trace effect.

POINT TO REMEMBER:

A solid ground plane provides most of the benefit of grounded guard traces.

5.7 NEAR-END AND FAR-END CROSSTALK

The crosstalk examples in Sections 5.1–5.6 use lumped-circuit analysis. This simple mutual inductive coupling model works well for many coupling problems but not for long lines.

This section treats coupling between two long distributed transmission lines. Such coupling involves both mutual inductance and mutual capacitance.

5.7.1 Inductive Coupling Mechanism

In this section we consider only the mutual inductive coupling. Section 5.7.2 discusses mutual capacitive coupling. B. L. Hart presents a somewhat more mathematical description of this same material.[2]

Figure 5.15 illustrates a typical crosstalk situation. The ends of this system are marked “near” and “far,” as is common in the language of long-wire crosstalk.

Wire A-B carries a signal whose magnetic fields induce voltages in wire C-D. Magnetic coupling (mutual inductance) normally acts like a transformer. Because the mutual inductance is distributed, it appears as a succession of small transformers connected between the two lines.

Assuming the coupling is small (it had better be), the transformers do not significantly affect the propagation of signals from A to B.

As a voltage step moves from A to B, at each coupling transformer a small blip of interference appears on the adjoining line. Each blip propagates both forward and backward along line C-D.

For the moment, let’s just consider the blip caused by transformer k. When a positive step arrives from A, the changing current induces a momentary voltage across transformer k, as shown in Figure 5.15. This voltage blip is the reaction of inductor k to a change in its current:

$$L_M = L\frac{dI}{dT} \qquad [5.16]$$

[2]B. L. Hart, *Digital Signal Transmission Line Circuit Technology,* Van Nostrand Reinhold, New York, 1988.

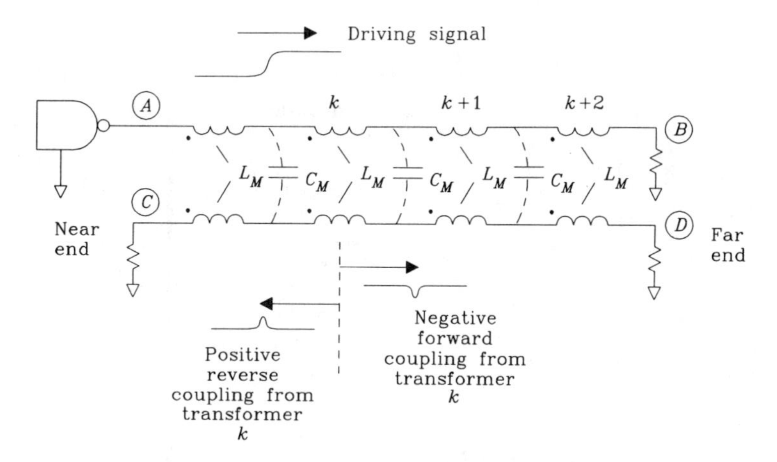

Figure 5.15 Mutual coupling between two long transmission lines.

The transformer reproduces this voltage blip on line *C-D,* with the polarity as marked.

The interesting thing about mutual inductive coupling is that the polarity at each end of transformer *k* is different. A positive blip of voltage goes down line *C-D* to the left (reverse coupling), while a negative blip goes to the right (forward coupling).

The reflection diagram in Figure 5.16 shows the blips from all transformers adding in a curious pattern. The negative (forward) blips all arrive together at the far end. The positive (reverse) blips arrive at different times, requiring a total of $2T_p$ to get to the near end.

Let's work with just the forward crosstalk coefficients for a moment. Each of the forward blips is proportional to the derivative of the input signal and to each mutual inductance L_M. Since all the forward blips arrive at the far end simultaneously, the total forward-blip size is proportional to the total mutual inductance between the two lines. If the line were stretched longer, the total mutual inductance would increase and so would the forward mutual inductive crosstalk.

Reverse mutual inductive coupling is different. The total amount of coupling (total area) is the same as for forward coupling, but it spreads out over a period of $2T_p$. In actual practice, all the reverse blips smooth out into one continuous blob of reverse coupling. The ideal reverse step response due to mutual inductive coupling is a rectangular function as shown in Figure 5.17.

If the line were stretched longer, the total mutual inductance would increase. The reverse coupling would respond by increasing its duration but not its height.

5.7.2 Capacitive Coupling Mechanism

Distributed mutual capacitance works almost the same way as distributed mutual inductance. The difference lies in the polarity of couplings.

When a voltage step passes one of the mutual capacitors in Figure 5.15, a small blip of interference appears on the adjoining line. Each blip propagates both forward and backward along the line *C-D.*

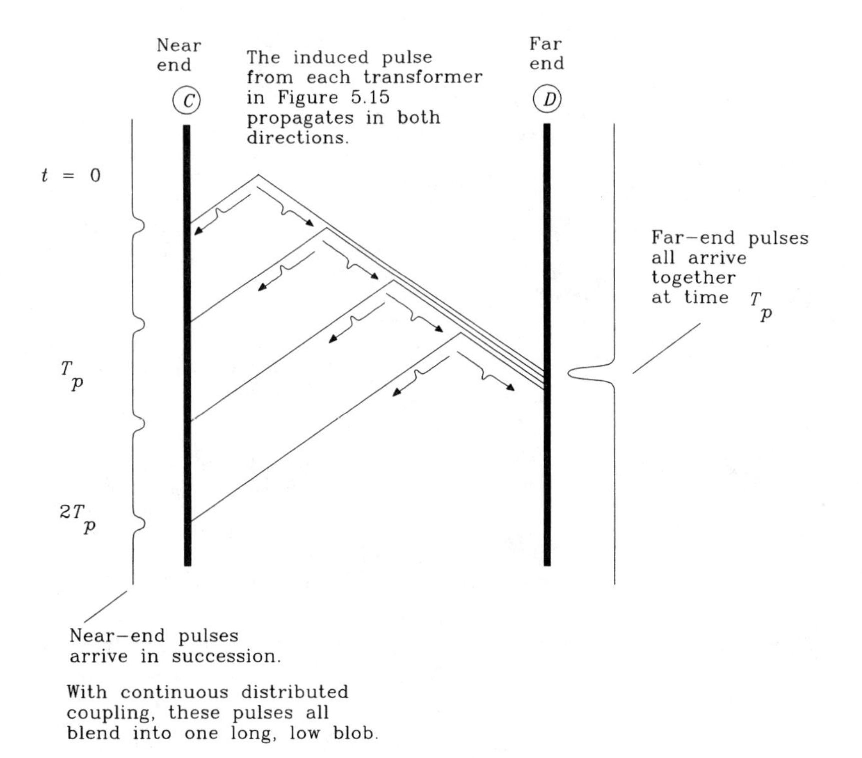

Figure 5.16 Reflection diagram showing mutual inductive coupling from the four transformers shown in Figure 5.15.

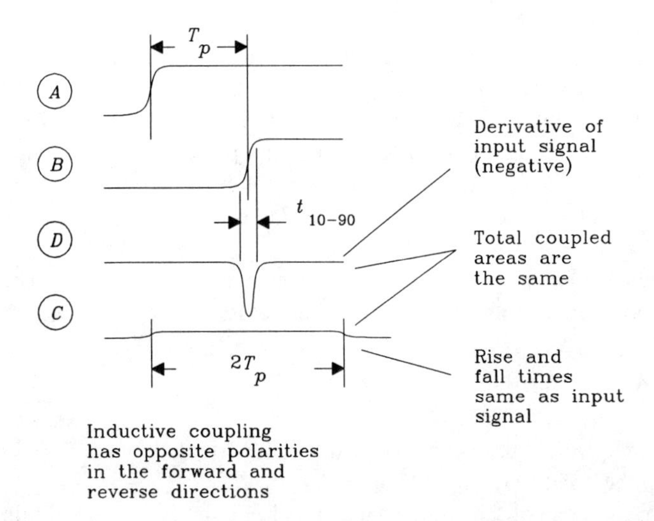

Inductive reverse coupling = $K[V(t) - V(t - 2T_p)]$

Figure 5.17 Forward and reverse mutual inductance coupling (distributed).

The polarities of mutual capacitive coupling are positive for forward and backward blips. Other than that, they behave the same as mutual inductive interference blips.

The mutual capacitance forward coupling looks like the derivative of the input signal. It gets larger for longer lines. Its polarity is positive (opposite of forward mutual inductive coupling).

The mutual capacitance reverse coupling area is the same as for forward coupling, but it spreads out over a period of $2T_p$. The ideal reverse step response due to mutual capacitive coupling is a rectangular function as shown in Figure 5.18.

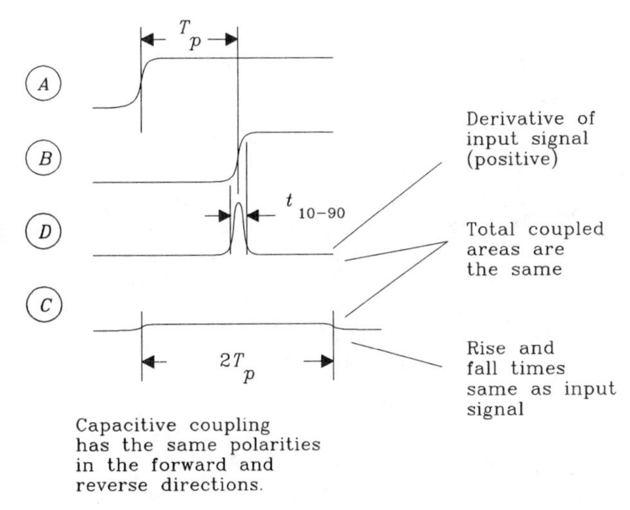

Capacitive reverse coupling = $K[V(t) - V(t - 2T_p)]$

Figure 5.18 Forward and reverse mutual capacitive coupling (distributed).

5.7.3 Combining Mutual Inductive and Mutual Capacitive Coupling

Under normal conditions, over a solid ground plane, the inductive and capacitive crosstalk voltages are of roughly equal size. The forward crosstalk components (voltage at *D*) cancel, while backward crosstalk components (voltage at *C*) reinforce.

Stripline circuits show particularly good balance between inductive and capacitive coupling and have tiny forward-coupling coefficients. Microstrips, for which the electric field lines responsible for crosstalk travel mostly through air instead of through the dielectric, have somewhat less capacitive crosstalk than inductive, leading to a small negative forward-coupling coefficient.

Over a slotted, hatched, or otherwise imperfect ground plane, the inductive crosstalk component is much larger than capacitive coupling and the forward crosstalk is large and negative. The forward crosstalk is never larger than the reverse crosstalk.

5.7.4 How Near-end Crosstalk Becomes a Far-end Problem

In Figure 5.15, the forward and reverse coupling signals are different. Each signal propagates to its respective end of cable *C-D* and extinguishes at the terminator.

Practical applications are often different from this model. In digital applications without source terminations, the device connected to the left end in Figure 5.19 is a low-impedance driver. Just like any other signal, reverse crosstalk, when it hits this driver, reflects. The reflection coefficient for a low-impedance driver is almost –1. This changes the reverse coupling from positive polarity to negative polarity and sends it back toward the far end.

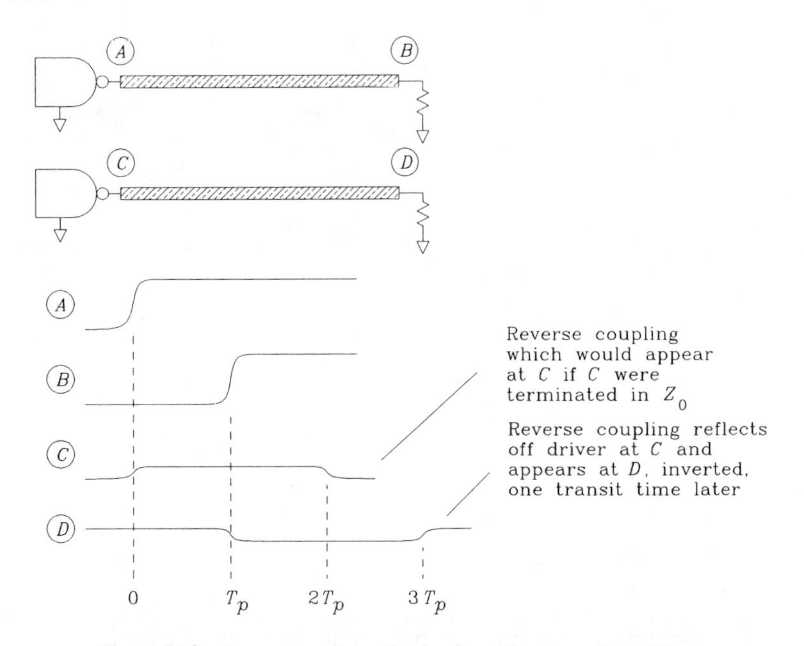

Figure 5.19 Reverse coupling reflecting from a low-impedance driver.

The signal seen at the far end *D* is a copy of the reverse coupling signal at *C*, delayed by one propagation time and inverted.

Because the mutual inductive and mutual capacitive portions of the forward coupling nearly cancel each other, the forward coupling is not visible when juxtaposed with this much larger reflected reverse coupling. When we measure crosstalk as defined by Figure 5.20, we are mostly measuring reflected reverse coupling.

EXAMPLE 5.2: REFLECTED REVERSE CROSSTALK

Figure 5.20 shows the measurement setup which produced the reflected reverse crosstalk waveforms in Figure 5.21.

The pulse generator drives trace *A-B* with a 2.5-V step input having a rise time of 880 ps. The waveform monitored at *A* appears in the top of Figure 5.21 at a scale of 1 V/division.

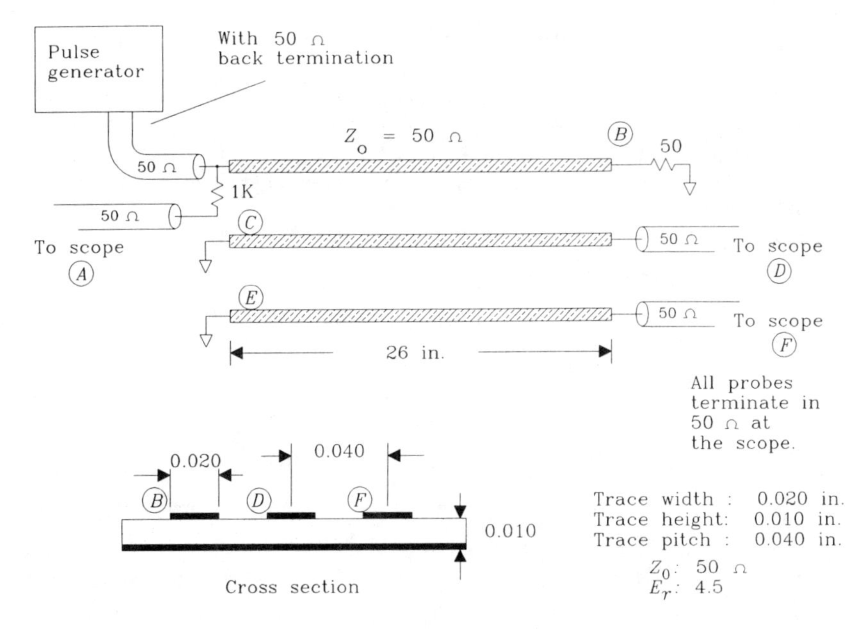

Figure 5.20 Setup for reflected reverse crosstalk measurement.

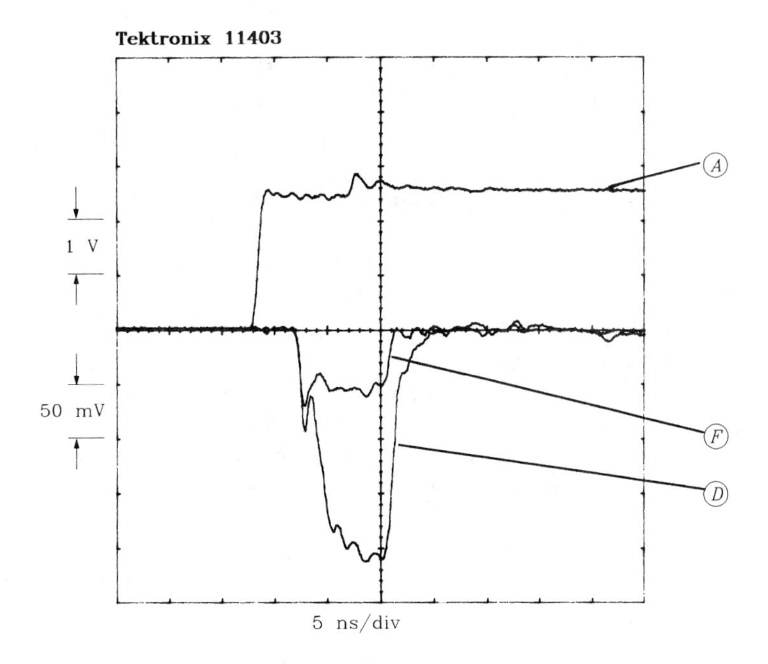

Figure 5.21 Reflected reverse crosstalk measurement.

Crosstalk signals at points D and F connect to regular coax probes and display in Figure 5.21 at 50 mV/division.

All probes are the same length. All probes terminate at the scope in 50 Ω.

The near ends of both pickup traces connect directly to ground, simulating a low-impedance driver.

Both crosstalk signals start together 4.5 ns after the initial rising edge:

$$T_p = 4.5 \text{ ns} \quad [5.17]$$

Both signals have a duration of 9 ns, and are of negative polarity.

$$\text{Crosstalk duration} = 2T_p = 9\text{ns} \quad [5.18]$$

The crosstalk signals measured at points D and F are

$$D = (4 \text{ divisions})(50 \text{ mV/division}) = 200 \text{ mV} \quad [5.19]$$

$$F = (1 \text{ division})(50 \text{ mV/division}) = 50 \text{ mV} \quad [5.20]$$

The crosstalk ratios (output divided by input) measured for these two geometries are

$$\frac{D}{A} = \frac{0.200}{2.5} = 0.08 \quad [5.21]$$

$$\frac{F}{A} = \frac{0.050}{2.5} = 0.02 \quad [5.22]$$

The crosstalk ratios predicted by Equation 5.2 ($K = 1$) for these geometries are

$$\frac{D}{A} \approx \frac{1}{1+(0.040/0.010)^2} = 0.059 \quad [5.23]$$

$$\frac{F}{A} \approx \frac{1}{1+(0.080/0.010)^2} = 0.015 \quad [5.24]$$

5.7.5 Characterizing Crosstalk Between Two Lines

Forward crosstalk is proportional to the derivative of the driving signal, and to line length. The constant of proportion depends on the balance between inductive and capacitive coupling. Once we measure this ratio for one known signal, modeling the response for other signals is trivial.

Modeling reverse crosstalk for fast-rising edges is equally simple. The reverse coupling looks like a square pulse, with rise and fall times comparable to the input signal and height proportional to the driving signal amplitude. The reverse coupling percentage is a physical constant determined by the line parameters which is independent of line length. The duration of the pulse is $2T_p$.

Reverse crosstalk for slow-rising edges is a little more involved. Once we have found the reverse coupling coefficient for fast edges, the reverse signal for any input can be found from

$$\text{Reverse coupling } (t) = \alpha_R\left[V(t) - V(t-2T_p)\right] \quad [5.25]$$

where t = time, s
$V(t)$ = driving waveform, V
α_R = reverse coupling coefficient for fast-edged signal
Tp = propagation delay of line, s

For lines longer than half the signal rise time, the reverse coupling has time to build up to its full value. The reverse coupling coefficient for such a line equals approximately

$$\alpha_R \approx \frac{1}{1+(D/H)^2} \qquad [5.26]$$

where D = separation between lines, in.
H = line height above ground, in.

For lines shorter than half the signal rise time, the reverse coupling ramps up and then back down, never reaching its steady-state maximum value.

5.7.6 Using Series Terminations to Reduce Crosstalk

A series terminator extinguishes reverse-coupled crosstalk at the near end. An end terminator attenuates the returning reflection of the main signal, the reverse coupling from which would be again pointed toward the far end. Using both terminators eliminates both sources of reverse coupling noise, improving overall crosstalk considerably.

The reduced coupling gained by combining series and end terminations lets us route parallel bus traces closer than would otherwise be practical.

POINTS TO REMEMBER:

Regarding long transmission lines:

Over solid grounds, inductive and capacitive crosstalk are equal. Forward-crosstalk components cancel, while reverse crosstalk reinforces.

Over a slotted or imperfect ground plane, the inductive coupling exceeds capacitive coupling, making forward crosstalk large and negative.

Forward crosstalk is proportional to the derivative of the input signal and to line length.

Reverse coupling looks like a square pulse, with a constant height and duration equal to $2T_p$. For short lines, reverse coupling does not climb to its full value.

Reverse crosstalk, when it hits a low-impedance driver, reflects toward the far end.

5.8 HOW TO STACK PRINTED CIRCUIT BOARD LAYERS

A *layer stack* for a printed circuit board specifies the arrangement of circuit board layers. It specifies which layers are solid power and ground planes, the dielectric constant of the substrate, and the spacings between layers. When planning a layer stack, also compute the desired trace dimensions and minimum trace spacing.

Manufacturing constraints heavily influence the layer stack. As a rule, the greater your circuit wiring density, the greater your production costs per square inch. This section sets out some basic rules of thumb for planning layer stacks.

5.8.1 Power and Ground Planning

Design the power and ground layers first. To plan a power and ground system, first establish the signal rise times, the number of signals, and the physical dimensions of the circuit board.

Included among the physical dimensions, make a guess as to the trace width. The trace width assumption is not particularly critical at this stage.

Next estimate the self-inductance and mutual inductance using solid, hatched, and fingers ground plane models. At this point it is usually clear which model suits the design. Remember that for the ground fingers model, all traces interact. For the hatched model, traces running along the same hatch grid interact. For the ground plane model, only adjacent traces interact.

If you will be using a solid ground plane, plan on using ground and power planes in pairs. The symmetric pairing of solid planes in a layer stack helps prevent warping in the circuit board. A board with a single plane, offset to one side, can warp noticeably.

Power planes may be used as low-inductance signal return-current paths just as ground planes. Assuming adequate bypass capacitors between power and ground (Chapter 8), transmission lines routed over a power plane operate as well as those routed over a ground plane. Transmission striplines routed between one power and one ground layer, or two power layers, also work.

5.8.2 Chassis Layer

Sometimes you will want to run a signal outside your digital system. For this application you may pick a low-speed or controlled rise-time driver. This is a good choice because it reduces external radiation, which helps with FCC problems.

If the ground for the driver connects to ordinary digital logic ground, the effective driver output equals its intended drive voltage *plus* any noise voltage present on the digital logic ground. See Figure 5.22.

Digital logic grounds are notorious for high-frequency noise voltages. Grounds carry fluctuating voltages caused by the action of many returning signal currents acting across their self-inductance. These high-frequency fluctuations are too small to cause trouble in the digital circuits, but plenty large enough to exceed FCC limits. Any wire connected to the digital logic ground which runs outside the cabinet almost always fails FCC tests.

Without other precautions, the controlled rise-time driver effectively picks up ground noise and broadcasts it outside the chassis.

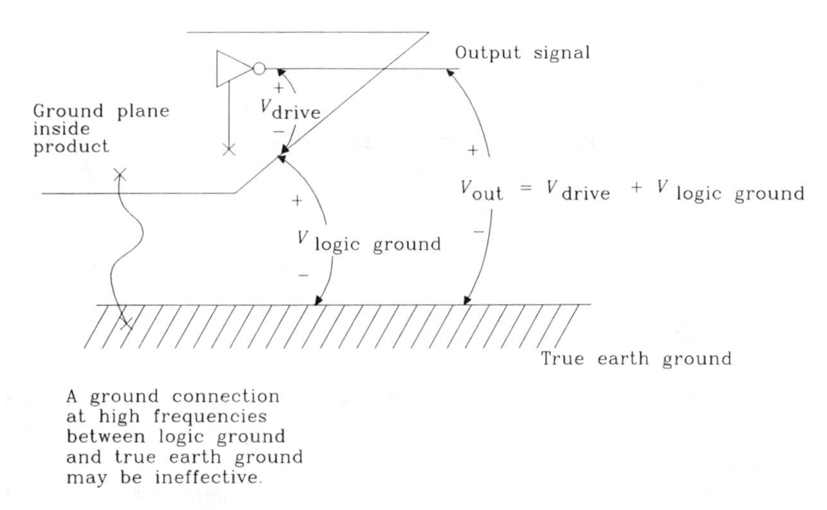

Figure 5.22 Using a controlled rise-time driver.

One solution to this problem adds a *chassis plane* to the layer stack. This plane stacks directly next to a ground plane, giving a very tight capacitive coupling between the two planes. At high frequencies, the two planes are effectively tied together. The chassis plane then screws, solders, or welds to the external chassis along one continuous axis near the controlled rise-time driver. At high frequencies, we have effectively shorted the digital ground plane to the chassis. This reduces the amount of digital ground noise at that point, also reducing noise carried by the controlled rise-time driver to the outside world.

Ordinary capacitors will not function as a short between chassis and digital logic ground, because they have too much lead inductance. Only the large, wide, parallel surface area between the chassis plane and digital ground plane has a low enough inductance to effectively short the two together.

With the chassis plane approach, the digital logic and external chassis remain electrically isolated at low frequencies. This may be desirable for safety or other reasons. If the isolation does not matter, simply short digital logic ground directly to the chassis without using a separate chassis layer. Make this connection by screwing, soldering, or welding the ground plane to the external chassis along one continuous axis near the controlled rise-time driver.

When using a chassis plane, counterbalance it in the layer stack with some other solid plane layer. For mechanical reasons, always lean toward using symmetric arrangements of planes in your layer stack.

5.8.3 Selecting Trace Dimensions

Squeezing traces tightly together increases the circuit packing density. Very dense designs require fewer circuit board layers. Since printed circuit board cost is proportional to the number of layers, as well as to the board surface area, we are tempted to always design using the fewest number of layers that will do the job.

Smaller, more closely spaced traces also yield more crosstalk and less power-handling capacity. This tradeoff among crosstalk, routing density, and power is critical to low-cost product design.

Let's deal with power-handling capacity first, because it is the simplest of the constraints.

The power-handling capacity of a printed circuit trace depends mostly on its cross-sectional area and the allowable temperature rise. For a given cross-sectional area, a trace's temperature rise above ambient is roughly proportional to the power it dissipates. Large temperature rises are unreliable and heat up nearby digital circuits. A conservative upper limit on trace heating inside digital products is 10° C.

Figure 5.23 relates maximum power-handling capacity to temperature rise. The horizontal axis in Figure 5.23 measures cross-sectional area in units of square inches.[3] The vertical axis of Figure 5.23 shows the allowable current for that trace at a given temperature rise.

For example, a 0.010-in.-wide trace of 1-oz copper (0.00135 in. thick) can safely pass 750 mA of current at a temperature rise of 10° C.

Power is rarely a serious constraint except for large power distribution buses. As thin-film technology, with extremely small trace cross sections, becomes more widely available, heating limitations may become more prevalent.

A second lower bound on trace width results from the manufacturing process. Table 5.1 lists the minimum trace widths attainable in various production processes.

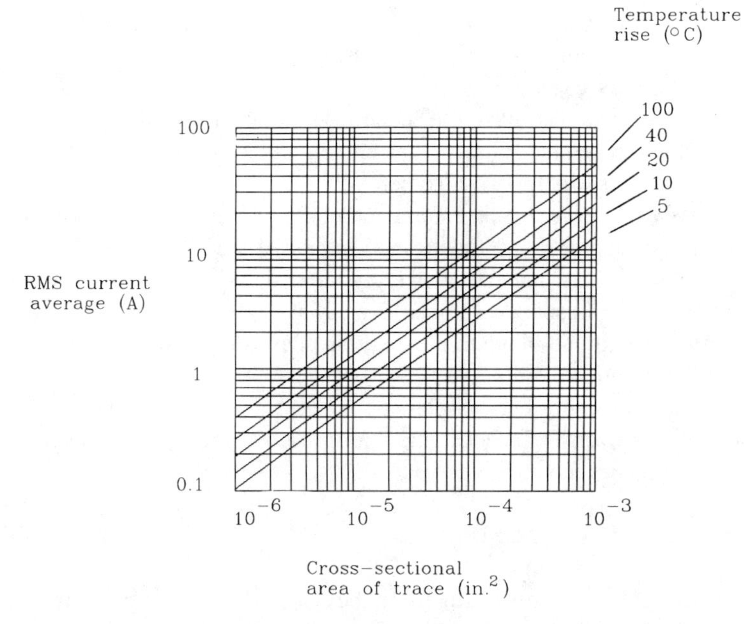

Figure 5.23 Current-carrying capacity of copper printed circuit traces.

[3]Example: A trace 0.010 in. wide, fabricated in a 1-oz-weight copper (0.00135 thickness), has a cross-sectional area of 1.35×10^{-5} in.^{2}

TABLE 5.1 MINIMUM LINE WIDTHS ATTAINABLE WITH VARIOUS PRODUCTION PROCESSES

Process	Minimum line width (in.)
Gold screened onto thick film substrate	0.010
Etched copper on epoxy board with plating	0.004
Etched copper on epoxy board with no plating	0.003
Gold evaporated onto thin film substrate and then etched	0.001

With any process the manufacturing yield will drop, and cost will go up, as one approaches the minimum attainable trace width. This factor prevents most designers from using the minimum attainable line width.

Other factors tend to increase line width. Poor control over the etching process can result in large line width variations. At low line widths, the percentage line width variation, which controls the percentage impedance tolerance, may be unacceptable. A need for accurate impedance control can force the use of lines much wider than the minimum attainable trace width.

Use the formulas for trace impedance in Appendix C to find a combination of trace width and height sufficiently large that, over expected variations in width and layer height, the impedance stays within your design range. Remember that you must also allot room in your impedance budget for variations in the electric permittivity of the substrate.

Considerations of power, cost, and impedance tolerance usually drive the selection of a particular trace width. Given the width, the impedance constraint sets the layer height.

Next, using the formula for crosstalk (see Section 5.7.5 and Equation 5.2), figure the minimum spacing between adjacent traces (center to center). This number is called the minimum *trace pitch.* The unused distance between traces is called the *trace separation.* The sum of the trace separation and the trace width equals the trace pitch.

POINTS TO REMEMBER:

As a rule, the greater your circuit wiring density, the greater your production costs per square inch.

Printed circuit board cost is proportional to the number of layers and to the board surface area.

Design the power and ground layers first.

For mechanical reasons, lean toward using a symmetric arrangement of ground and power planes in your layer stack.

Smaller, more closely spaced traces yield more crosstalk.

5.8.4 Routing Density Versus Number of Routing Layers

With more layers, we can spread the traces out farther. That makes routing easier and reduces the risk of crosstalk problems. Unfortunately, the cost of multilayer printed circuit boards is proportional to the number of layers times the surface area. Using more layers costs more.

With fewer layers, we must use a narrower trace pitch, which can also cost extra. Not only that, at a fine enough pitch we run the risk of too much crosstalk.

Determining the least cost number of layers for a board is a matter of experience and guesswork. The central issue involves estimating the trace pitch required by N connections routed on a certain-size board, using M layers. Knowing the trace pitch, we can find out how much the board will cost and at the same time model the crosstalk.

Trace pitch is determined from wiring density. One useful model for wiring density is called Rent's rule, after the IBM engineer who popularized it. Rent noticed that most large square boards, when divided into quadrants, reveal half their wiring going between quadrants and half staying within each quadrant. Further subdividing of each quadrant, reveals the same distribution. If, upon traveling between two quadrants, we assume (this is very hypothetical) a wiring length on average equal to the spacing between quadrants, we arrive at a total average wiring length equal to three-eighths of a board side.

Knowing the average wire length and the number of wires we can compute the total board surface area occupied by those wires using any trace pitch. This knowledge, in the form of Equation 5.27, shows the trace pitch required to route N connections on a fixed-size board using M layers.

Of course, if we have some other information about the routing requirements, such as large buses or other structures, we should use it. Given no other information, we can attempt to calculate the required trace spacing on a board using Rent's idea:

$$p_{\text{avg}} = \frac{(XY)^{1/2}}{N} 2.7M \qquad [5.27]$$

where N = number of connections (assumed distributed according to Rent's rule)
p_{avg} = average trace pitch, in.
X = board width, in.
Y = board height, in.
M = number of routing layers

For example, on an 8 in. × 12 in. board having 800 interconnections routed on four layers, we need a trace pitch, on average, of 0.132 in. This means that if the board is pretty much covered with DIP through-holes, we will need one trace routed between almost every pin.

Don't count on filling more than half the spaces between pins. In the above example we should plan on more layers or plan on using double-track routing (two traces between each pin).

For through-hole boards, the average pitch predicted by Equation 5.27 and the minimum pitch required are very different. Use the minimum pitch determined from crosstalk considerations to determine if you need double- or triple-track routing between pins. Use the average pitch predicted by Equation 5.27 to determine how many available routing tracks we need.

Inner layers of surface-mounted boards may have more routing space available than DIP boards. The total number of vias is roughly the same, but the vias are smaller in surface-mount designs since IC pins don't have to stick through them. The average and minimum pitch for inner layers of surface-mount boards may be similar.

Routing up to four tracks between pins is possible on inner layers of epoxy circuit boards, but this may lead to severe crosstalk problems.

Extra routing space on a board can be gained by leaving large spaces between chips, but this takes more total surface area. Most designers choose to go with more layers instead.

If crosstalk is a problem, ensure that the layout squeezes traces together only to fit between pins and then immediately spreads them back out again on their way to the next chip. This requires a lot of hand adjustment, but even modest increases in trace separation reduce crosstalk.

With any luck, we can find some number of circuit board routing layers that yield acceptable crosstalk and won't cost too much.

POINTS TO REMEMBER:

Don't count on filling more than half the spaces between through-hole pins.

When all else fails, use Rent's reasoning to figure the average trace length.

5.8.5 Classic Layer Stacks

Figures 5.24–5.26 illustrate three classic layout stacks for 4, 6, and 10 layers. These stacks are designed for use with the ordinary epoxy multilayer fabrication process described below. Beyond 10 layers, designers usually incorporate additional ground planes to isolate sets of routing layers.

These stacks are appropriate for high-speed computer products embedded within well-shielded card cages. If you are planning systems that must pass FCC, VDE, Tempest, or other mandated rules for electromagnetic emissions without the benefit of a well-shielded card cage, these simple stacks will be inadequate for your purposes.

In each figure, the terms *horizontal routing* and *vertical routing* refer to the orientation of traces on that layer (horizontal or vertical). Traces on each layer traditionally lie

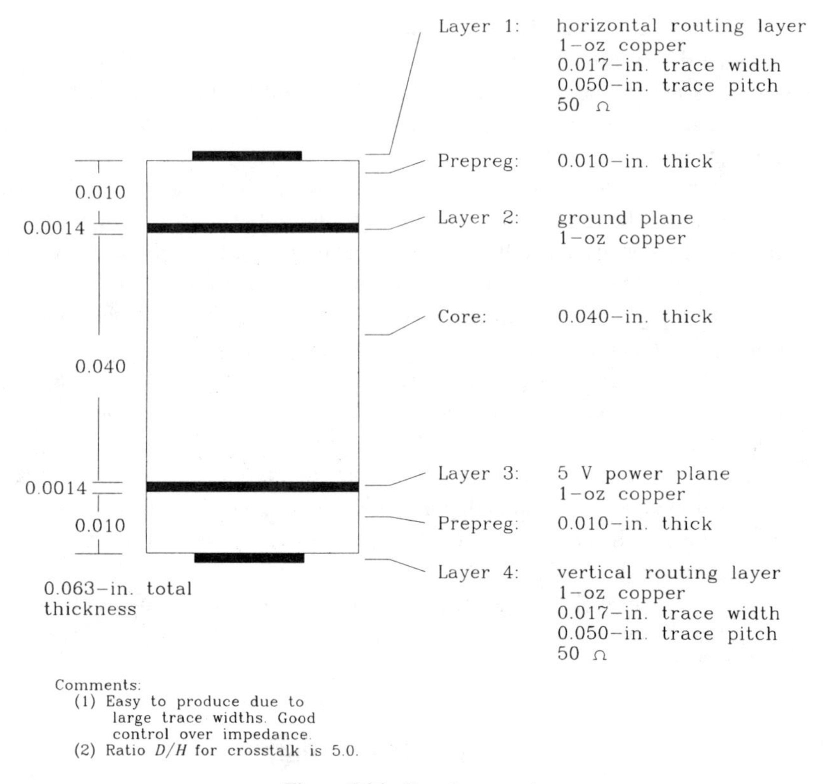

Figure 5.24 Four-layer stack.

parallel to each other, with every other layer built at right angles to the one below it. Very few traces cut diagonally across a layer or make large 90-degree bends. This increases the routing efficiency.

Power and ground layers in Figures 5.24–5.26 are marked with thick solid lines. Trace layers show the proportional line width and trace height.

The notations *core* and *prepreg* refer to materials used in the substrate lamination process. The following paragraphs briefly describe one commonly used process for building circuit cards. If you want to tightly control trace-to-ground spacings, you need to know about core and prepreg layers.

The multilayer buildup process starts with a set of raw two-sided laminate layers coated with copper on each side. Surfaces destined to become inner layers are etched in this form. Surfaces destined to become outer layers are left fully copper-plated. These etched laminates are called cores. The thickness between opposing layers on a core depends on the thickness of the original laminate.

The cores are then stacked together, with a sheet of prepreg epoxy material placed between each pair of cores. This sheet melts into an epoxy glue when heated and pressed. The thickness of the prepreg sheet determines the spacing between core layers. The prepreg cures into a hard epoxy layer having the same dielectric constant as the core layers. Core and prepreg layers alternate.

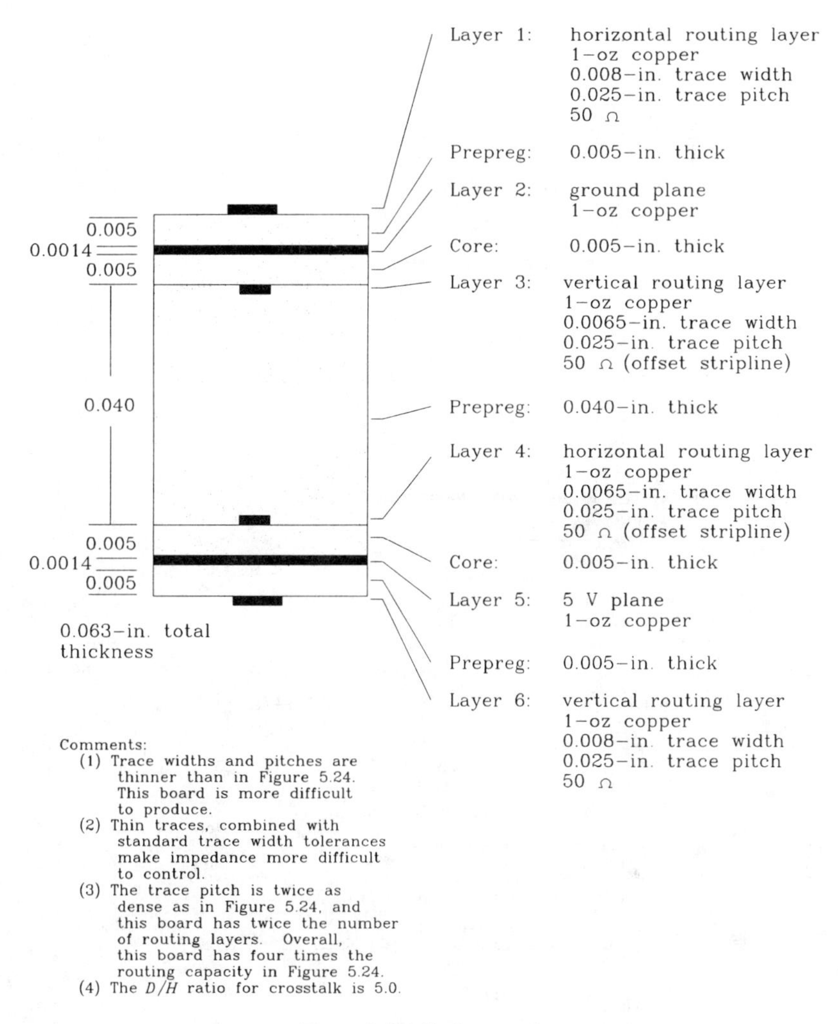

Figure 5.25 Six-layer stack.

Because the prepreg partially melts in the process, traces on opposite sides of the prepreg tend to sink into the melted, gluey prepreg material. Accurate analysis of trace-to-ground separation takes into account the fact that the spacing between facing sides of two traces on either side of a prepreg layer will be reduced by twice the trace thickness as they sink into the prepreg. Ground layers do not sink into the prepreg.

Manufacturers sometimes form the outer layer from one side of a core, and in other cases they form it from a metal foil pressed onto a top layer of prepreg. In either case, the outer layer is a solid copper sheet (no etching yet).

After the prepreg cures, holes are drilled through the assembly. The drilling exposes the various copper layers and pads which it penetrates in the inner layers, but they are at this point not connected electrically.

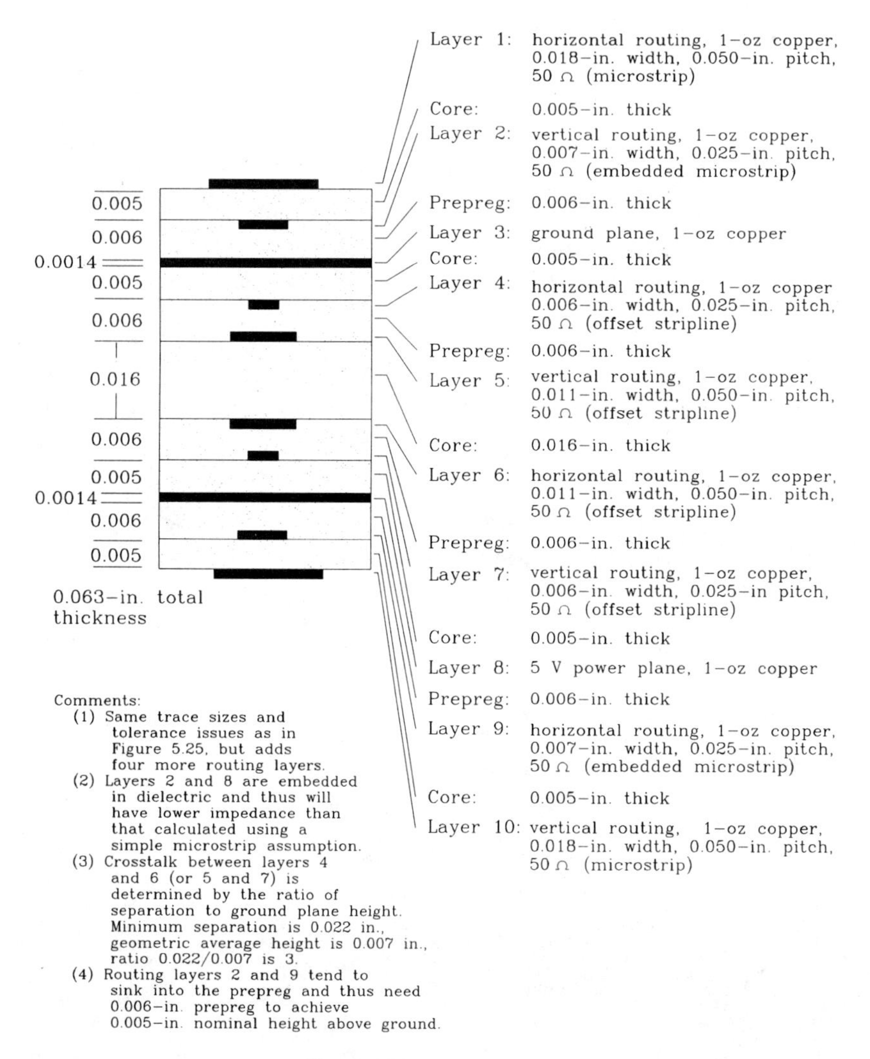

Figure 5.26 Ten-layer stack.

The plating step covers the inside surfaces of the holes and the outer surfaces of the board simultaneously. To save plating material and time, most manufacturers mask off the outer surfaces except around holes and the outer-layer traces. After plating, the outer-layer traces come out a little thicker than the raw copper sheet. It is this additional thickness which causes greater uncertainty in the finished line widths of outer layers compared to inner layers.

The last step etches away unwanted copper on the outer layers, leaving a finished board. The board is then tinned (optional), coated with solder mask, and silk-screened on both sides.

POINTS TO REMEMBER:

Core and prepreg layers alternate.

Outer layers, if plated, have greater trace width variation than inner traces.

Traces on routing layers tend to sink into the prepreg mixture. Their thickness doesn't add to the total board thickness.

The thickness of solid ground plane layers always adds to the total board thickness.

5.8.6 Extra Hints for High-speed Boards

For the very-highest-speed boards, place power and ground layers directly together. This positioning maximizes their capacitive coupling, reducing power supply noise.

Use plenty of extra ground planes (not power planes) to isolate sets of routing layers. Sprinkle around lots of ground vias, connecting together the many ground planes. Returning signal currents will flow through these ground vias as they jump from layer to layer following the contorted path of each signal trace.

Had we used a mix of power and ground planes for isolating routing layers, instead of just ground planes, the return currents, which always flow in the nearest plane, would have had to traverse many bypass capacitors as they jumped between ground and power planes. This is not a good idea, because any currents we send through the bypass capacitors will cause voltages across them. These voltages radiate very effectively from the power and ground planes, adding to our radiated noise problems.

POINTS TO REMEMBER:

At the highest speeds, keep ground and power planes directly adjacent.

Use extra ground planes, not power planes, to isolate routing layers.

Terminations

When does a system need terminating resistors? According to Chapter 4, there are two cases: the long reflection case and the short ringing case.

When a line is long, meaning the cable length exceeds one-sixth of the electrical length of a rising edge, the cable needs terminators. Without terminators, reflections at either end of a long cable render signal transmission impossible. Chapter 4, Section 4.3, shows how to precisely determine the effect of signal reflection. Section 4.3.5 proposes a simple mathematical test for determining the duration of reflections when a line is left not terminated.

When a line is short, it may still need terminators if it drives a capacitive load. Section 4.1 analyzes highly inductive circuits which are capacitively loaded, showing the effect of high-Q ringing. The ringing phenomenon in short lines has the same practical effect as reflections in long lines.

Resistive terminators can cure problems with either reflections or ringing.

This chapter addresses three main topics:

- Comparison of end versus series termination
- Selection of appropriate terminating resistors
- Crosstalk among terminating components

6.1 END TERMINATORS

When using end terminations, each driving gate connects directly to its transmission line, with the terminator located at the receiving end (see Figure 6.1). End-terminated lines have these properties:

(1) The driving waveform propagates at full intensity all the way down the cable.

(2) All reflections are damped by the terminating resistor.

(3) The received voltage is equal to the transmitted voltage.

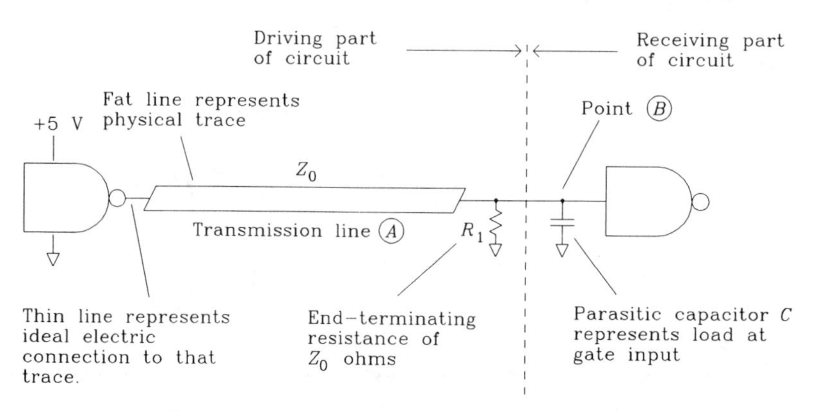

Figure 6.1 Calculating the rise time of an end terminator.

6.1.1 Rise Time of an End Terminator

We may deduce the signal rise time of an end-termination circuit by intuitive reasoning or by more detailed mathematics. We shall attempt the intuitive approach first and then double check using detailed math.

Our intuitive approach splits the circuit in Figure 6.1 into two parts. The left part, or driving part, is composed of the driving gate, the transmission line, and the terminating resistor. We can model the Thevenin equivalent driving impedance for this part of the circuit as the impedance of a long transmission line Z_0, in parallel with the terminating resistor (also Z_0). The net effect, for short-term events, is a drive impedance of $Z_0/2\ \Omega$.

The right part, or receiving part, includes only the receiving gate, modeled in Figure 6.1 as a capacitor. This capacitive model is appropriate for most CMOS, TTL, or ECL situations. Recognizing this circuit as a simple *RC* filter, we know the *RC* time constant:

$$RC \text{ time constant} = \frac{Z_0}{2} C \qquad [6.1]$$

Using the formula from Section 3.1 for the 10–90% rise time of an *RC* filter,

$$T_{\text{term}} = 2.2 \frac{Z_0}{2} C = 1.1\, Z_0 C \qquad [6.2]$$

Given an incoming signal with a rise time of T_1, we combine it with the rise time of the termination circuit T_{term} to find the resulting actual rise time at point *B*:

$$T_B = \left(T_{\text{term}}^{\;2} + T_1^{\,2}\right)^{1/2} \qquad [6.3]$$

When a line is long compared to a rising edge, the impedance of its output is effectively Z_0. If we shorten the length of a transmission line down to a size comparable to the length of its rising edges, the impedance of that line, as seen at *B*, goes down. Eventually, for very short lines, the driving impedance at *B* is just equal to the output impedance of the driving gate and we get a faster rise time at point *B*.

Next let's try a math-intensive method of estimating rise time. Recall that Equation 4.61 from Chapter 4 models the overall response of a transmission line:

$$S_\infty(w) = \frac{H_X(w)A(w)\left[R_2(w)+1\right]}{1 - R_2(w)R_1(w)H_X{}^2(w)} \qquad [6.4]$$

If the length of the transmission line exceeds the length of a rising edge, we can ignore any reflections that may bounce off the end terminator. This is reasonable because end reflections will not have time to travel to the source, and then rebound back to the far end again before the effects of the first rising edge have completed their course. We may have late reflections, but they do not influence the shape of the initial rising edge. Mathematically, to force zero reflections we set the reflection factor $R_1(w)$ in Equation 6.4 to zero. We may then simplify Equation 6.4:

$$S_\infty(w) \approx H_X(w)A(w)\left[R_2(w)+1\right] \qquad [6.5]$$

Our next simplification assumes the drive impedance is very low compared to the transmission line characteristic impedance, and so $A(w)$ is unity. Further assume the line is not so long that it disperses the signal, so the magnitude of $H_X(w)$ is unity. Incorporating these simplifications, we have

$$S_\infty(w) \approx R_2(w) + 1 \qquad [6.6]$$

Finally, rearranging the definition of $R_2(w)$ from Equation 4.53 and adding 1:

$$S_\infty(w) \approx \frac{2Z_L(w)}{Z_L(w)+Z_0(w)} \approx \frac{2}{1+\dfrac{Z_0(w)}{Z_L(w)}} \qquad [6.7]$$

Next, model the characteristic impedance function $Z_0(w)$ as a constant resistance having value Z_0. Finally, note that $Z_L(w)$ is composed of the parallel combination of a terminating resistor (also equal to Z_0) and a capacitor of value C. Combining the parallel components,

$$\frac{1}{Z_L(w)} = \frac{1}{Z_0} + jwC \qquad [6.8]$$

And substituting into Equation 6.7,

$$S_\infty(w) \approx \frac{2}{1+Z_0\left[\left(1/Z_0\right)+jwC\right]} \qquad [6.9]$$

$$= \frac{1}{1+jw\left[\left(Z_0/2\right)C\right]} \qquad [6.10]$$

Equation 6.10 is the response of a simple RC filter having a time constant of $(Z_0/2)C$ seconds. This corroborates our intuitive model.

The rise time of an end-teriminated circuit, when capacitively loaded, is half that of a series-terminated line driving the same load (see Section 6.2.2).

6.1.2 DC Biasing of End Terminator

The termination circuit in Figure 6.1 rarely appears in TTL or CMOS circuits because of the large drive current required in the HI state. When the driving gate in Figure 6.1 switches its output to V_{CC}, it must supply a current of V_{CC}/R_1 to the terminating resistor. When the driving gate switches its output to ground, no output current flows. Assuming we are using typical 65-Ω transmission lines, the current required for a 5-V drive signal is 5/65 = 76 mA. Very few drivers can source that much current.

Compare this drive requirement with the drive capabilities of TTL, which sources much more current when driving LO than when driving HI, or CMOS which sources equal amounts of current in both directions.

Figure 6.2 shows a popular terminating arrangement called the split termination. In this arrangement, the parallel combination of R_1 and R_2 equals Z_0, the characteristic impedance of transmission line A. The ratio R_1/R_2 controls the relative proportions of HI and LO drive current. Figure 2.10 presents equations for converting this split termination into a single resistor and Thevenin equivalent voltage source.

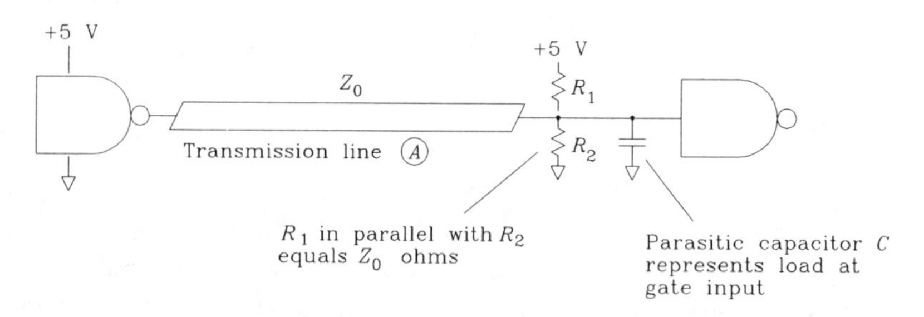

Figure 6.2 Split termination.

If R_1 equals R_2, the HI and LO drive current requirements are the same. This setting is appropriate for the HCMOS digital logic family.

If R_2 exceeds R_1, the LO current requirement exceeds the HI current requirement. This setting is appropriate for TTL and HCT families.

The selection of values for R_1 and R_2 is best done graphically. The selection is controlled by three constraints:

(1) The parallel combination of R_1 and R_2 must equal Z_0.

(2) We must not exceed $I_{OH\ \max}$ (maximum high-level output current).

(3) We must not exceed $I_{OL\ \max}$ (maximum low-level output current).

In this example, we use the following convention for output current: Current entering the driver (sink current) is positive, while current leaving the driver (source current) is negative. A TTL or CMOS gate sinks (positive) current in the low state and sources (negative) current in its high state. An ECL gate sources (negative) current in both states.

The first constraint is easily expressed in the admittance domain. Let the variables Y_1 and Y_2 stand for the admittances of resistors R_1 and R_2, respectively:

$$Y_1 = \frac{1}{R_1} \qquad Y_2 = \frac{1}{R_2} \tag{6.11}$$

We will solve for values of Y_1 and Y_2 and then invert them to find R_1 and R_2 as the last step. The advantage of this approach is that it makes our constraint equations linear.

The first constraint is graphed in Figure 6.3:

$$Y_1 + Y_2 = \frac{1}{Z_0} \tag{6.12}$$

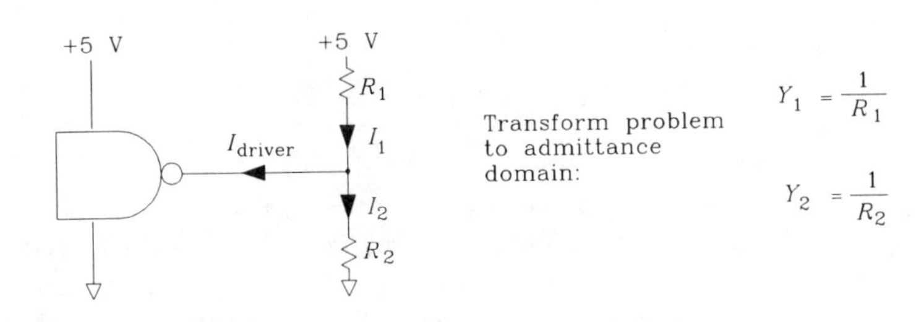

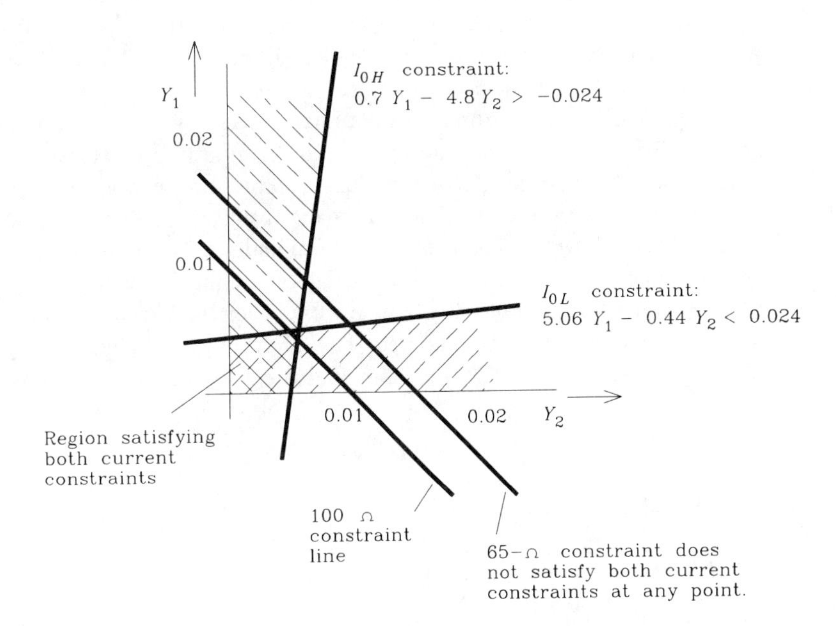

Figure 6.3 End-termination constraints.

This appears as a diagonal line on the constraint graph (Figure 6.3). All valid combinations of Y_1 and Y_2 lie on this line.

Derive an equation for the second constraint by noting that the current into the driver equals the current flowing in R_2 minus the current flowing in R_1. These two currents depend on voltages V_{CC}, V_{EE}, and the driver output voltage. For generality we use the symbol V_{CC} for the more positive power supply voltage and V_{EE} for the more negative voltage. Often, one or the other of these voltages is zero.

Constraint (2) is calculated using the required HI state driving voltage:

$$(V_{CC} - V_{OH})Y_1 - (V_{OH} - V_{EE})Y_2 > I_{OH\,\max} \tag{6.13}$$

The direction of the inequality in Equation 6.13 may appear backward, but it isn't. We expect both sides of Equation 6.13 to be negative (the driver normally sources current). Equation 6.13 asks that the actual drive current required be less negative (i.e., more positive) than the maximum limit I_{OH}. The value of $I_{OH\,\max}$ should enter Equation 6.13 as a negative number.

Constraint (3) is calculated using the LO state output voltage:

$$(V_{CC} - V_{OL})Y_1 - (V_{OL} - V_{EE})Y_2 < I_{OL\,\max} \tag{6.14}$$

The value of $I_{OL\,\max}$ is a positive number for TTL and CMOS. For ECL the value for $I_{OL\,\max}$ is zero because ECL circuits cannot sink any current.

All three constraints appear in Figure 6.3, calculated for a 74HC11000 NAND gate. The output voltages and current limits are all evaluated at the maximum power supply input voltage of 5.5 V (that is usually the worst case). Constraint (1) appears twice, once for a characteristic impedance of 65 Ω and again for 100 Ω. The 100-Ω constraint line passes into the good zone of both current constraints near the value ($Y_1 = 0.05$, $Y_2 = 0.05$). This corresponds to resistance values of $R_1 = 200$, $R_2 = 200$.

The 65-Ω constraint line does not intersect the region allowed by both current constraints. There is no combination of split terminating resistors that will work. The 74HC11000 cannot adequately drive a 65-Ω end-terminated transmission line.

Sometimes an end terminator is made using only one resistor leading to a fixed intermediate voltage provided solely for terminations. The above procedure for designing split terminations also works for finding a good termination voltage.

First design a split terminating network. Then transform the split termination into a Thevenin equivalent voltage source. The Thevenin equivalent source impedance will always turn out to be Z_0. The Thevenin equivalent source voltage will be

$$V_{\text{terminate}} = \frac{R_1 V_{EE} + R_2 V_{CC}}{R_1 + R_2} \tag{6.15}$$

Use this value as the termination voltage.

6.1.3 Other Topologies Used with End Terminators

The bifurcated line in Figure 6.4 cannot be terminated properly. Regardless of where we place terminators, signal energy from the driver still reflects off the juncture at A, causing ringing.

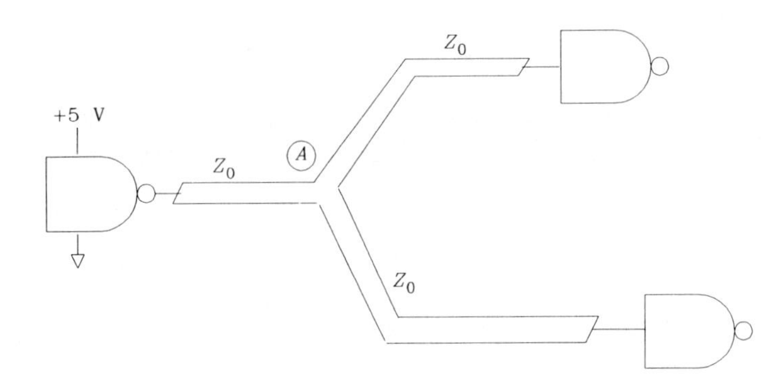

Figure 6.4 Bifurcated line.

The bifurcated line in Figure 6.5 differs from that in Figure 6.4 and can be terminated correctly. The characteristic impedance of each of the bifurcated wires in Figure 6.5 is twice the impedance of the source wire. This trick is accomplished by making the bifurcated wires skinnier than the main feed wire. An end terminator of $2Z_0$ appears at the ends of each bifurcated segment. The input impedance of each bifurcated section, as seen at point *A*, is $2Z_0$. The characteristic impedance of the driving section Z_0 matches perfectly at point *A* with the impedance created by the two bifurcated transmission lines in parallel, each having characteristic impedance $2Z_0$. Few systems use this technique owing to the difficulty of fabricating lines of widely varying impedance on one board.

Using an end terminator, each transmitted step progresses once down the line and then stops. No reflections occur.

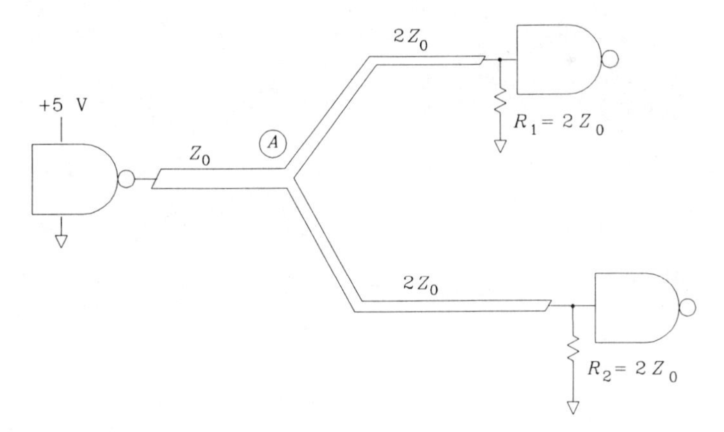

Figure 6.5 Bifurcated line with matching trace impedances.

Because a delayed reproduction of the input signal appears at each point along an end-terminated line, we can attach receivers to the line at any point. This configuration is called a *daisy chain*. The receivers in Figure 6.6 each see delayed versions of the original input signal.

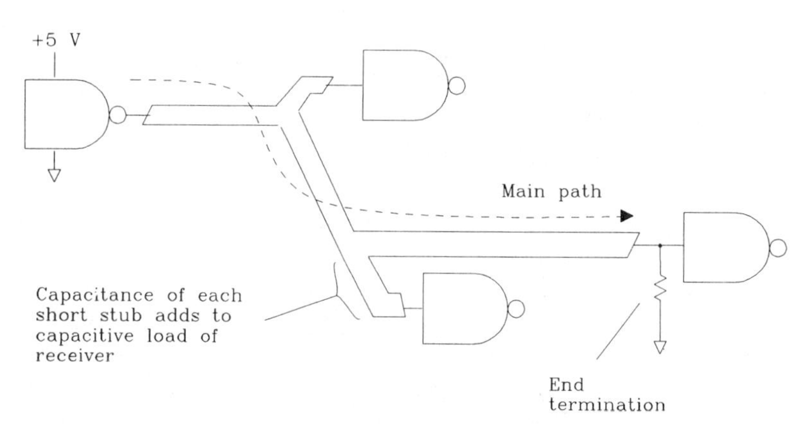

Figure 6.6 Daisy-chain configuration with end termination.

Keep the connecting stubs for each receiver short compared to the length of a rising edge. This avoids reflections from the bifurcation points, as explained above. Short stubs (and their associated receiver capacitances) act like simple capacitive loads which degrade the signal rise time as described in Section 4.4.2. If the stubs occur at regular intervals, the approximations in Section 4.4.3 may be useful.

The ideal arrangement for end terminators places the terminating resistor beyond the last receiver, with no side branch or stub. (See Figure 6.7.)

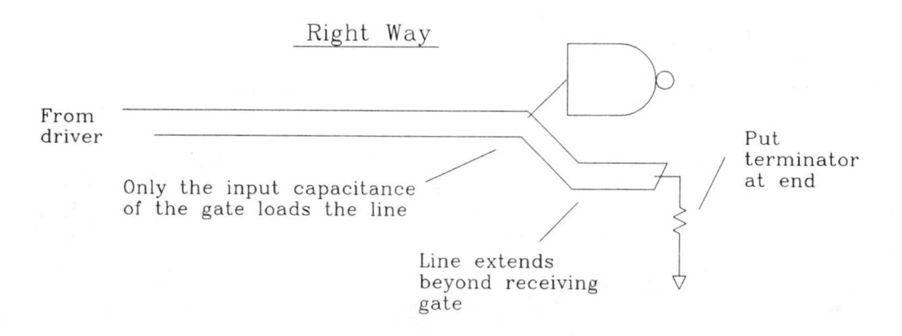

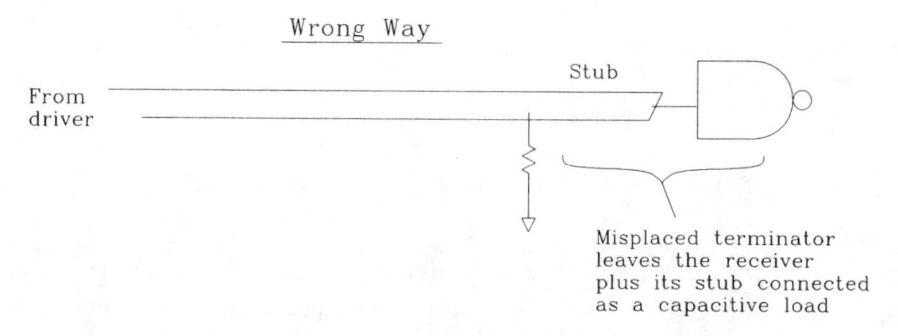

Figure 6.7 Detail of an ideal end-terminator placement.

6.1.4 Power Dissipation in End Terminators

The power dissipated in a terminating load is a function of its HI and LO operating voltages, the various power supply voltages, and the load impedance. Load dissipation is inversely proportional to the terminating impedance. Higher-impedance transmission lines dissipate less power in their terminators.

The equations for power dissipation inside the driving circuit appear in Section 2.2.6. The total power dissipated in the load resistors in Figure 6.3 may be calculated (assuming equal amounts of time are spent in HI and LO states) by Equation 6.16:

$$P_{\text{load}} = \frac{(V_{\text{HI}} - V_{EE})^2 + (V_{\text{LO}} - V_{EE})^2}{2R_2} + \frac{(V_{CC} - V_{\text{HI}})^2 + (V_{CC} - V_{\text{LO}})^2}{2R_1} \qquad [6.16]$$

POINTS TO REMEMBER:

The rise time of an end-terminated circuit, when capacitively loaded, is half that of a series-terminated line driving the same load.

Most TTL or CMOS logic gates can't source enough current to drive end terminators.

You can daisy-chain receivers on an end-terminated line.

6.2 SOURCE TERMINATORS

The source termination scheme connects each driving gate through a series resistor to its transmission line. The value of the series resistor plus the output impedance of the driving gate should equal Z_0, the characteristic impedance of the transmission line. The reflection coefficient at the source will then be zero. See Figure 6.8.

Source terminated circuits have these properties:

(1) The driving waveform is cut in half by the series-termination resistor before it begins propagating down the line.

(2) The driving signal propagates at half intensity to the end of the line.

(3) At the far end (an open circuit) the signal reflection coefficient is +1. The reflected signal is half-intensity. The half-sized reflection plus the original incoming half-sized signal together bring the signal at the receiving end to a full level.

(4) The reflected signal (half-sized) propagates back along the line toward the source, where it damps out at the source termination.

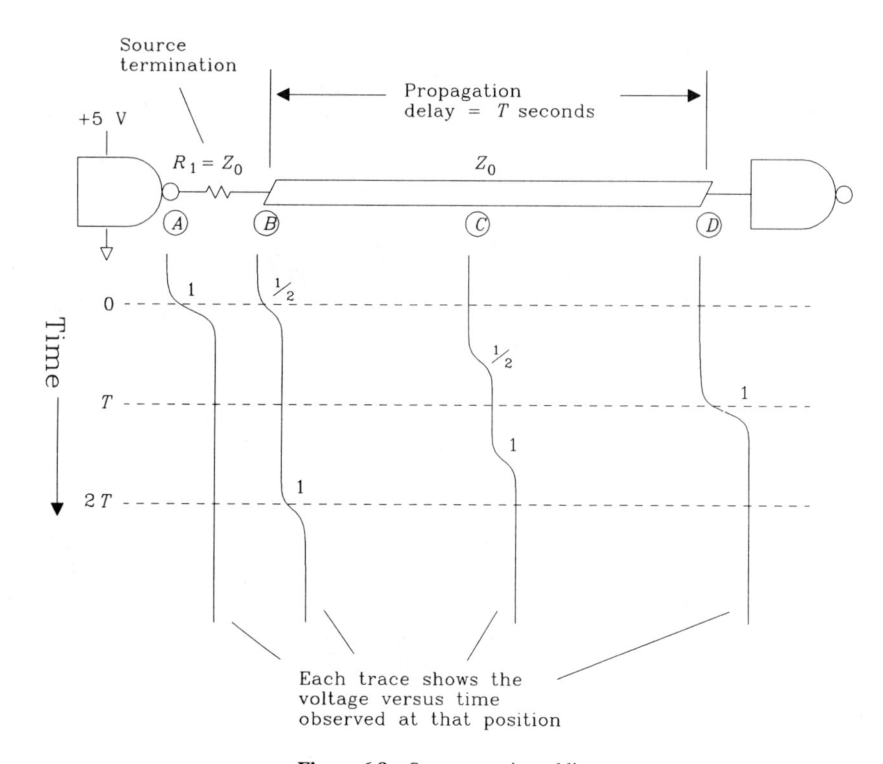

Figure 6.8 Source-terminated line.

(5) After the end reflection returns to its source, the drive current drops to zero where it remains until the next transition. In fast systems, the next transition starts before the end reflection returns.

6.2.1 Resistance Value of Source Termination

An ideal driver has zero output impedance. Practical drivers have a small resistive output impedance. An ECL output has about 10 Ω of output resistance in both high and low states. When forming a source terminated transmission line, the driver output impedance plus the source terminating resistor must match the line impedance. Source-terminating resistors somewhat smaller than the line characteristic impedance are therefore common.

Both TTL and CMOS circuits have different output impedances in their HI and LO states (see Example 2.1). There are no correct values for source-terminating resistors when using TTL or CMOS drivers; there are only compromises.

6.2.2 Rise Time Of Source Termination

At any point along the line, looking back toward the source, we see a drive impedance equal to Z_0. When driving a capacitive load, we therefore get a response which looks like a simple *RC* low-pass filter with an *RC* time constant of

$$RC \text{ time constant} = Z_0 C \tag{6.17}$$

Using the formula from Section 3.1 for the 10-90% rise time of an *RC* filter,

$$T_{10-90} = 2.2 Z_0 C \tag{6.18}$$

This rise time is twice as long as the rise time of an end-terminated circuit with the same transmission line impedance and same load.

6.2.3 Flatter Step Response of Source Termination

It is easier to eliminate reflections at the source than at the far end of a transmission line in typical digital circuits. The source typically has a resistive output impedance (plus a little inductance). On the other hand, a receiver at the far end usually has a parasitic capacitive load. The mismatch effects caused by the capacitive load when attempting end termination are often worse than those caused by driver inductance when attempting source termination, especially when driving multiple loads. Source terminations often yield a more nearly zero reflection coefficient than end terminators, leading to a flatter overall frequency response.

It's worth figuring out which form gives better results with your logic family.

6.2.4 Drive Current Required by Source Termination

The composite input impedance of a source-terminated line includes both the transmission line characteristic impedance Z_0 and the source-termination resistor. Their sum is nearly twice the characteristic impedance. The worst-case drive current required when a gate switches equals $\Delta V/2Z_0$. This worst-case drive current requirement lasts only as long as the round-trip propagation delay of the cable, after which the drive current falls to zero. For source-terminated signals that are infrequently switched, the average drive current is small, although the peak drive current is $\Delta V/2Z_0$.

Contrary to popular wisdom, end-terminated lines are no more difficult to drive than source-terminated ones. End-terminated transmission lines require exactly the same maximum drive current as source-terminated lines *if the end terminator is biased halfway between logic levels*. The input impedance of an end-terminated line is Z_0 (half that of a source-terminated line), but the maximum voltage difference between logic output and halfway bias point is only half the logic swing. The resulting current is $\Delta V/2Z_0$.

Note that if the end-termination bias point is offset from zero, we can decrease one direction of current drive while increasing the other. With source terminations, both polarities take the same amount of current.

Even though the maximum drive current requirements are the same for end-terminated and source-terminated lines, the *average current* required by a source-terminated line, for slow-moving signals, is lower. In fast systems, the next transition on a source-terminated line may start before the end reflection has returned to its source. This fast condition can require maximum drive power continuously.

6.2.5 Other Topologies with Source Termination

The daisy-chain topology does not work with source terminators. All loads must connect at the end of the line. A load-connected midline will see a waveform like that shown in Figure 6.8 at point *C*.

6.2.6 Power Dissipation in Source Terminators

The assumptions of Section 2.2.6 do not work well for estimating power dissipated in the driving circuit. This happens because the current waveform pulled by the source-terminated load abruptly shuts off after one round-trip propagation delay $2T$. We must develop a better model.

From the time the driver switches until one round-trip delay later when the line comes to rest uniformly at the new drive voltage, the source-termination resistor has a voltage of $\Delta V/2$ impressed across it. During this period, the source-terminating resistor dissipates a total energy equal to:

$$E = 2T\left(\frac{\Delta V}{2}\right)^2 \frac{1}{R} \qquad [6.19]$$

where ΔV = difference between HI and LO logic levels, V
T = propagation delay of transmission line, s
$2T$ = round-trip delay of transmission line, s
R = source-terminating resistor, Ω

Multiply this energy per pulse by the pulse rate to get a rough idea of the power dissipation. This approximation works only if the pulse interval is greater than twice the propagation delay. If the pulse interval is shorter, then assume a worst case of $\Delta V/2$ across resistor R at all times.

$$\text{Power} \approx \frac{(\text{Pulse frequency})T\,\Delta V^2}{2R} \qquad [6.20]$$

where ΔV = difference between HI and LO logic levels, V
T = propagation delay of transmission line, s
R = source-terminating resistor, Ω

This amount of power is less than the power dissipated in an end-terminated load for the same conditions of drive voltage and transmission line impedance.

POINTS TO REMEMBER:

Source terminators have a slower rise time and usually smaller residual reflections than end terminators.

Do not daisy-chain receivers on lines having source terminators.

Subtract from the ideal source-termination value the output impedance of your driver.

At low-pulse repetition rates, source terminators dissipate little power.

The peak drive power for a source-terminated line and an end-terminated line (biased at the halfway point) are the same.

6.3 MIDDLE TERMINATORS

Sometimes an engineer ties together a big hair-ball network of gates without thinking about terminations. The length of a rising edge may be much shorter than the size of the network. This problem is exacerbated by tristate drivers, for which there is no well-defined source or destination.

Intuition tells us that each transmitted step will rattle around in the wiring for quite some time before settling down. Section 4.3.5 provides a quick way of estimating the settling time of a straight section of wire. In this case we have a jumble of interconnected wires, which usually rattles at least as long as the longest contiguous length in the bunch.

If devices connected to the network require monotonically rising edges, we are in trouble. There is in general no way to fix such a problem, short of slowing down the rising edges (or filtering the received signal).

If the network is sampled in time, we can arrange for the sampling time to wait until the net settles after each transition before sampling it. In the sampled case, we just need to reduce the settling time, not eliminate it.

There are at least four approaches to this problem:

(1) Add a source terminator to every driver.
(2) Add an end terminator at each receiver.
(3) Add a shunt termination in the middle of the network.
(4) Add series resistance between every juncture of branches.

Option (1) is well defined, takes little power, provides a little bit of damping, and reduces the settling time.

Option (2) requires a lot of drive power but works well in star configurations. A star configuration has a discrete wire leading from each active circuit to a single point where all wires tie together. Reflections are confined to the segment between the source and the central connection.

Combining options (1) and (2), while wasting even more power, is a perfect solution to the star problem. Unfortunately, that configuration attenuates each signal as it passes through the central star. There are no reflections, but the received signal levels are very small.

We don't know why people use option (3). It just lowers the impedance of the central part of the network, where it is already too low.

Option (4) attenuates the signal at every juncture. Using the circuit in Figure 6.9, the signal attenuates by one-half as it passes through each juncture. This damps out reflections quickly (the round-trip attenuation is one-fourth) but also cuts down the signal level severely when the signal goes through many junctures. Constraining the system to perhaps no more than three series junctures, we can easily arrange receivers sensitive enough to tolerate that much attenuation.

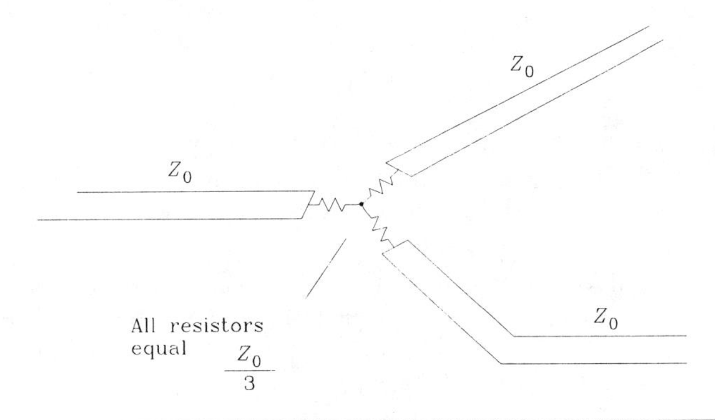

Figure 6.9 Attenuating juncture for hair ball networks.

> **POINT TO REMEMBER:**
>
> Middle terminations can improve system step response only at the expense of signal attenuation.

6.4 AC BIASING FOR END TERMINATORS

Capacitors are sometimes incorporated in end-termination circuits to reduce the quiescent power dissipation. Consider the two circuits in Figure 6.10. The time constant R_1C is chosen very large compared to the signal clock time.

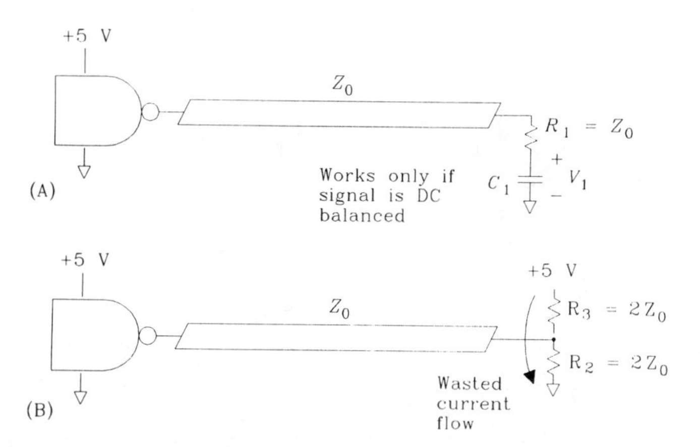

Figure 6.10 Power-saving advantage of capacitive termination.

If we can guarantee that the drive circuit spends half its time in each state (we call such a circuit *DC-balanced*), the average value accumulated on capacitor C_1 will be halfway between the HI and LO voltages. Resistor R_1 will then have a voltage magnitude of $\Delta V/2$ continuously impressed upon it. The power dissipated in R_1 will be

$$P_{R_1} = \frac{(\Delta V/2)^2}{Z_0} = \frac{(\Delta V)^2}{4Z_0} \qquad [6.21]$$

where ΔV = difference between HI and LO logic levels, V
Z_0 = value of terminating resistor, Ω

By contrast, in the split terminator one or the other resistor always has the full ΔV across it, but each resistor is twice as big as R_1 and so the average power dissipation is

$$P_{R2+R3} = \frac{(\Delta V)^2}{2Z_0} \qquad [6.22]$$

Equation 6.22 shows twice the power of Equation 6.21. The additional wasted power dissipation flows from V_{CC} directly to ground through R_2 and R_3.

From the perspective of the driving circuit, the two terminations are indistinguishable. Power dissipation in the driving circuit is the same for both cases. Only the dissipation in the terminating resistors differs.

6.4.1 DC Imbalance in the Capacitive Termination

If the signal in Figure 6.10A stays in the HI state for too long, then capacitor C_1 will charge up to a full HI value. When the gate switches LO, the full ΔV will appear across R_1. The drive current $\Delta V/R_1$ is twice as much as the current required when capacitor C_1 is balanced halfway between HI and LO voltages.

If the driver can source the full amount of $\Delta V/R_1$, then why don't we just tie one end of R_1 to either ground or V_{CC}? If it can source that much current, then forget about split or capacitive terminations. If, on the other hand, the driver cannot source so much current, then we would need DC-balanced signals in order for a capacitively terminated line to operate properly.

Sometimes designers will compromise by making the value of C_1 so small that the R_1C_1 time constant is quite short. Their hope is that C_1 is large enough to help damp out reflections but small enough not to stress their driver with the double-current requirement. This reduces average power dissipation in the driver (so it won't overheat) but does not address the driver's inability to source full logic levels under the load conditions imposed by the initial charge on the capacitor. Even with a small C_1, the rising edge of each step may not meet specifications until C_1 charges.

With a short time constant, even if the signal doesn't meet its V_{OL} and V_{OH} specifications until the capacitor discharges (or charges), at least the result will be monotonic. That tradeoff can be useful, especially when terminating interrupt lines or clock signals.

6.4.2 End Terminators for Differential Lines

Given two signals which are complementary (a differential pair), we can connect their end-terminating resistors together onto a single capacitor. This provides a power-saving end terminator with a guaranteed correct voltage always present on C_1. See Figure 6.11.

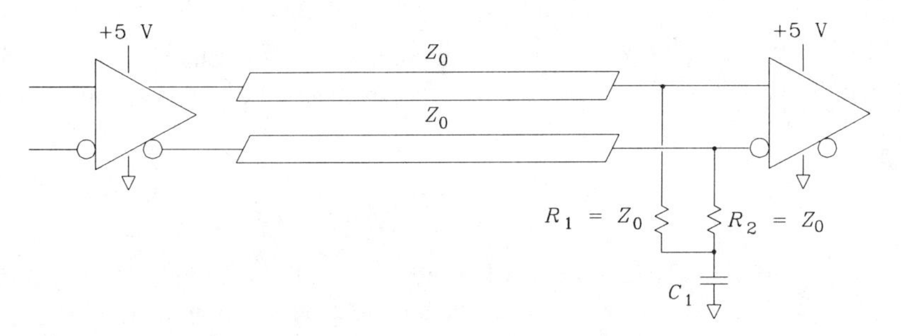

Figure 6.11 Termination of a differential pair.

> **POINT TO REMEMBER:**
>
> Combination RC circuits can terminate DC-balanced lines with no wasted quiescent power.

6.5 RESISTOR SELECTION

6.5.1 Accuracy in Terminating Resistors

A terminating resistor should reduce or eliminate unwanted reflections on a transmission line. It can perform this function only when its resistance value matches the characteristic impedance of the transmission line.

Add the uncertainty in terminating resistance to the uncertainty in transmission line impedance to calculate the total worst-case terminating mismatch. The resulting total is divided by 2 to find the expected reflection percentage (a consequence of Equation 4.53). The transmission line impedance is usually less certain than the terminating resistor value. Knowing, for example, that the transmission line is likely to vary by ±10%, most designers would specify terminating resistors with a 1% tolerance.

If signal fidelity is of the utmost importance, consider using both source *and* end terminators. This procedure cuts the received signal level in half but reduces reflections dramatically. Any reflecting signal must bounce off both source and destination ends, squaring the effective reflection coefficient. The tolerance required for termination matching at either end is greatly relaxed. This approach is used extensively in microwave circuitry to improve gain flatness over wide frequency ranges. In digital electronics, this double-termination technique is used only in conjunction with *line receiver* components capable of discriminating reduced-size receive signals.

6.5.2 Power Dissipation in Terminating Resistors

Regardless of your operating speed, always calculate the expected *worst-case* power dissipation in each terminator. Do not assume your circuits will operate at 50% duty cycle for this calculation.

For example, each of the split terminating resistors used in Figure 6.12 will dissipate a worst-case power of

$$P_{\text{worst}} = \frac{(5\text{ V})^2}{100\ \Omega} = 0.25\text{ W} \qquad [6.23]$$

Standard 1/8-W resistors will overheat in this application at room temperature. Larger 1/4-W resistors may overheat at elevated temperatures. Check with the manufacturer to see if the resistor can safely dissipate the full 1/4-W at your maximum expected ambient temperature. The power-handling capability of many resistors declines at elevated ambient temperatures.

Follow the manufacturer's suggested mounting and heat sinking instructions carefully. Resistor bodies have a thermal resistance rating measured in degrees Celsius rise per watt, just like integrated-circuit packages (see Chapter 2). Resistors can usually tolerate much higher operating temperatures than integrated circuits, especially if made of ceramic materials.

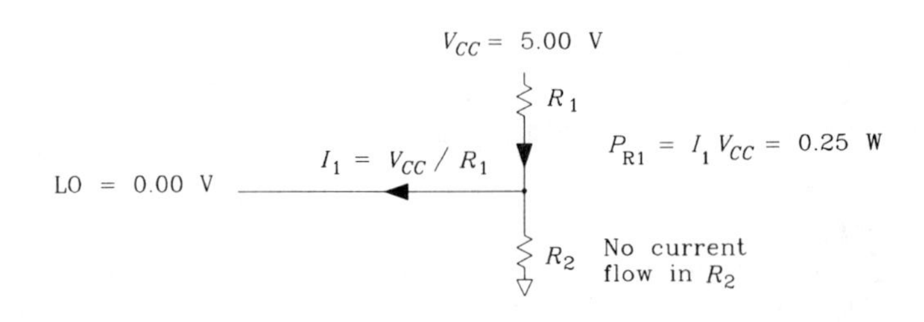

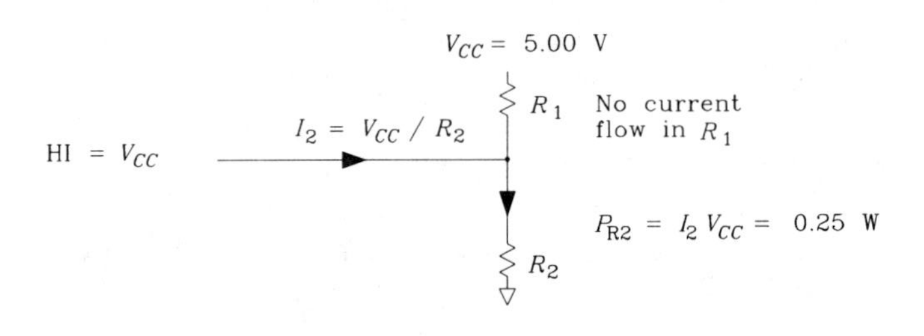

Figure 6.12 Power calculation for split termination with the worst-case LO or HI signal.

Unlike integrated-circuit packages, resistors can be mounted in two different physical configurations. The vertical mount illustrated in Figure 6.13 has a lower thermal resistance in still air than the horizontal mount.

The consequence of overheating a resistor is that its resistance value may drift, causing reflections. In extreme cases a resistor will crack open, unterminating your carefully designed circuit.

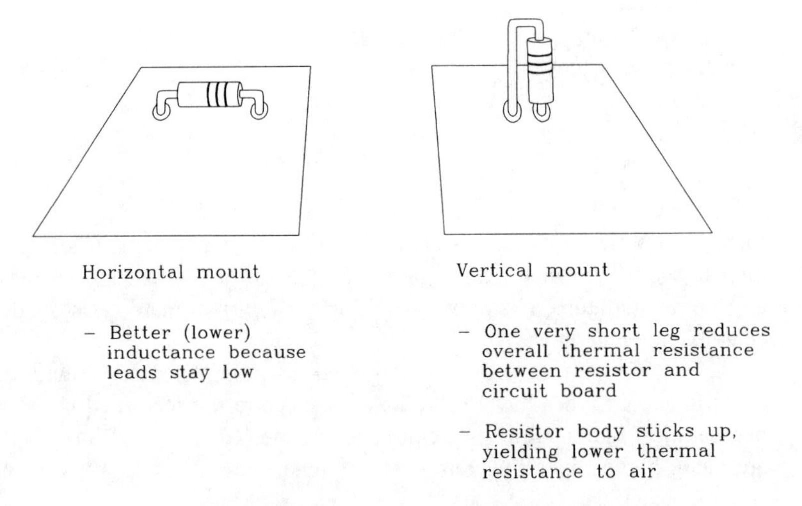

Figure 6.13 Two ways to mount axial resistors.

6.5.3 Series Inductance of Terminating Resistors

Assuming you have selected a resistor value, a tolerance, and a power rating, the next most important factor is the parasitic series inductance. Every resistor has a parasitic series inductance. Each resistor's inductance depends on its internal construction, external lead type, and the mounting configuration. The wiring inductance of your printed circuit board must also be included as part of the overall inductance in a series with every terminating resistor.

The effect of series inductance is a function of operating frequency. For digital signals, we will analyze the inductive effect at the knee frequency (see Equation 1.1). Using Equation 1.1 to relate rise time to frequency, we can compute the inductive reactance magnitude seen by a rising edge directly as a function of the operating rise time:

$$\left|X(T_r)\right| = \frac{\pi L}{T_r} \qquad [6.24]$$

where T_r = rise time of digital signals, s
$\left|X(T_r)\right|$ = magnitude of inductive reactance as seen by rising edge T_r, s
L = inductance, H

Parasitic series inductance causes mismatch in a termination just like an error in the terminating resistance value. Expressing the magnitude of inductive reactance as a percentage of the terminating resistance value, every 1% of reactance causes 1/2% of reflection. When the quantity $|X(T_r)|$ equals 10% of the terminating resistance value, the reflection is 5%.

Table 6.1 shows the results of laboratory measurements taken on three different resistor types. The first two types are 2.2-Ω axial carbon-film resistors. The last type is a surface-mounted 0-Ω resistor measuring 0.120 in. long and 0.060 in. wide. The larger 1/4-W axial body has more inductance than the 1/8-W axial body.

These measurements are critically affected by lead length. For Table 6.1, each axial resistor was horizontally mounted with its leads pulled down tightly at each end and soldered as close as possible to the resistor body.

TABLE 6.1 TYPICAL SERIES INDUCTANCE OF RESISTORS

Resistor type	Series inductance (nH)
1/4-W axial	2.5
1/8-W axial	1.0
1/8-W 1206, surface-mount	0.9

EXAMPLE 6.1: Effect of Inductance of Terminating Resistor

Let's use 1/8-W axial resistors to terminate a digital signal having a rise or fall time of 1 ns.

Signal rise or fall time	1 ns
Transmission line impedance	50 Ω
Inductance of 1/8-W axial	1 nH

We will use a split termination of 100 Ω going to +5 V and 100 Ω to ground. In this configuration, the ratio of inductive reactance magnitude to resistance for each resistor is the same. For the general case of split terminations, just use the ratio of inductive reactance magnitude to resistance for the smaller of the two resistors.

Calculate the inductive reactance magnitude:

$$|X(T_r)| = \frac{\pi(1 \text{ nH})}{1 \text{ ns}} = 3.14 \qquad [6.25]$$

Find the ratio $|X(T_r)|/R$:

$$\frac{|X(T_r)|}{100\ \Omega} = 3.14\% \qquad [6.26]$$

The reflection due to this inductance will be 1.5%.

The use of a split terminator in Example 6.1 cut the expected reflection in half, compared to a single terminating resistor of 50 Ω having the same inductance. Putting resistors in parallel is in general a good way to make a very-low-inductance structure.

The measurements in Table 6.1 were taken with the series inductance measuring jig depicted in Figure 6.14. This jig has a 4.3-Ω source impedance, and the source waveform is a step. When testing a pure inductance, we expect to measure an inductive spike from the jig, having a total area equal to (Chapter 1)

$$\text{Spike area} = \frac{L}{R_S}\Delta V \qquad [6.27]$$

where ΔV = voltage step size, V
L = inductance under test, H
R_S = test jig source resistance, Ω

When testing a pure resistance, we expect to measure a step output having a final value of

$$\text{Final value} = \frac{R_1\,\Delta V}{R_1 + R_S} \qquad [6.28]$$

where R_1 = resistance under test, Ω
R_S = test jig source resistance, Ω

When testing an unknown series combination of resistance and inductance (a physical resistor), we expect to see in the output a superimposition of a step waveform due to the resistance plus an inductive spike due to the inductance. Both step and spike waveforms are clearly visible in Figure 6.15.

When analyzing such an output, first use your knowledge of the test jig source resistance and the final output value to solve for the unknown resistance. Alternately, measure the unknown resistance with a DC ohmmeter. Knowing the resistance of the device under test, we can scale a copy of the test jig open-circuit output waveform to rep-

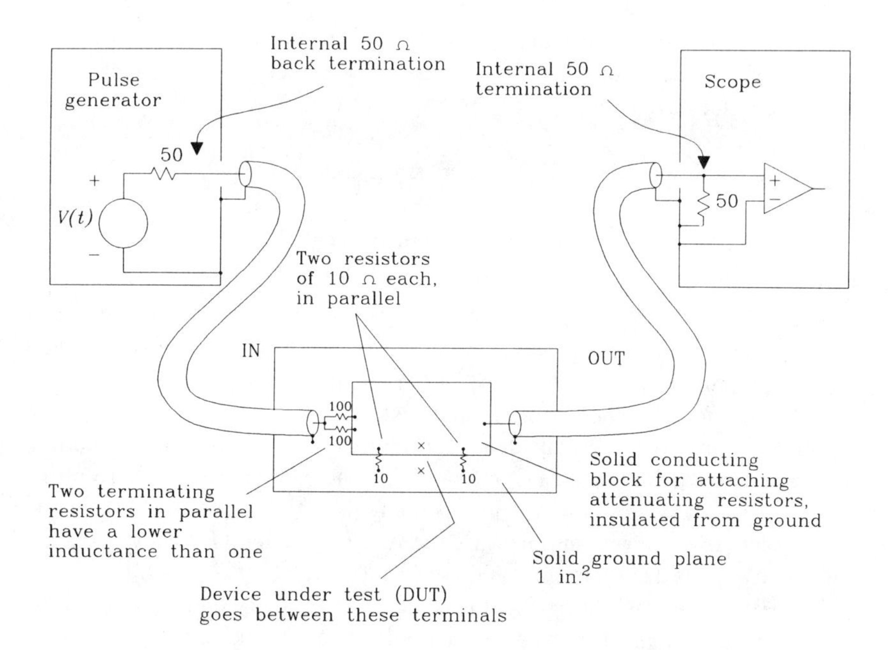

Figure 6.14 A 4.3-Ω lab setup for measuring the inductance of resistor packages.

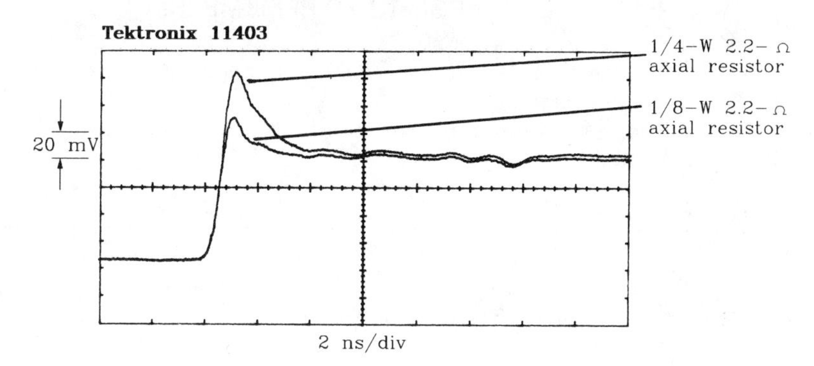

Figure 6.15 Output of a 4.3-Ω test setup for 1/4- and 1/8-W resistor bodies.

resent the theoretical output we should get due only to a resistance of this value. Subtract this theoretical output from the actual measured waveform. The difference should consist only of the inductive spike.

We can then use Equation 6.29 to solve for the unknown inductance. Equation 6.29 is similar to Equation 6.27, except that we have accounted for the effect of the self-resistance of the device under test.

$$L = (\text{spike area})\frac{R_1 + R_S}{\Delta V} \qquad [6.29]$$

where L = inductance under test, H
R_1 = resistance under test, Ω
R_S = test jig source resistance, Ω

When subtracting the theoretical waveform from the actual measured waveform, it is important to use a scaled copy of the test jig open-circuit waveform, not an ideal rectangular step. Because of the limited rise time of your measuring equipment, the difference in area between an ideal step and the rounded corner of your actual test jig waveform may introduce a significant error into your experimental result.

You may store the test jig open-circuit output, scale it, subtract it from the actual measured waveform, and measure the resulting area quite easily using a digital oscilloscope such as the Tektronix 14000 series.

Use the smallest practical resistor value for this experiment. For a fixed amount of inductance, the measured spike area is inversely proportional to resistance. This makes the spike very difficult to see at high resistances.

Be aware that some metal-film resistors use serpentine patterns etched in the metal film to construct high-value resistors. High-value resistors sometimes have markedly higher inductance than low-resistor values. Resistors from any family in the range 10–100 Ω usually have the same physical topology.

POINTS TO REMEMBER:

Specify both a resistance value tolerance and a power rating on terminating resistors.

Parasitic inductance in terminating resistors causes unwanted reflections.

6.6 CROSSTALK IN TERMINATORS

The adjacent terminating circuits in Figure 6.16 cross-couple signal energy between circuit traces. This cross-coupling can be much worse than the natural crosstalk which occurs between adjacent transmission lines.

This section presents the results of practical measurements of terminator cross-coupling along with some heuristics for predicting crosstalk in terminating circuits.

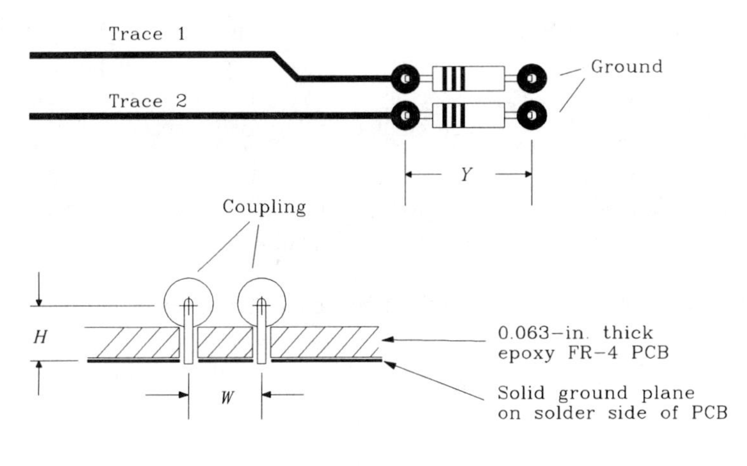

Figure 6.16 Configuration of terminating resistors.

Crosstalk in terminations comes from both mutual inductive and mutual capacitive coupling. The inductive coupling is usually larger. The total coupling is the sum of both inductive and capacitive parts, both of which couple proportionally to the derivative of the applied input signal.[1] For our purpose, we need only work with one overall coupling coefficient, without worrying whether it is magnetic (inductive) or electrostatic (capacitive) coupling.

$$\text{Noise voltage} = \frac{K}{R}\frac{\Delta V}{T_{10-90}} \qquad [6.30]$$

where noise voltage = peak crosstalk coupled onto trace 2

K = cross-coupling coefficient (units for K are ohm-seconds, which are the same as henries)

R = resistance, Ω

ΔV = driving signal step size, V

T_{10-90} = driving signal rise time, s

6.6.1 Crosstalk From Adjacent Axial Resistors

Inductive coupling between adjacent through-hole terminating resistors generally follows the rule of Equation 6.30. We may estimate the crosstalk coefficient using this handy approximation:

$$K = \left(5.08\times10^{-9}\right)Y\frac{1}{1+(W/H)^2} \qquad [6.31]$$

[1]The coupled noise voltage on trace 2 will be a pulse when we apply a step signal to trace 1.

where Y = length of resistors between through-holes, in.
H = centerline height above ground plane, in.
W = separation between resistor centerlines, in.
K = crosstalk coupling coefficient

Figure 6.17 plots several measured values for K along with the calculated values. The measured values (points) were determined from actual crosstalk measurements taken with a physical sample and then using Equation 6.30 to infer values of K. The calculated values (solid line) were derived from Equation 6.31 using the same length (0.400) and height (0.108) as in the physical sample, but differing separations.

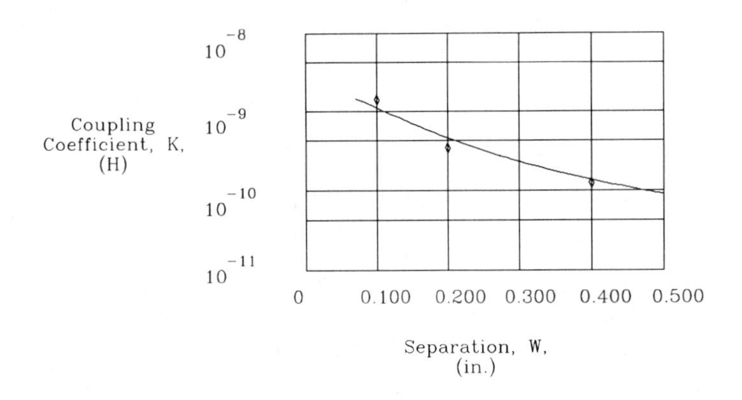

Same layout as in Figure 6.16
1/4-W resistor bodies
Body length, 0.400 in.

Centerline height above surface of board	0.045 in.
Distance, surface of board to ground	0.063 in.
Total centerline height	0.108 in.

Figure 6.17 Measured and calculated values for the coupling coefficient between two terminating resistors.

If the layout staggers the resistors, as depicted in Figure 6.18, then substitute their overlap length for the parameter Y in Equation 6.31.

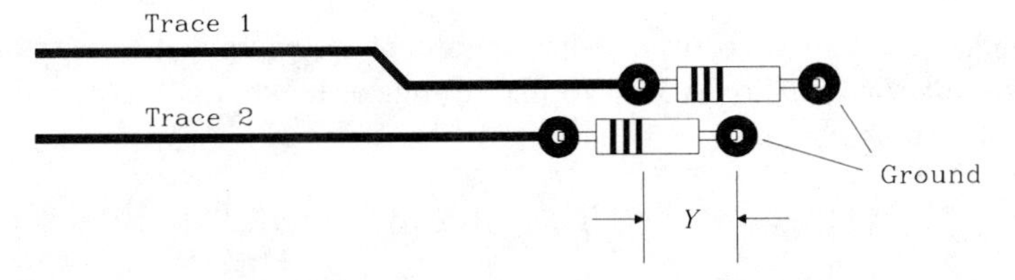

Figure 6.18 Staggered terminating resistors showing overlap length (parameter Y.)

6.6.2 Crosstalk From Adjacent Surface-Mounted Resistors

Surface-mounted resistors, being naturally closer to the circuit board, can exhibit dramatically lower crosstalk coefficients than axial components. To get maximum benefit from this effect, bury a ground plane layer near the circuit board outer surface directly underneath the surface-mounted parts. This decreases parameter H in Equation 6.31, lowering the crosstalk.

6.6.3 Crosstalk From SIP Terminating Resistors

These parts can behave well or poorly, depending on their internal wiring. Figure 6.19 shows the common current path shared by terminators in the single ground pin design. This common current path introduces lots of mutual inductance between the resistors in that package.

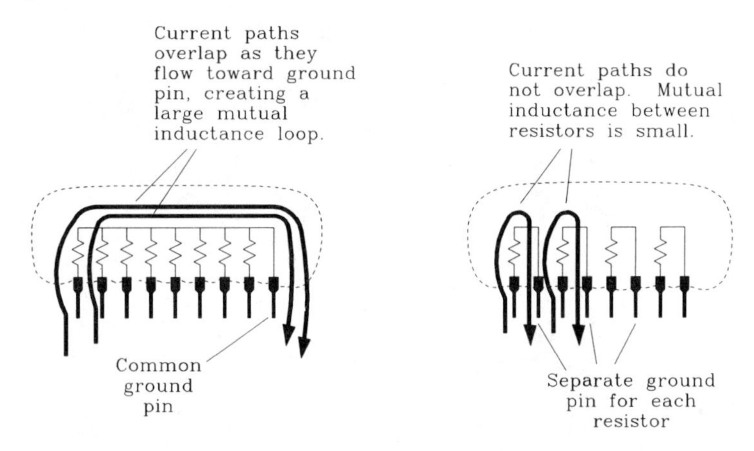

Figure 6.19 Two styles of SIP terminating resistors.

Table 6.2 lists typical coupling coefficients for resistors in 0.1-in. spacing SIP packages. Package SIP-A contains seven resistors in an eight-pin package, with one common ground pin at the end. Resistor 7 in package SIP-A is the furthest from the ground pin. Package SIP-B contains four resistors in an eight-pin package with an independent

TABLE 6.2 COUPLING COEFFICIENTS IN SIP TERMINATING NETWORKS

Package	From	To	Coupling coefficient
SIP-A	7	6	8250.0 ps-Ω (worst)
SIP-A	7	1	2050.0 ps-Ω (best)
SIP-B	4	3	95.0 ps-Ω (worst)
SIP-B	4	1	8.0 ps-Ω (best)

ground for each package. All resistors are 50 Ω. The independent SIP-B package performs almost a factor of 100 better than the common ground network.

Use Equation 6.30 to convert these coupling coefficients into coupled noise voltages.

POINT TO REMEMBER:

The physical layout of terminating resistors affects crosstalk between signal paths.

7

Vias

The term *via* commonly refers to a hole in a printed circuit board. A via can be used for mounting a through-hole component or for routing traces between layers. The only difference, from our point of view, is that during assembly a through-hole via has one leg of a component soldered into it, while a trace-routing via remains empty. Electrically, the two types of vias behave similarly, as described in the following sections.

7.1 MECHANICAL PROPERTIES OF VIAS

When vias are too big, there isn't any room left to route signal traces. Obviously we want small vias, but how small? Even when vias are already small, making them even smaller will permit more traces (higher routing density) on the board. Designers concerned with overall product size are inescapably forced toward smaller and smaller vias.

At smaller sizes, vias have less parasitic capacitance. This means they work better at high speeds. For the highest speed work, tiny vias are mandatory.

Of course, small vias cost more to produce. This dynamic of higher cost for better performance is central to engineering design; vias are no exception. So far, we have three rules concerning vias:

- Smaller vias take up less room.
- Smaller vias have less capacitance.
- Smaller vias cost more.

Don't underestimate the importance of properly sizing vias. The remainder of Section 7.1 discusses tradeoffs between density and cost. Sections 7.2 through 7.4 examine speed issues.

7.1.1 Finished Via Diameter

Let's start at the beginning, with via diameter. Later sections discuss the size of pads surrounding a via and then the gap between pads left over for routing traces between them.

A through-hole via must accommodate a physical component lead. The finished hole's diameter must exceed the size of the component lead inserted into it. The excess diameter required for good soldering on typical boards ranges from 0.010 to 0.028 in., depending on the solder process. There isn't much we can do to shrink the diameter of a through-hole via.

The correct finished diameter for trace-routing vias is more difficult to determine. The minimum attainable diameter for trace-routing vias is constrained by drilling and plating technology.

Smaller holes require smaller drill bits, and smaller drill bits break more often than big, sturdy ones. Your board fabricator would be very happy if all holes were at least 0.050 in. in diameter. Unfortunately, such a large diameter would seriously limit the routing density.

Small holes also take longer to drill. For big holes, a drill shop sets up stacks of several boards, drilling them through in one pass. Tiny drill bits cannot penetrate a deep stack without wandering off center (the slender bit actually bends as it penetrates). Small holes must be drilled in smaller batches, increasing production time.

Electroplating action will not penetrate a deep, skinny hole. Holes deeper than six times their diameter will not plate uniformly. At a standard board thickness of 0.063 in., this limits the minimum hole diameter to about 0.010 in., depending on the care with which the plating shop adjusts its equipment and on the yield required.

All these factors raise the cost of small vias. When talking with a printed circuit board manufacturer about costs, be sure to separate the discussion of hole drilling and plating capability from line etching capability. The two subjects go together, but not exactly. What you need is a chart showing the cost per drilled hole as a function of hole diameter, and another chart showing cost per square inch of circuit board area as a function of line width. Combine your two charts with the information below to pick the best combination of hole size, trace width, and number of layers for your application. Most board fabricators charge proportionally to the number of layers.

What are reasonable limits for hole size? Military Specification MIL-STD-275E lists three categories of acceptable tolerance data for hole size: preferred, standard, and reduced producibility (you can thank the U.S. government for this catchy phrase). The preferred specifications are easiest (and cheapest) to manufacture. The reduced producibility specifications are much harder to meet and usually cost extra. A related document, IPC-D-300G (Interconnections Packaging Circuitry Standard), lists similar information for commercial products with slightly different numbers. Tables 7.1–7.3 include a minimum sampling of information from MIL-STD-275. Both sets of standards are summarized nicely by R. H. Clark.[1]

[1]Raymond H. Clark, *Printed Circuit Engineering,* Van Nostrand Reinhold, New York, New York, 1989.

TABLE 7.1 MIL-STD-275E HOLE DIAMETER

	Preferred	Standard	Reduced producibility
Minimum hole diameter*	$T/3$	$T/4$	$T/5$

*T is the board thickness.

TABLE 7.2 MIL-STD-275E HOLE TOLERANCES

	Preferred	Standard	Reduced producibility
Plating allowance*	0.0028	0.0021	0.0014
Plated hole diameter tolerance†			
Holes 0.015—0.030 in.	0.008	0.005	0.004
Holes 0.031—0.061 in.	0.010	0.006	0.004
Hole alignment allowance‡			
Board <12 in.	0.009	0.006	0.004
Board >12 in.	0.012	0.009	0.006
Required annular ring			
Inner layer	0.008	0.005	0.002
Outer layer	0.010	0.008	0.005

*Not part of MIL-STD-275E. Standard plating for digital boards is 1 oz (0.0014 in.). For fine line fabrication, some manufacturers use 1/2-oz (0.0007 in.). Plating allowances for hole diameters are twice the plating thickness.

†Includes allowance for variation in plating thickness.

‡Sum of hole location tolerance and master pattern (etching) accuracy as listed in MIL-STD-275E.

TABLE 7.3 MIL-STD-275E MINIMUM AIR GAP

	Preferred	Standard	Reduced producibility
Air gap for wave solder*	0.020	0.010	0.005

*This is the air gap required to prevent solder bridging. Larger gaps are requirerd by UL, CSA, and TUV safety regulations for protection against high-voltage arcing.

7.1.2 Required Via Pad Size

Every via requires additional space for a pad, and for clearance around the pad, on the printed circuit board surface. The pad electrically connects the plated interior of the via to traces on the surface (or in the interior) of a printed wiring board.

The correct size for the pad surrounding a via is determined principally by four factors. Table 7.2 lists typical values for these parameters.

- Plating allowance
- Hole diameter tolerance
- Hole alignment allowance
- Required annular ring.

A via must be drilled before it can be plated. The plating process, which coats the interior of the holes to make them conductive, also builds up the wall thickness by about 0.001 or 0.002 in. The diameter of the finished hole may then be 0.002–0.004 in. smaller than the drilled size. The difference between the drilled hole size and the finished hole size is the *plating allowance*. The plating allowance is twice the maximum plating thickness. Figure 7.1 illustrates the relationship between finished hole size and drilled hole diameter. Don't worry about *variations* in plating thickness; they show up as part of the hole diameter tolerance. The plating allowance just accounts for the nominal plating thickness.

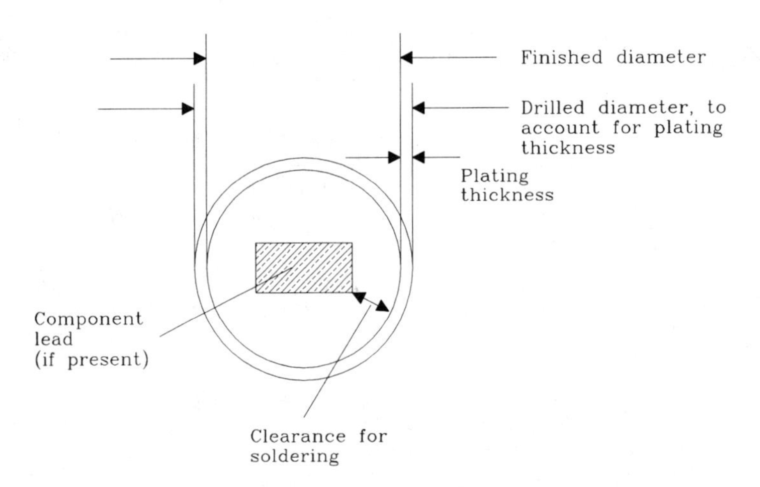

Figure 7.1 Finished hole diameter versus drilled diameter.

No manufacturer can drill holes perfectly. They always insist on a *hole diameter tolerance*. A hole diameter with tolerance is usually expressed like this: **0.032±0.003 in.** The hole diameter tolerance introduces two constraints.

We must oversize the nominal hole diameter slightly so that the smallest hole leaves enough room for a component lead and satisfies the depth-to-breadth ratio required for electroplating. This oversize adds to the plating allowance.

On the other hand, the maximum-sized hole must not obliterate its pad as it is drilled. Pads are drawn oversized to avoid obliteration.

The *hole alignment allowance* accounts for mechanical slop in the drilling machine. The drilling machine attaches to special reference holes provided in the board. The copper patterns etched on the board align with these same reference holes. Mechanical processes being what they are, neither alignment is perfect. The manufacturer can quote a hole **alignment** allowance stating how far his holes will be from the nominal etched pad centers. This alignment tolerance includes both drilling and etching alignment errors.

In reference to Figure 7.2, the donut-shaped copper ring left after drilling through a pad is called an *annular ring*. If the hole lands off-center, the annular ring may appear dangerously thinned or broken through on one side. The condition is called *breakout*. Severe breakout, if it occurs on the trace side of a pad, can jeopardize electric contact between the trace and the interior surface of its via. The *required annular ring* specifies the minimum amount of copper pad we want surrounding a via under worst-case drilling conditions. If your layout program paints a bulge on the trace side of each via, you can get away with a zero or slightly negative annular ring requirement (see Figure 7.3) in commercial products. This practice is not acceptable for military or high-reliability products.

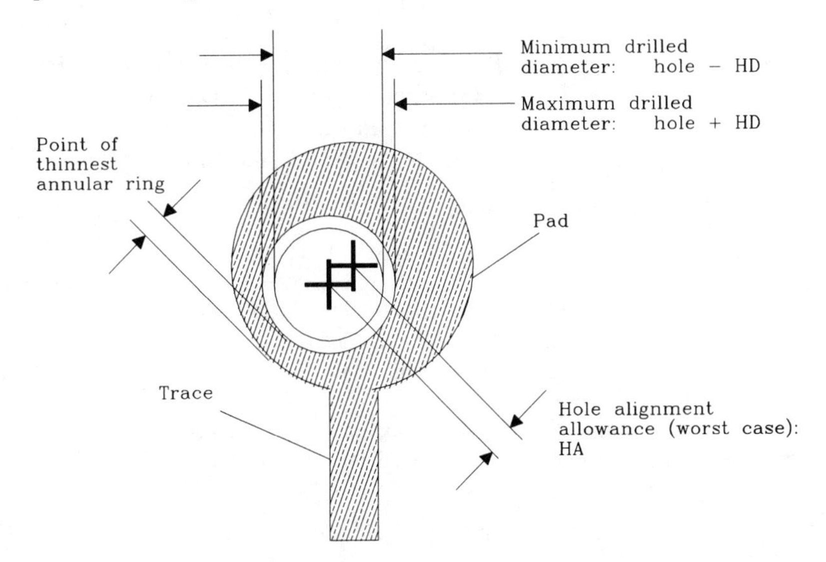

Figure 7.2 Annular ring surrounding a hole drilled in a pad.

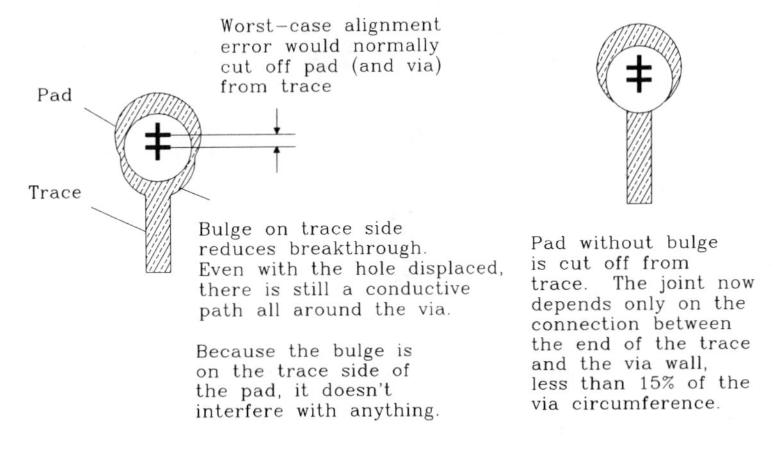

Figure 7.3 Bulge on a pad used to bolster an annular ring at a trace juncture.

The minimum pad diameter may be calculated:

$$\text{PAD} = \text{FD} + \text{PA} + 2(\text{HD} + \text{HA} + \text{AR}) \quad [7.1]$$

where PAD = minimum pad diameter, in.
FD = required minimum finished hole diameter, in.
PA = plating allowance, in.
HD = hole diameter tolerance, in.
HA = hole alignment allowance, in.
AR = annular ring required, in.

The correct nominal drilled hole diameter is

$$\text{HOLE} = \text{FD} + \text{PA} + \text{HD} \quad [7.2]$$

where HOLE = correct nominal drilled hole diameter, in.
FD = required minimum finished hole diameter, in.
PA = plating allowance, in.
HD = hole diameter tolerance, in.

EXAMPLE 7.1: Pad Size Calculation

Let's design pads for a 0.063-in. epoxy FR-4 printed circuit board.

Our manufacturer tells us there is one price for big holes and a premium of 30% on holes from 0.015 to 0.020 in. The smallest size available is 0.015 in. The hole diameter tolerance from this manufacturer is not so good. They need a tolerance of ± HD = 0.003 in.

The plating thickness is 1 oz (0.0014 in.), and so the plating allowance is PA = 0.0028 in. Call it 0.003 in.

We decide to specify a minimum finished diameter of FD = 0.015 in., instructing the manufacturer to drill the hole to a diameter of 0.021 ± 0.003 in. We avoid the hole drilling premium.

$$\text{HOLE} = \text{FD} + \text{PA} + \text{HD} = 0.015 + 0.003 + 0.003 = 0.021 \quad [7.3]$$

Next we ask about the hole alignment allowance, HA = 0.002, and select our required annular ring, AR = 0.005 in.

The pad size should be

$$\begin{aligned} \text{PAD} &= \text{FD} + \text{PA} + 2(\text{HD} + \text{HA} + \text{AR}) \\ &= 0.015 + 0.003 + 2(0.003 + 0.002 + 0.005) \quad [7.4] \\ &= 0.038 \text{ in.} \end{aligned}$$

The pad diameter needed to give us enough annular ring for this example is close to twice the size of the finished hole diameter. This is typical for tiny holes.

7.1.3 Clearance Requirements: Air Gap

The space between copper features on a printed circuit board is called an *air gap*. This historical term harks back to the days when all wiring was done by hand between metal connection posts. A minimum air gap was originally required to prevent electric arcing between high-voltage terminals. On modern printed circuit boards the gap between copper features is embedded in circuit board material or solder mask, but we still call it an air gap.

Newly designed boards include a specification for all nominal pad and trace sizes. From these specifications we can calculate the air gap between nominal features. At low voltages we need only a very tiny air gap to prevent arcing, almost never a source of failure on digital boards. Failures in digital printed circuit boards more often come from solder bridging.

Imperfections in the etching process cause solder bridging. These imperfections result in ragged edges, bumps, or nodules of copper hanging on the sides of traces and pads. Such imperfections cause adjacent copper features to draw nearer to each other than normal. At the points of closest approach, solder used in the assembly process is likely to bridge between the features. The minimum safe clearance which prevents solder bridging depends on these factors:

- Precision of the etching process
- Assembly method
- Required yield.

Etching precision is controlled by your board manufacturer. Ask what the line width tolerance is for the process (see Section 4.5.1.4 on typical line width tolerances). Always subtract the line width tolerance from the nominal air gap when figuring the worst-case clearance. This accounts for worst-case swelling of both features. Each feature protrudes only one-half the line width tolerance toward the other, so you need to subtract only once.

Wave soldering and reflow soldering are the two major types of assembly processes. The wave-soldering method is more prone to solder bridging than reflow soldering. Through-hole boards always use wave soldering. Surface-mounted boards may use reflow soldering, wave soldering, or both.

The required yield depends on your manufacturing volume and the depth of your pocketbook. At very low volumes, you may choose to have an assembly rework person visually inspect each board and manually clear any solder bridges. This manual procedure is not practical if you are building 100,000 units. For high volumes it is much better to spend extra design effort locating and fixing clearance problems.

Both etching imperfections and solder bridges are random phenomena. Increasing the air gap reduces their probability but never completely eliminates them. Finding the right balance between your desire for packing density and your need for high manufacturing yields takes time and practice.

7.1.4 Trace-Routing Density Versus Via Pad Size

The cost of a printed circuit board is roughly proportional to the number of layers. The number of layers required depends on the wiring density of each layer.

Wiring density is in turn influenced by the way traces sneak between vias. Most boards are so filled with vias they look like Swiss cheese. Long traces must often squeeze through the spaces between adjacent vias. The number of traces that we can squeeze between a pair of adjacent vias is called the number of *tracks*. A one-track board can fit only one trace between a pair of vias. A double- or triple-track board can fit two or three traces between vias. A multilayer board often supports more tracks on the inner layers than on the outer layers. Solder bridging cannot occur on inner layers, allowing us to reduce the required air gap and squeeze through more tracks.

Wiring density is measured in units of *trace pitch*. Trace pitch equals the spacing between parallel trace centers. Trace pitch is also the inverse of the number of parallel traces per inch. Trace pitch usually refers to the minimum spacing between parallel trace centers. In the context of this section we discuss the effective, or average, trace pitch.

There are usually so many holes in a board that most trace positions are blocked by vias. If we line up a row of adjacent vias, the maximum number of traces that will fit between them is equal to the number of vias times the number of tracks. This is much less than the theoretical number of traces that would fit in the same space given no vias. A board with lots of vias limits the effective trace pitch to

$$\text{Effective trace pitch} = \frac{\text{via spacing}}{\text{tracks}} \qquad [7.5]$$

When designing a new board, be aware that minor adjustments in the pad annular ring requirement, via spacing, or trace line width tolerance can easily make the difference between one or two or even three tracks. This dramatically increases the wiring density and may save board layers. On the other hand, sacrifices in annular ring requirements and minimum air gaps will directly reduce the manufacturing yield.

Designers usually set a fixed minimum separation between vias, keeping them on a grid pattern of minimum separation positions. That way they can add a new via at any open grid position without having to move other vias. The via positioning grid is usually set at 0.100 in. for through-hole designs, corresponding to the pin separation of DIP packages. For surface-mount designs, the via positioning grid can be different. IPC-D-300G asks that you use a grid of 0.100, 0.050, or 0.025 in. where possible.

POINTS TO REMEMBER:

The finished diameter of routing vias depends on drilling and plating technology. Smaller holes cost more.

Pad sizes are determined by drilling tolerances and the annular ring requirement. The annular ring controls breakout.

The minimum air gap is determined by line width tolerances and nominal pad positions. The air gap controls solder bridging.

Sacrificing pad size and air gap increases tracks but reduces yield.

7.2 CAPACITANCE OF VIAS

Every via has parasitic capacitance to ground.[2] Vias being physically small structures, they behave very much like lumped circuit elements. We can predict, within an order of magnitude, the amount of parasitic capacitance for a via:

$$C = \frac{1.41\,\varepsilon_r T D_1}{D_2 - D_1} \qquad [7.6]$$

where D_2 = diameter of clearance hole in ground plane(s), in.
D_1 = diameter of pad surrounding via, in.
T = thickness of printed circuit board, in.
ε_r = relative electric permeability of circuit board material
C = parasitic via capacitance, pF

When the pad size approaches the clearance hole diameter, pads pick up a lot more capacitance. If your ground clearance holes must remain small to maintain ground continuity, shrink or eliminate the pads on ground layers. For trace-routing vias, it doesn't matter if you get some breakout on the plane layers.[3]

The primary effect of via capacitance is that it slows down, or degrades, the rising edges of digital signals.

Equation 7.6 assumes there is a pad on every layer. Some designers omit pads on layers not connecting to traces, slightly reducing the parasitic capacitance. In many practical situations, the parasitic capacitance is so low anyway that it isn't worth worrying about.

If you must know beforehand the capacitance of a via, measure it using a physical model. When building a physical model, make use of the scaling principle for capacitance:

> A scale model of a via or trace has X times the capacitance of a real via, where X is the model scale.

For example, Figure 7.4 shows a simple pad model constructed of aluminum foil and cardboard. This is a 100:1 scale model of a routing via in a surface-mount design. The central tube, representing the inside surface of the plated through-hole, is 1.6 in. in diameter. The pad at each end of the tube is 2.8 in. in diameter. The ground plane has a 5.0-in. clearance cutout. These dimensions result in a measured capacitance value of 11.0 pF. Scaling down by 100, the actual capacitance in air would be 0.11 pF. Remember that the actual capacitor will be embedded in FR-4, with a relative permeability of 4.7, so the via capacitance will be close to 0.5 pF. It is much easier to accurately measure the relatively large capacitance of 11 pF than the capacitance of the actual finished via. Besides, making mock-ups is fun.

Let's cross-check this measured value of capacitance with Equation 7.6:

$$C = \frac{(1.41)(4.7)(0.063)(0.028)}{0.050 - 0.028} = 0.53\text{ pF} \qquad [7.7]$$

[2](Assuming your high-speed circuit board has at least one ground plane layer.)

[3]C. F. Coombs explains that completely removing the pad can in some circumstances lead to a short circuit between the ground plane and trace routing via.

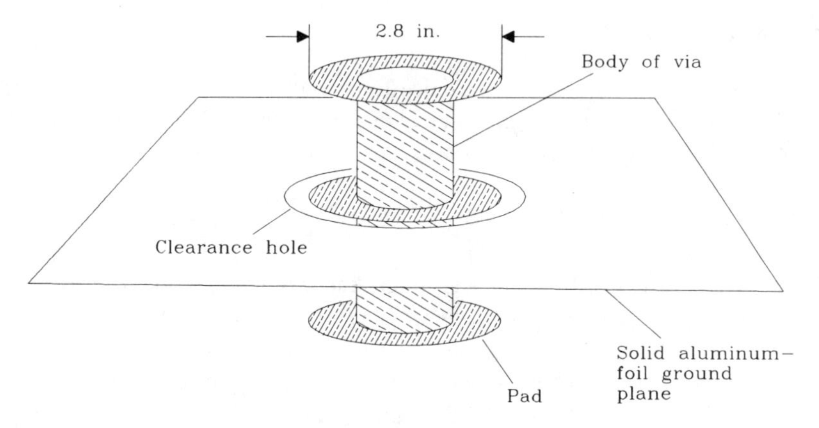

Figure 7.4 A 100:1 model of a via.

Don't expect the formula to be so close all the time.

How much will this via affect a 50-Ω transmission line? In reference to Equation 4.76, the 10–90% rise-time degradation contributed by the finished via will be

$$\begin{aligned} T_{10-90}(\text{step response}) &= 2.2C(Z_0/2) \\ &= (2.2)(0.5)(50/2) \\ &= 27.5\text{ ps} \end{aligned} \qquad [7.8]$$

Twenty-seven picoseconds is a tiny interval of time, indeed.

If you must make many pad capacitance predictions, invest in electromagnetic field modeling software.[4] These packages can (with enough computer resources) accurately model the inductance and capacitance of three-dimensional structures.

> **POINTS TO REMEMBER:**
>
> Via capacitance is a measurable, but small, effect.
>
> A scale model of a via or trace has X times the capacitance of a real via, where X is the model scale.

7.3 INDUCTANCE OF VIAS

The inductance of vias is more important than their capacitance to digital designers. Every via has parasitic series inductance. Vias being physically small structures, they behave very much like lumped circuit elements. The primary effect of series via induc-

[4]Try Quantic Laboratories, in Winnipeg, Canada, or Quad Design of Camarillo, California.

tance is that it degrades the effectiveness of power supply bypass capacitors. This can defeat your whole power supply filtering strategy.

The purpose of a bypass capacitor is to short together, at high frequencies, two power planes. If we imagine an integrated circuit connected between power and ground planes at point *A* in Figure 7.5 with a perfect surface-mounted bypass capacitor connected at point *B*, we expect the chip to see zero high-frequency impedance between the V_{CC} and ground planes at its attachment point. This is not, however, the case. The finite inductance of each attachment via used to connect the capacitor to the V_{CC} and ground planes introduces a small but measurable inductance. The magnitude of this inductance is approximately

$$L = 5.08h\left[\ln\left(\frac{4h}{d}\right)+1\right] \qquad [7.9]$$

where L = inductance of via, nH
h = length of via, in.
d = diameter of via, in.

Because Equation 7.9 involves a logarithm, changing the via diameter does little to influence the inductance. A big change may be effected by changing the via length.

Using Equation 1.15, we can find the inductive reactance of the example via in Section 7.3 to a rising edge speed of 1 ns. First compute the inductance:

$$h = 0.063 \text{ (length of via, in.)}$$
$$d = 0.016 \text{ (diameter of via, in.)}$$
$$T_{10-90} = 1.00 \text{ (rising edge speed, ns)}$$

$$L = (5.08)(0.063)\left[\ln\frac{4(0.063)}{0.016}+1\right]$$

$$= 1.2 \text{ nH} \qquad [7.10]$$

$$X_L = \frac{\pi L}{T_{10-90}} = 3.8\ \Omega \qquad [7.11]$$

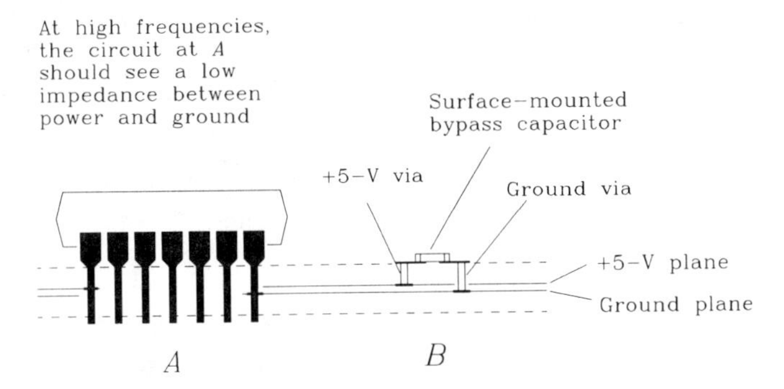

Figure 7.5 Mechanical arrangement of a bypass capacitor.

A value of 3.8 Ω may not be sufficiently low to shunt off high-frequency currents from the chip. Also remember that a bypass capacitor usually connects by a via at one end to the ground plane, and by a via at the other end to the +5-V plane, doubling the impact of via inductance. Mounting the bypass capacitors on the side of the board nearest the power and ground planes can help reduce this effect. Last, any traces leading between the capacitor pads and the vias add more inductance. These traces should always be extra fat.

It is possible to achieve very low impedances between power and ground using multiple bypass capacitors. As a rough guide for digital products, assume that the power and ground planes are perfect conductors having zero inductance. We need worry only about the inductance of our bypass capacitors, their associated traces, and the vias. Within a certain radius, all the bypass capacitors will act as if connected in parallel, lowering the power-to-ground impedance. The effective *radius* within which this effect works is equal to $l/12$, where l is the electrical length of a rising edge. All capacitors within the *diameter* of $l/6$ act in concert as a lumped circuit. Section 1.3 lists the speed of electromagnetic wave propagation in various media for the purpose of determining the electrical length l of a rising edge.

A rising edge of 1 ns propagating in FR-4 material has a length of about $l = 6$ in. No benefit will be derived in this example from a capacitor grid spaced further apart than $l/12 = 0.5$ in.

Power supply bypassing gets progressively more difficult as rise times get shorter. When the rise time shrinks, the size of the effective radius goes down. The number of capacitors within the effective radius shrinks with the *square* of the rise time. Compounding the problem, as rise time shrinks, digital knee frequency rises (see Equation 1.1), forcing the inductive reactance of each via *up*. The net result is that a particular configuration of bypass capacitors that works at one frequency is *eight times less effective* when we halve the rise time. Experience gathered working with one speed range translates easily into a new speed range with this scaling principle.

POINTS TO REMEMBER:

Via inductance degrades the shunting capability of bypass capacitors.

An array of bypass capacitors is more effective than a single bypass capacitor.

Power supply bypassing gets progressively more difficult as rise times shrink.

7.4 RETURN CURRENT AND ITS RELATION TO VIAS

In multilayer boards with more than one ground plane, we must consider carefully the issue of where ground return current flows.[5]

[5]Thanks to W. Michael King of Costa Mesa, Calif., for pointing out this effect.

Figure 5.2 illustrates the basic principle of return-current flow: *high-speed returning signal currents follow the path of least inductance.*

If we imagine that Figure 5.2 has more than one ground plane, there appears to be a choice as to which ground plane will carry the returning signal current. The solution to this puzzle (the least inductance path) is that return signal current flows along whichever ground plane is closest to the signal wire, following a path directly underneath the signal trace.

Still referring to Figure 5.2, let the ground pin of the gate at *A* penetrate through a stack of ground planes, connecting to each. Do the same for the ground connection at resistor *B*. As drawn, the signal trace is adjacent to the top ground plane, which carries all the returning signal current.

Now modify the circuit by burying the signal-routing trace between two internal ground layers. The return current is now shared between the two internal ground layers. Most of the return current flows on whichever ground is closest.

Because both the gate *A* and resistor *B* have connections to every ground plane, returning signal current can easily flow to the internal layers. The inductance of this modified ground return path is comparable to the inductance of the original path because it has a similar topology.

Next we will establish a connection between inductance and electromagnetic radiation: *The inductances being equal, we know the total magnetic flux generated by the two paths will be equal. Therefore we conclude that electromagnetic radiation from the two configurations will be equal.*

An interesting consequence of this connection is that an inner-layer trace radiates no more or less than an outer-layer trace. This is particularly true for traces near the edge of a board. The ground planes, which lie parallel to the axis of magnetic flux generation, offer almost no magnetic shielding.

Now let's add a nasty modification to the basic circuit. Let the trace from *A* go halfway to *B* along the top layer and then use a via to drop it down to an inner layer. Complete the path to *B* on a buried trace between two internal ground planes. Where does the ground return-current flow?

At the point where the signal jumps from layer to layer, there is no way for return signal current to make the jump! We have provided no connections between ground planes except at locations *A* and *B*. The ground return current must therefore do something other than follow along closely underneath the signal trace, something which invariably involves *more* inductance than the original path. We have discovered one mechanism whereby unrestricted use of vias creates additional electromagnetic interference. Not only do we radiate more, but by diverting the return signal current from its intended path, we introduce more crosstalk.

A number of solutions to the return signal current plane jumping problem come to mind, listed in decreasing order of effectiveness.

(1) Arrange the board so return currents for high-speed traces never have to jump between planes. Do this by restricting each trace to remain on whatever layer it starts.

(2) Restrict traces to remain on *either side* of whichever ground plane they start out nearest. This rule allows the use of naturally grouped horizontal and vertical routing layer pairs. It works almost as well as solution (1).

(3) Provide ground vias next to every signal via explicitly for the purpose of letting return currents jump between layers.

(4) Make sure there are plenty of ground vias everywhere. Regardless of where a signal via occurs, its return current won't have to divert far to find a place to jump layers.

Do not use guard traces to provide a nearby return-current path. This idea looks good on paper but doesn't work in practice. First of all, guard traces don't do anything until they get very close to the signal trace. Once guard traces are placed close enough to serve as an effective ground return path, they are close enough to mess up (lower) the trace impedance. Third, to provide a low enough impedance to make any difference, guard traces must be very, very wide.

After committing to using a solid ground plane, guard traces cause nothing but trouble.[6]

[6]We have seen designers insist on using guard traces during layout only to remove them at the last minute. Their temporary guard traces force other traces away from high-speed lines during routing, thus reducing crosstalk problems.

8

Power Systems

Power systems in modern digital machines serve two essential purposes:

- Provide stable voltage references for exchanging digital signals
- Distribute power to all logic devices.

This chapter examines how power systems provide stable voltage references and distribute power.

8.1 PROVIDING A STABLE VOLTAGE REFERENCE

Figure 8.1 illustrates the voltage reference problem as it occurs in *single-ended* logic systems. Logic gate A generates output voltage V_1 which propagates along wire B to the input of gate C. Gate C must determine whether the incoming logic level is one or zero. To accomplish this feat, gate C uses a differential amplifier, comparing the incoming voltage against its internal reference voltage R. We normally don't think about a gate input incorporating a differential amplifier, but it does. The topology of this differential amplifier causes our voltage reference problems.

The internal reference voltage usually connects to some combination of the power input terminals. No matter which terminals we pick, the same basic problem appears.[1] For our examples here we assume the reference is a fixed offset above ground. Including the effect of noise source N, the voltage received at the input of the differential amplifier is

$$\text{Differential input} = V_1 - N - R \qquad [8.1]$$

[1]Referring to the most positive power terminal as V_{CC} and the most negative as V_{EE}, the major logic families use the following reference voltages: CMOS, weighted average of V_{CC} and V_{EE}; TTL, fixed offset above V_{EE}; ECL, fixed offset below V_{CC}.

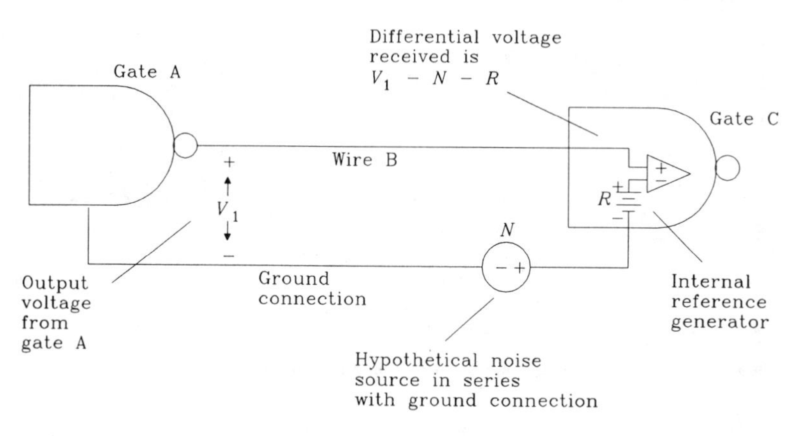

Figure 8.1 Voltage reference used with single-ended logic.

Any noise which induces a voltage difference between the ground terminals of gates A and C (like N) shows up directly at the differential amplifier as if added to the incoming signal voltage. Noise voltage N reduces the *noise margin*[2] for gate C.

What causes noise voltages between grounds? The most common cause involves returning signal currents. Whenever gate A sends a signal to gate C, the outgoing signal current returns to gate A along the power distribution wiring. The returning signal current, acting across the inductance of the ground wiring, causes noise voltages like N. Returning signal currents between any two gates, not just gates A and C, can also generate ground noise that interferes with reception at gate C. Figure 8.2 illustrates common-path noise generation. Such noise voltages are called *common-path noise* voltages.

Common-path noise voltage is the product of a returning signal current and ground impedance. To ensure low common-path noise, we must have low-impedance

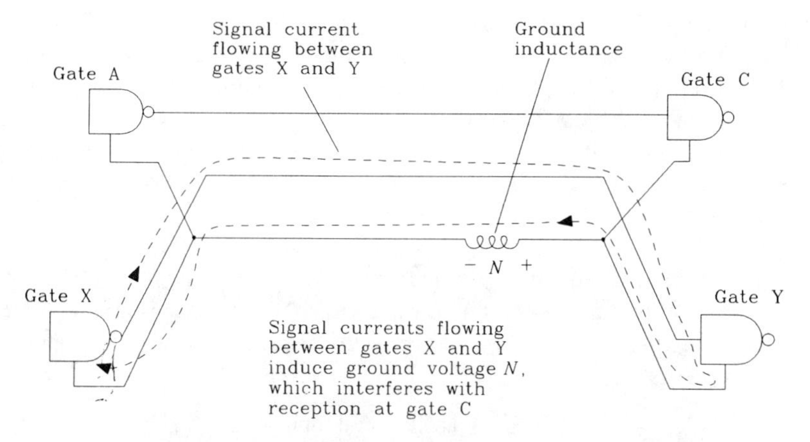

Figure 8.2 Common-path noise caused by a ground connection.

[2]Noise margin is a measure of the safety margin remaining between the worst-case transmitted logic levels and the logic levels which the receiver is guaranteed to receive.

ground connections between gates. This principle becomes our first power system design rule:

Power Rule 1. Use low-impedance ground connections between gates.

Are there structures with sufficiently low inductance to avoid problems with common-path noise? Yes. As a practical matter, a solid ground plane (even one filled with small holes) presents a remarkably low inductance to returning signal currents.

Common-path noise is related to the mutual inductive coupling described in Chapter 5. Both effects involve inductive coupling between returning signal current loops. Common-path noise differs in that we ascribe it to the lumped inductance of a particular circuit element or wire. The discussion in Chapter 5 encompasses situations in which return currents flow in nearby but separate paths which interact only through shared magnetic fields.

Low ground inductance alone does not solve the common-path noise problem. Figure 8.3 illustrates the point that even if every gate uses a perfect ground connection, common-path inductance in the power wiring can still cause trouble. Remember that in the HI state, a gate's output voltage depends on the voltage at its power terminal. Any changes in the power voltage, caused by returning signal currents flowing in the power wiring, directly affect the output voltage. The impedance between power pins on any two gates should be just as low as the impedance between ground pins. This is our second power system design rule:

Power Rule 2. The impedance between power pins on any two gates should be just as low as the impedance between ground pins.

Note that in Figure 8.3 returning signal current flows through the power supply battery. Apparently, to maintain stable transmitted signal levels, the impedance of the battery must be very low, as well as the impedance of both the ground and power

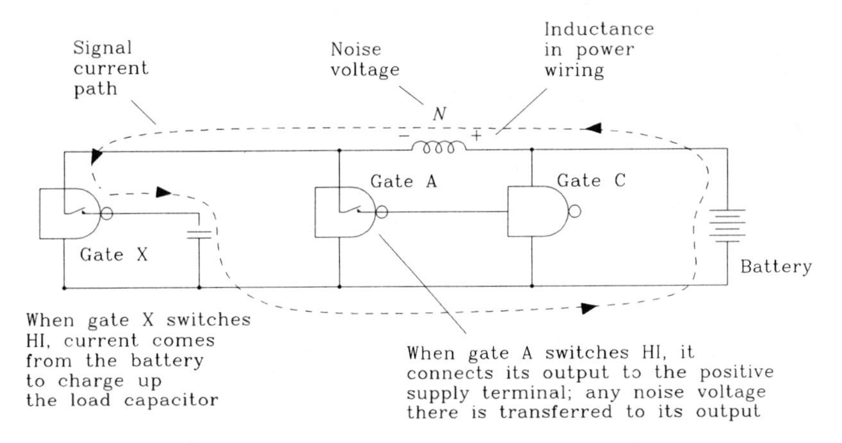

Figure 8.3 Common-path inductance in power wiring.

connections. In Figure 8.3 the only path between power and ground is through the battery. In a practical power system design there are other components which provide this low-impedance path. However it is accomplished, there must be some low-impedance path between power and ground. This is our third power system design rule.

Power Rule 3. There must be a low-impedance path between power and ground.

Any power system that satisfies our three power system design rules will have low common-path noise, and it will also distribute power everywhere at a uniform voltage. The properties of providing a stable reference voltage, having low common-path noise, and maintaining everywhere a uniform power distribution voltage are inseparable. Techniques that help with one property also help with the others.

The power system in Figure 8.4 satisfies all three rules. It does this by first providing a single ground plane which carries all returning currents. Then *bypass capacitors* are added at each gate between power and ground. The power wiring can be arbitrary. Let's check the three power design rules for this configuration:

(1) Between grounds, we have a ground plane connection.
(2) Between power terminals, we have in series the impedance of one capacitor, then the ground plane, and finally a second capacitor.
(3) At each gate, there is a bypass capacitor from power to ground. We can measure from any power to any ground point and see a low impedance.

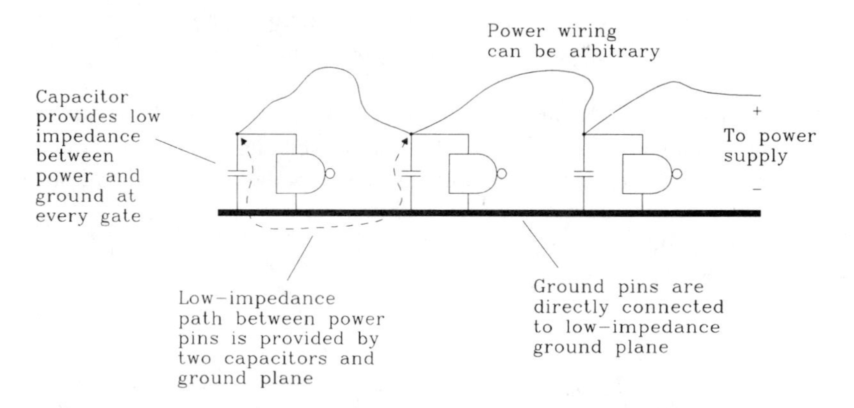

Figure 8.4 Single-plane power system.

The biggest disadvantage of the single-plane approach is the impedance of its bypass capacitors, which may not be low enough. Section 8.3 discusses the tradeoffs involved in selecting a good bypass capacitor.

A better approach (Figure 8.5) uses separate copper planes for power and ground. This guarantees an almost perfect performance between either the power terminals or ground terminals of any two gates. When the planes lie very close to each other, they share a lot of mutual capacitance. This capacitance has a very low impedance at high frequencies,

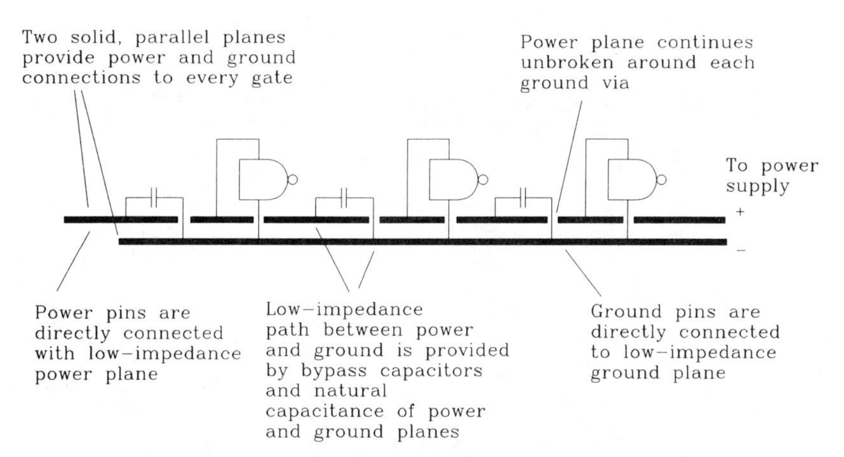

Figure 8.5 Power and ground plane system.

allowing high-frequency currents to cross back and forth easily between planes. At lower frequencies, discrete bypass capacitors at each gate help to short together power and ground.

Let's check the three power design rules for this configuration:

(1) Between grounds, we have a ground plane connection.
(2) Between power terminals, we have a power plane connection.
(3) Between power and ground, we have a combination of bypass capacitors and the natural capacitance of the power and ground planes.

Before leaving this section, take a moment to examine Figure 8.6. The *differential transmission* configuration provides a built-in return-current path for every signal wire. Not only that, but every signal carries its own reference voltage! Note that the differential amplifier in the receiver connects to neither power terminal. Differential transmission is an excellent way to manage communication between gates which do not share good power and ground connections.

Differential transmission uncouples the problem of distributing power from the problem of providing stable reference voltages.

POINTS TO REMEMBER:

Three power system design rules:

(1) Use low-impedance ground connections between gates.
(2) The impedance between power pins on any two gates should be just as low as the impedance between ground pins.
(3) There must be a low-impedance path between power and ground.

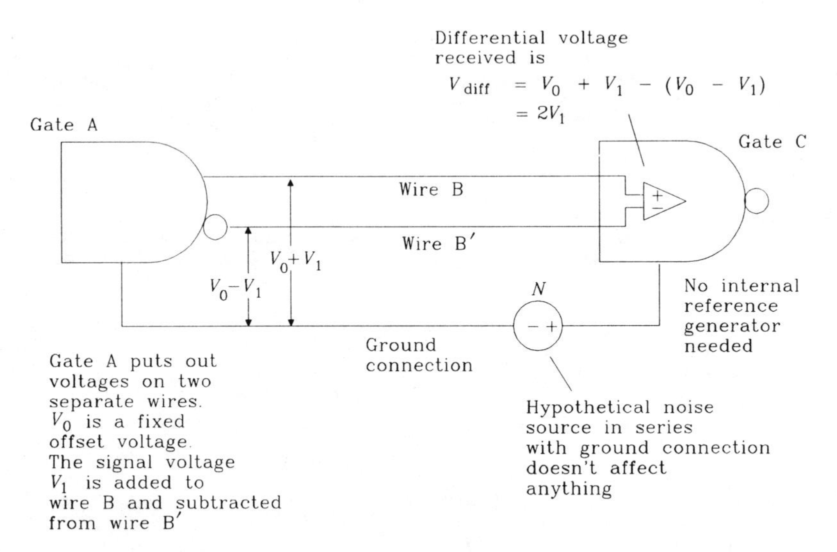

Figure 8.6 Differential signal transmission between gates.

8.2 DISTRIBUTING UNIFORM VOLTAGE

Power supplies sold for use in digital electronics have very, very low output impedances. As measured directly across their output terminals, power supplies generally satisfy rule (3). Circuits mounted directly across the power supply output terminals fully benefit from the power supply's low output impedance.

Circuits mounted anywhere else must connect to a power supply by wires, cables, or circuit board traces. The relatively large inductance of this wiring, termed *power distribution wiring*, raises the naturally low output impedance of most power supplies. Measured at the end of a power distribution cable, DC regulation may be very good, but high-frequency impedance will be too large.

In an effort to circumvent the problems created by power distribution wiring inductance, designers commonly place a single large bypass capacitor on each printed circuit card. This capacitor connects in parallel to the power supply. In the frequency range where wiring inductance starts becoming a problem, the bypass capacitor provides a low impedance between power and ground. At some even higher frequency, the bypass capacitor loses its effectiveness as a result of the inductance of its mounting leads.

To fix inadequacies in the large bypass capacitor, designers include an array of other smaller bypass capacitors on the card. The capacitor array picks up where the big bypass capacitor left off. The array has a total capacitance less than the big bypass capacitor but a much better series inductance.

Working together, the power supply, its wiring, the big bypass capacitor, and the small bypass capacitor array provide a low-impedance power source for every logic device across the entire operating frequency range. The combination of power distribution wiring, a big bypass capacitor, and an array of little bypass capacitors is called a *multilayered power distribution* system.

Sections 8.2.1–8.2.5 build up the theory behind multilayered power distribution systems. Section 8.2.6 describes how to measure the performance of a completed power distribution system.

8.2.1 Resistance of Power Distribution Wiring

The wires leading from a power supply to the logic it feeds may include appreciable resistance. This resistance induces a voltage drop across the wiring proportional to the operating current. If this voltage drop is too high, it might cause the supply voltage at the logic gates to fall outside their specified operating range.

The resistance of wiring is easy to calculate, as is the expected operating current. You should always determine beforehand whether wiring resistance will present a problem.

If wiring resistance is a problem, use a bigger wire. Resistance being proportional to the inverse square of diameter, only a 40% increase in diameter cuts the resistance by one-half.

Many new regulated power supplies include a provision for remote sense wires. Once connected, these sense wires inform the supply of the output voltage *as measured at the far end of its distribution wiring*. The supply can then take corrective action to adjust for the wiring resistance. Such a supply will include a specification for the maximum amount of wiring voltage drop it can accommodate. A specified drop of 1/2 V is typical. With such a supply, we don't have to use low-resistance cable.

POINT TO REMEMBER:

Sense wires correct for resistance in power distribution wiring.

8.2.2 Inductance of Power Distribution Wiring

Inductance in power wiring presents a much harder problem than resistance. Rapidly changing currents, acting across the inductance of power distribution wiring, induce voltage shifts between the supply and the logic it feeds. These voltage shifts are more sudden and far larger than shifts introduced by wiring resistance.

Unfortunately, sense wire circuitry cannot respond quickly enough to correct for wiring inductance.

There are three approaches to dealing with the power wiring inductance problem:

- Use lower-inductance wiring.
- Use logic immune to power supply noise.
- Reduce the size of changing power supply currents.

Since inductance is a logarithmic function of diameter, it is almost impossible to reduce wiring inductance by simply using a bigger wire. Equation 8.2 expresses the inductance of two parallel power distribution wires (power and ground):

$$L = 10.16X \ln\left(\frac{2H}{D}\right) \quad [8.2]$$

where X = length of wire, in.
H = average separation between wires, in.
D = wire diameter, in.
L = inductance, nH

According to Equation 8.2, even ridiculously large wires have too much inductance. Wide, flat parallel structures work much better as distribution wiring than round wires. The lowest-inductance distribution wiring uses multiple parallel flat ribbons, with power and ground on alternating layers. Equation 8.3 expresses the inductance of a stack of parallel flat ribbons.

$$L = 31.9\frac{XH}{W(N-1)} \quad [8.3]$$

where X = length of ribbon, in.
H = separation between ribbon plates, in.
W = width of ribbon, in.
N = number of plates (count 2 for single power and ground; count 3 for double ground and one power, etc.)
L = inductance, nH

Differential transmission (see Figure 8.6) is practically immune to power supply fluctuations. For communication between circuit cards where there is no cheap way to provide low-impedance power distribution, differential drivers and receivers excel. The cost and extra space required for differential transmission is often less than the cost and space needed for improved power distribution cabling.

A final approach to reducing the impact of power supply wiring inductance involves reducing the magnitude of changing currents. Note that we use the word *changing* currents. We cannot reduce the average current flow through power wiring, but we can surely reduce the *rate of change* of current. The next section shows how board-level bypass capacitors accomplish this goal.

POINTS TO REMEMBER:

It is almost impossible to reduce wiring inductance by simply using a bigger wire.

Wide, flat parallel structures work much better as distribution wiring than round wires.

Differential transmission (see Figure 8.5) is practically immune to power supply fluctuations.

8.2.3 Board-Level Filtering

Let's see how bad the wiring inductance problem can get. We will calculate the maximum *dI/dt* in the circuit in Figure 8.7. Then we will multiply that *dI/dt* by the wiring inductance to estimate the power supply noise voltage.

The circuit in Figure 8.7 drives a big capacitive load. The power supply current requirement at the power pin of gate A consists of a big spike every 100 ns. The current spikes correspond to the times when gate A drives the capacitive load HI. The current path for the HI drive operation appears as a dashed line.

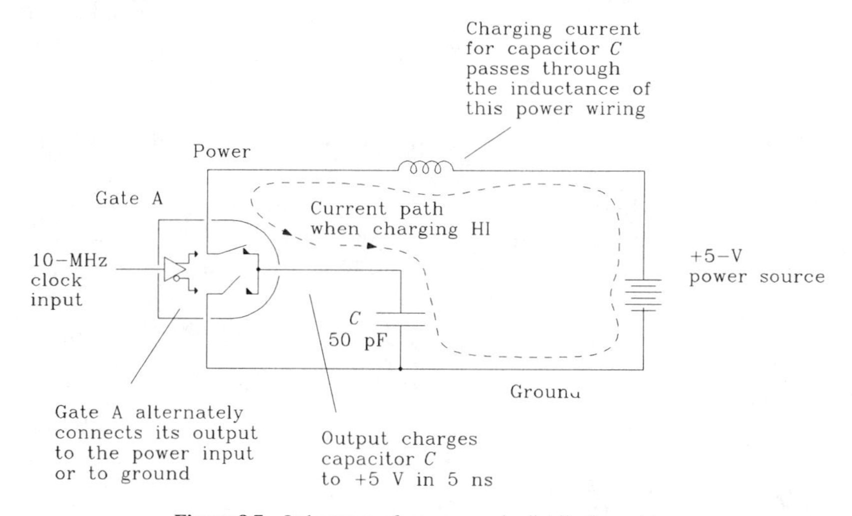

Figure 8.7 Inductance of power supply distribution wiring.

Figure 8.7 indicates that the HI drive current circulates through the power supply and through the power wiring inductance. The rise time of gate A is 5 ns, and so we can compute the maximum *dI/dt* of this drive current using Equation 2.42:

$$\text{Max}\frac{dI}{dt} = \frac{1.52\ \Delta V}{\left(T_{10-90}\right)^2} C_1 = 1.5 \times 10^7\ \text{A/s} \qquad [8.4]$$

$$\Delta V = 5\ \text{V (drive voltage)}$$
$$T_{10-90} = 5\ \text{ns (drive rise time)}$$
$$C_1 = 50\ \text{pF (load capacitance)}$$

Next we need to calculate the inductance of the power supply wiring, using Equation 8.2

$$L = 10.16X \ln\left[\frac{2H}{D}\right] = 164\ \text{nH} \qquad [8.5]$$

where $X = 10$ in. (length of wire)
$H = 0.1$ in. (average separation between wires)
$D = 0.04$ in. (wire diameter, 18 AWG)
$L =$ inductance, nH

Multiplying the maximum dI/dt by the inductance, we find the peak noise voltage:

$$\text{Noise} = \left(1.5\times10^{7}\right)\left(164\times10^{-9}\right) = 2.5\text{ V} \qquad [8.6]$$

This is a ridiculous answer! Do we really get that much noise?

Actually, we have an even worse problem. Equation 8.6 doesn't work properly because of a flawed assumption. In Equation 8.5 we *assumed* a rise time equal to 1 ns. In this circuit, the power supply inductance is so large that when gate A tries to drive HI, the power supply input at the card will droop near zero, slowly rising as capacitor C_1 is charged through the inductance of the power supply wiring. When the power supply droops, gate A may no longer function, or it may break into oscillation.

The solution to this power supply droop problem is to install a bypass capacitor as shown in Figure 8.8. If the impedance of capacitor C_2 is lower than the impedance of the power wiring, changing currents will flow through it instead of through the wiring. The power droop experienced by gate A when switching HI will then be a function of the impedance of capacitor C_2, not the power wiring.

Current flow in the power wiring in Figure 8.8 is smoothed by capacitor C_2 into a continuous average value. We have successfully reduced the rate of change of current flowing in the power wiring. This is a major result. We have hit on the idea of building a staged power distribution system. The power supply provides low impedance at low frequencies. The local bypass capacitor provides low impedance at higher frequencies.

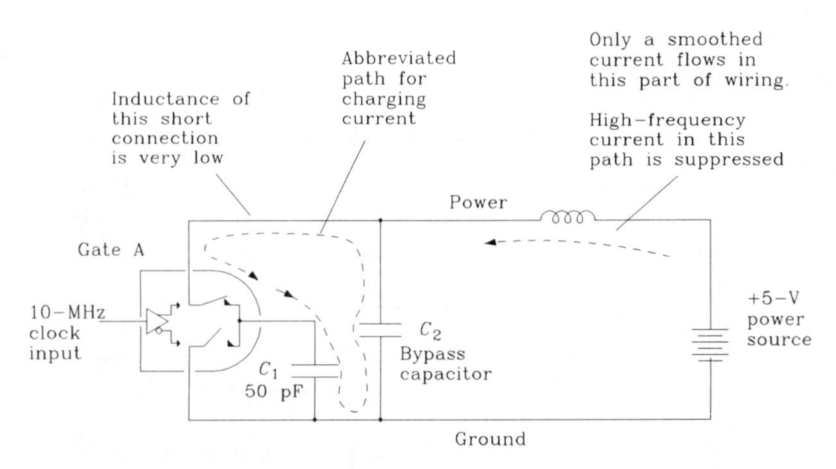

Figure 8.8 Bypass capacitor smoothes current flowing in power wiring.

Determining the correct value of bypass capacitance involves several steps:

(1) Figure the maximum step change in supply current (ΔI) expected on the board. We don't know when the gates will switch, so assume that in the worst case they all switch together at some fixed frequency.

(2) Figure the maximum amount of power supply noise your logic can tolerate (ΔV). Derate that value to leave a safety margin.

(3) The maximum common-path impedance we can tolerate is $X_{max} = \Delta V/\Delta I$. If we are using solid power and ground planes, we can allocate the entire X_{max} impedance to the connection between power and ground. Otherwise we must split it up, allocating part to ground connections, part to power connections, and part to the connection between power and ground.

$$X_{max} = \frac{\Delta V}{\Delta I} \quad [8.7]$$

(4) Figure the inductance of the power supply wiring L_{PSW}. Combine that with the maximum allowable impedance X_{max} to find the frequency below which the power supply wiring is adequate. If the gates all switch together at this frequency, we will get power supply noise less than ΔV.

$$F_{PSW} = \frac{X_{max}}{2\pi L_{PSW}} \quad [8.8]$$

(5) Below frequency F_{PSW} the power supply wiring is fine. Above frequency F_{PSW} we need a bypass capacitor to take over. Find the value of capacitance that has impedance X_{max} at frequency F_{PSW}. Use a bypass capacitor at least that big.

$$C_{bypass} = \frac{1}{2\pi F_{PSW} X_{max}} \quad [8.9]$$

EXAMPLE 8.1: Board-Level Bypass Capacitor

Liz has designed a CMOS circuit card having 100 gates each switching 10-pF loads in 5 ns. The power supply inductance is 100 nH. Find the right value of bypass capacitor.

$$\Delta I = NC\frac{\Delta V}{\Delta t}$$

$$= 100(10\text{ pF})\frac{5\text{ V}}{5\text{ ns}} \quad [8.10]$$

$$= 1\text{ A (worst case peak while charging all loads)}$$

$$\Delta V = 0.100\text{ V (from noise margin budget)} \quad [8.11]$$

$$X_{max} = \frac{\Delta V}{\Delta I} = 0.1\ \Omega \quad [8.12]$$

$$L_{PSW} = 100\text{ nH} \quad [8.13]$$

$$F_{PSW} = \frac{X_{max}}{2\pi L_{PSW}} = 159\text{ kHz} \quad [8.14]$$

$$C_{bypass} = \frac{1}{2\pi F_{PSW} X_{max}} = 10\ \mu\text{F} \quad [8.15]$$

Bypass capacitors in the range of 10–1000 μF are common on digital printed circuit cards.

The naturally low output impedance of a power supply and its associated wiring work to prevent power supply noise at frequencies up to F_{PSW}. Above frequency F_{PSW}, local bypass capacitors prevent power supply noise. At some even higher frequency F_{bypass}, the bypass capacitors will stop working. What causes that to happen and what to do about it are the subject of the next section.

> **POINTS TO REMEMBER:**
>
> A power supply provides low impedance at low frequencies.
> Local bypass capacitors provide low impedance at higher frequencies.

8.2.4 Local Filtering at Individual Integrated Circuits

Every printed circuit card needs a relatively large bypass capacitor to counteract the inductance of power distribution wiring. A single perfect bypass capacitor on each card could completely solve the distribution problem.

Unfortunately, no capacitor is perfect. Every discrete capacitor has some finite series lead inductance L_{C2}, which causes its impedance to go up, not down, at very high frequencies. Whether lead inductance is a problem or not depends on the digital knee frequency F_{knee} (see Equation 1.1) and the impedance X_{max} which you must attain.

We can calculate the highest frequency at which a given bypass capacitor is effective:

$$F_{bypass} = \frac{X_{max}}{2\pi L_{C2}} \qquad [8.16]$$

A properly sized bypass capacitor will be effective between frequencies F_{PSW} and F_{bypass}. Hopefully, there is a big gap between the two frequencies.

EXAMPLE 8.2: Highest Effective Frequency of a Bypass Capacitor

From Example 8.1, assume the 10-µF capacitor has a series inductance of L_{C2} = 5 nH. We were working to achieve an X_{max} of 0.1 Ω.

Find the maximum frequency at which it is effective.

$$F_{bypass} = \frac{X_{max}}{2\pi L_{C2}} = 3.18 \text{ MHz} \qquad [8.17]$$

This capacitor is effective from 159 kHz (see Example 8.1) to 3.18 MHz, a range of about 16:1.[3]

One big bypass capacitor allowed us to reach the frequency F_{bypass}. To guarantee low impedance above F_{bypass}, we need another capacitor with lower series inductance.

[3]Other factors, such as equivalent series resistance discussed in Section 8.3.2, can cause a capacitor to be ineffective.

The best way to get very low inductance is to parallel a lot of small capacitors. Sprinkle the parallel array of bypass capacitors around the circuit card.

Three factors will dominate the impedance between power and ground:

- At low frequencies, by the inductance of power distribution wiring
- At middle frequencies, by the impedance of a card-level bypass capacitor
- At high frequencies, by the impedance of a distributed capacitor array.

The procedure below steps though the design of a bypass capacitor array. This procedure looks a lot like the procedure in Section 8.2.3. The difference is that in the last section we fixed the inductance of the power supply distribution wiring, while here we design to fix the series inductance of our local bypass capacitor.

(1) We want the system to work up to F_{knee}. Calculate how much inductance we can tolerate at that high frequency (see Equation 1.1 for a definition of knee frequency):

$$L_{tot} = \frac{X_{max}}{2\pi F_{knee}} = \frac{X_{max} T_r}{\pi} \quad [8.18]$$

(2) Look up (or measure) the series inductance of the bypass capacitors you plan to use, L_{C3}. A typical series inductance value for surface-mounted capacitors with very short, fat vias is 1 nH. A typical series inductance value for through-hole bypass capacitors is 5 nH. Use this value to figure how many bypass capacitors are needed to meet the total inductance goal.

$$N = \frac{L_{C3}}{L_{tot}} \quad [8.19]$$

(3) The total array capacitance must have an impedance less than X_{max} at frequencies down to F_{bypass}. Use this fact to calculate the total array capacitance.

$$C_{array} = \frac{1}{2\pi F_{bypass} X_{max}} \quad [8.20]$$

(4) Calculate the capacitance of each element in the array.

$$C_{element} = \frac{C_{array}}{N} \quad [8.21]$$

EXAMPLE 8.3: Capacitor Array

Let's use numbers from Examples 8.1 and 8.2. The bypass capacitor is a 10-μF capacitor having a series inductance of 5 nH. We want to achieve $X_{max} = 0.1\ \Omega$.

$$X_{max} = 0.1\ \Omega \qquad \text{(from last section)} \quad [8.22]$$

$$T_r = 5\ \text{ns} \quad [8.23]$$

$$L_{tot} = X_{max} \frac{T_r}{\pi} = 0.159\ \text{nH} \quad [8.24]$$

$$L_{C3} = 5 \text{ nH} \qquad \text{(using through-hole capacitors)} \tag{8.25}$$

$$N = \frac{L_{C3}}{L_{\text{tot}}} = 32 \qquad \text{(number of caps required)} \tag{8.26}$$

$$F_{\text{bypass}} = 3.18 \text{ MHz} \qquad \text{(from Example 8.2)} \tag{8.27}$$

$$C_{\text{array}} = \frac{1}{2\pi F_{\text{bypass}} X_{\text{max}}} = 0.5\ \mu\text{F} \tag{8.28}$$

$$C_{\text{element}} = \frac{C_{\text{array}}}{N} = 0.016\ \mu\text{F} \tag{8.29}$$

We require an array of 32 capacitors each having a capacitance of 0.016 μF and a series inductance of 5 nH or less.

POINT TO REMEMBER:

The best way to get very low inductance is to parallel a lot of small capacitors.

8.2.5 Capacitance of Power and Ground Planes

Parallel power and ground planes provide a third level of bypass capacitance. The power-to-ground-plane capacitance has zero lead inductance and no ESR (see Section 8.3). It helps reduce power and ground noise at extremely high frequencies. The capacitance between power and ground planes is

$$C_{\text{power plane}} = \frac{0.225\, \varepsilon_r A}{d} \tag{8.30}$$

where ε_r = relative electric permeability of insulator (use 4.5 for FR-4 epoxy circuit boards)

A = area of shared power-ground planes, in.^2

d = separation between planes, in.

$C_{\text{power plane}}$ = capacitance between planes, pF

Power and ground planes separated by 0.01 in. of FR-4 have a capacitance of 100 pF/in.^2

Figure 8.9 plots, as a function of frequency, the impedance of the various elements involved in the power system. Note that Figure 8.9 includes the effect of parasitic series resistance, also called *equivalent series resistance* (ESR), for capacitors C_2 and C_3. The ESR effect is described in Section 8.3.

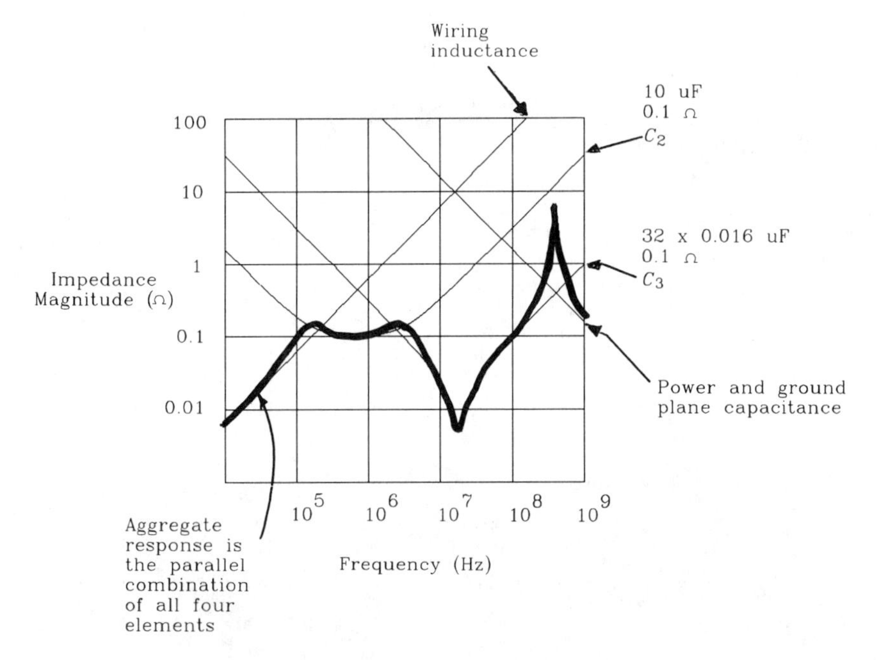

Figure 8.9 Impedance magnitude of a bypass capacitor design.

> POINT TO REMEMBER:
>
> Power and ground planes separated by 0.01 in. of FR-4 have a capacitance of 100 pF/in.^{2}

8.2.6 A Test Jig for Measuring the Step Response of a Power Distribution System

The test jig in Figure 8.10 applies a small current step to the power system and looks to see what the reaction is. The output impedance of this probe arrangement is 25 Ω (50 from the pulse generator in parallel with 50 from the scope).

Set your pulse generator to the same rise time expected in the actual system. Set the output step size (when loaded by the 50-Ω input impedance of your scope) to 5 V. The output current step will then be 5 V/25 Ω = 0.2 A. Scale the measured step response by a factor of $\Delta I/0.2$ to determine the response of the power system to a current step of ΔI amps.

When working with a finished system, apply this test with the power turned on. Disconnect all the clocks to stop the local logic from operating. This reduces the noise

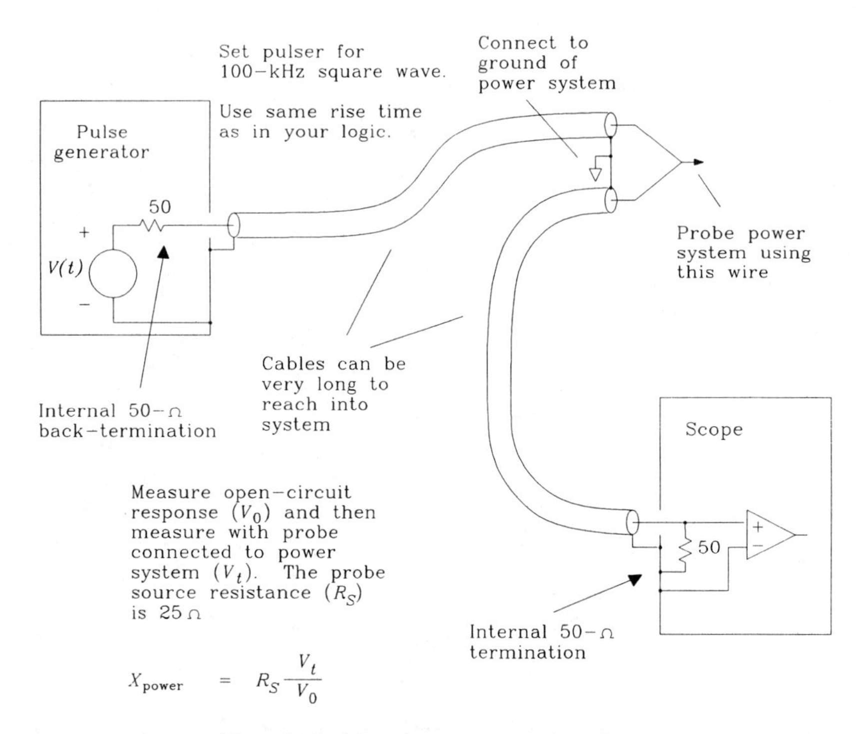

Figure 8.10 Measuring power system impedance.

level on the board so you can accurately measure the very small signals created by this test setup.

If you cannot turn off the clocks, then try using a digital scope. The Tektronix 11403 can perform averaging to pull a weak signal out from much larger random noise. To use this feature, connect the pulse generator trigger output to the scope as a trigger input. Leave the test circuit connected to the scope but disconnect the pulse generator signal output.[4] Then invoke the scope averaging feature. It should now begin averaging the power supply noise in your circuit synchronously with the pulse generator trigger signal. Since the pulse generator trigger is not synchronized to the circuit under test, the averaged signal value will be zero.

Next connect the pulse generator signal output to the test circuit and observe the averaged power system step response.

> **POINT TO REMEMBER:**
>
> A simple test jig measures power supply step response.

[4]Leave the pulse generator trigger output connected to the scope.

8.3 EVERYDAY DISTRIBUTION PROBLEMS

If your power system suffers from one of the symptoms listed below, try these helpful hints. This is not a complete compendium of solutions but should help get you started.

8.3.1 Random ECL Errors in a Combined TTL-ECL System

Combining TTL and ECL in one system without considering the system design consequences is not a good idea.

TTL circuitry induces more noise on the power line than ECL, while ECL is more sensitive to power fluctuations. A typical symptom is that the ECL experiences random errors.

Hints:

- First make sure that TTL and ECL signals stay clear of each other. This solves direct crosstalk problems. Maintain a separation at least equal to eight times the trace height above ground.
- If you are using +5 V for TTL and –5.2 V for ECL, you have a leg up. The power systems are already separated. Assuming there exists a solid ground plane, TTL noise has little chance of leaking into the ECL system. If there is no solid ground plane, get one. A competent layout house can add a ground plane with little effort. Then lay out the design again with the ground plane in place to see if it works.
- Some designers use +5 V for both TTL and ECL. This is not the optimal working voltage for ECL circuits, but they function. If possible, cut off the clocks in the TTL area to determine whether noise leaking from the TTL section causes ECL errors.
- To reduce noise leakage, physically separate the TTL and ECL sections of the design. Then cut the +5-V plane (but not ground) in two, bifurcating the printed circuit card into separate areas for TTL and ECL. The main power entrance to the board should be on the TTL side. Leave the ground plane intact. Make sure no long signal traces cross the boundary between the two +5-V areas. Then connect a 1-μH inductor of sufficient current-carrying capacity between the two +5-V planes. This will limit the amount of TTL noise that can enter the ECL system.
- For maximum reliability, use differential signal transmission between the two sections.

8.3.2 Too Much Voltage Drop in Distribution Wiring

When sourcing power to multiple cards over long wires, there is often no one correct location for attaching the power supply remote voltage sense wires.

If the power distribution wiring has too much resistance, the local voltage at each card will differ.

Ideas:

- Distribute DC power in raw form and re-regulate locally on each card. This requires local regulation circuitry on every card. Use +8-V distribution with a linear regulator on each card. Use +40-V distribution with a switching power supply regulator on each card.
- Distribute DC power in regulated form but at high voltage. This reduces the power drop associated with each wire as a result of the lessened power currents. Then build a DC-DC converter on each card. If the DC-DC regulator is sufficiently stiff (i.e., low natural output impedance), you will not have to re-regulate.
- Distribute high-voltage alternating current in regulated multiphase form with flat-topped waveforms. Use (at least) two transformers locally at each card to rectify the alternating current. With proper design, the flat-topped waveforms will overlap in such a way that little output filtering capacitance is required. A high-frequency flat-topped waveform does not require very large transformers. A mechanism similar to a car alternator can be manufactured to supply flat-topped waveforms. Use the field winding to regulate the output voltage and put a big flywheel on the shaft to ride through brief power outages.

8.3.3 Power Glitches When Plugging in Cards

Some systems must allow users to insert and remove circuit cards while in operation. When a card plugs into a working backplane, it draws a huge surge of current as its local bypass capacitor charges to full potential. This current draws mostly from other bypass capacitors on other cards. The result is an inevitable glitch on the power bus.

Hints:

- Use the smallest bypass capacitor practical on each card. Attach a giant bypass capacitor (or an array of giant capacitors) directly to the backplane. This works only if the lead inductance of the local bypass capacitor plus power distribution wiring (including the connector) on each card is much larger than the lead inductance of the giant-backplane capacitance.
- Put some intentional inductance in series with the power pin on each circuit card. Attach a giant bypass capacitor (or an array of giant capacitors) directly to the backplane. This works better than the previous idea because we have increased the inductance of each card.
- Use an active circuit to apply power slowly to each card. Build the active circuit on each card, using a large switching FET. The FET slowly applies power, decreasing the dI/dt and thus reducing the power supply glitch. A 10-μs charge time cleans up most problems.
- A slow-start FET switch, also called a soft-start circuit, often introduces too much voltage drop. To fix the voltage drop, use two power pins on the card.

Design the power pins to contact at different times as the card is inserted into its slot. The first pin connects to the FET slow-start circuit and charges the card to about 4.5 V. The second pin completes a direct connection to the power bus, fully charging the card to +5 V. The second glitch is unprotected, but of a magnitude one-tenth of what it would have been otherwise.

8.3.4 EMI Radiating From the Power Distribution Wiring

Changing currents in the power wiring easily radiate away from a digital product. This electromagnetic radiation can exceed federally mandated limits.

Hints:

- Use better bypass capacitors to limit the changing currents leaking from each card.
- Put a common mode choke in series with the power distribution wiring to limit common mode currents flowing in the wiring.
- Route the wires closer together, limiting the area for magnetic radiation.
- Cover the power distribution wires with a solid metal shield attached to chassis ground at each end.

POINTS TO REMEMBER:

Combining TTL and ECL in one system without considering the system design consequences is not a good idea.

If the power distribution wiring has too much resistance, the local voltage at each card will differ.

When a card plugs into a working backplane, it draws a huge surge of current as its local bypass capacitor charges to full potential.

Changing currents in the power wiring easily radiate away from a digital product.

8.4 CHOOSING A BYPASS CAPACITOR

Bypass capacitors are riddled with imperfections.

Every capacitor includes a parasitic series inductance, called the *lead inductance*, *package inductance*, or *mounting inductance*. The consequences of this inductance are described in Section 8.2.

Every capacitor also includes a parasitic series resistance, called the *equivalent series resistance* (ESR). It acts like the lead inductance to defeat the effectiveness of a capacitor. ESR is a real-valued impedance (not imaginary like inductance) and is not a strong function of frequency. It acts just like an ordinary resistor bonded in series with the capacitor.

Every bypass capacitor is temperature-sensitive. The dielectric properties can change significantly with temperature, leading to wide swings in capacitance.

Every bypass capacitor blows up or shorts out if exposed to too high a voltage.

The following sections detail these imperfections.

8.4.1 Equivalent Series Resistance and Lead Inductance of a Capacitor

Equivalent series resistance acts like a resistor in series with a capacitor. Lead inductance acts like an inductor in series with the same capacitor. Together, they degrade a capacitor's effectiveness as a bypass element. The full equation for capacitor impedance as a function of frequency is

$$X(f) = \left[\text{ESR}^2 + \left(\frac{-1}{2\pi f C} + 2\pi f L \right)^2 \right]^{1/2} \qquad [8.31]$$

where ESR = equivalent series resistance, Ω
C = capacitance, F
L = lead inductance, H
$X(f)$ = impedance magnitude, Ω at frequency f, Hz

Equation 8.31 calculates the plots in Figure 8.9 for capacitor C_2 and the array C_3. Figure 8.9 assumes an ESR of 0.1 Ω for C_2 and for each element of the C_3 array, a total circuit board area of 10 in.^{2}, and a 0.01-in. FR-4 dielectric between power and ground.

Figure 8.9 shows a resonance in the bypass circuit at about 300 MHz, which occurs because of the lead inductance of the capacitor array and the capacitance between the power and ground planes. Since the digital knee frequency (see Equation 1.1) in this design is down at 100 MHz, there is nothing to worry about. If your digital knee frequency is higher, try using a surface-mounted capacitor array. Their lower inductance raises the resonance frequency and shrinks its amplitude.

ESR does not always appear on a manufacturer's data sheet, but it is extremely important. Do not believe what the salesperson tells you about this specification; get it in writing.

To measure ESR, use the same measurement setup shown in Figure 6.14 for measuring the inductance of a terminating resistor.

When we put a bypass capacitor C across the device under test (DUT) terminals in Figure 6.14, we expect a nice, clean RC rise time. With a big source resistance, perhaps 1000 Ω, that is precisely what we will get. With the small source resistance of the test setup in Figure 6.14, we get a much different picture. The RC rise time quickens, and the effects of lead inductance and ESR become exaggerated. By examining the first few nanoseconds of step response, we can directly measure the effects of lead inductance and ESR. Source resistances on the order of 1 Ω and speeds in the nanosecond range are common in digital bypassing applications, and so this is a reasonable way to look at our bypassing components.

Figure 8.11 plots the actual step response from a 0.1-μF bypass capacitor. The response is plotted at both 10 ns/division and 2 ns/division. Both plots superimpose both the open-circuit response of the test jig and the response to the capacitor under test.

The step response shows three distinct features: a spike, a step, and a slow overall ramp. By properly interpreting these features we can determine the lead inductance, the ESR, and the capacitance of the device under test.

(1) During the first 2 ns there is a short spike. The spike results from the action of lead inductance. We can use the area under the spike to estimate the lead inductance.

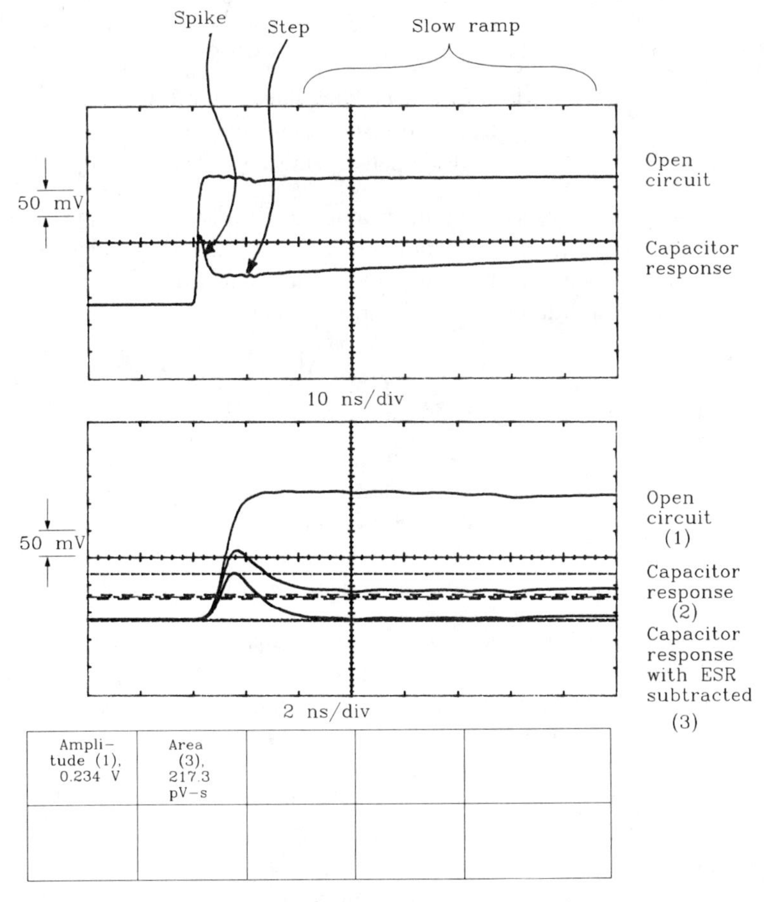

Amplitude (1), 0.234 V	Area (3), 217.3 pV–s			

Figure 8.11 Step response of a bypass capacitor.

$$L = \frac{R_S A}{\Delta V} \quad [8.32]$$

where R_S = source resistance of test jig, Ω
A = area under spike (see notes below), V-s
ΔV = open-circuit step voltage of test jig, V
L = lead inductance, H

(2) Directly after the spike, the waveform is relatively flat and offset above zero. This formation is caused by the ESR of the capacitor. At this moment the capacitor has not yet started to charge. A good model for the capacitor at this moment includes only the ESR tied directly to ground. The resistor divider formed by the source impedance of the test jig working into the ESR of the capacitor causes a voltage here roughly proportional to the ESR.

$$\text{ESR} = \frac{R_S X}{\Delta V - X} \quad [8.33]$$

where R_S = source resistance of test jig, Ω
X = measured step voltage after spike, V
ΔV = open-circuit step voltage of test jig, V

(3) After the step takes hold, it ramps up slowly. This is the effect of the capacitor slowly charging. The charge rate *dV/dt* is equal to the charging current divided by the capacitance. The charging current is equal roughly to the test circuit open-circuit voltage divided by the source resistance.

$$C = \frac{\Delta V - X}{R_S (dV / dt)} \quad [8.34]$$

where R_S = source resistance of test jig, Ω
X = measured step voltage after spike
ΔV = open-circuit step voltage of test jig, V
dV / dt = charge rate of ramp, V/s
C = capacitance, F

When watching the spike, remember that both the lead inductance and the ESR are active at that moment. If you calculate the ESR first, you can then subtract out its effect when measuring the area under the spike. The three traces shown on a scale of 2 ns/division in Figure 8.11 are the test jig open-circuit waveform, the raw measured response, and a copy of the raw measured response minus a scaled copy of the open-circuit waveform. This subtraction accounts for the effect of ESR. The subtraction and area measurement were carried out using a Tektronix Model 11403 digital oscilloscope.

The response in Figure 8.11 indicates a lead inductance of 4 nH, an ESR of 1.1 Ω, and a capacitance of 0.072 μF.

POINTS TO REMEMBER:

Lead inductance acts like an inductor in series with a capacitor.
ESR acts like a resistor in series with a capacitor.
Together they degrade a capacitor's effectiveness as a bypass element.

8.4.2 Relation of Capacitor Performance to Packaging

Go and buy many capacitors of equal capacitance and voltage rating, using the same dielectric, from different manufacturers. Surprisingly, they come in a very wide variety of shapes and sizes.

For large-valued capacitors (10 μF and up), smaller packages have higher series inductance and ESR than larger packages. Don't plan on buying the miniature package size without first checking to see if the ESR and lead inductance are acceptable.

For small bypass capacitors we can't tell much by looking at the package.

Capacitor performance varies a lot. Table 8.1 lists typical ESR and lead inductance values for various sample capacitors purchased through a typical electronic distributor. Items 1–5 are grab bag parts, representative of what your purchasing department will probably buy unless you specify otherwise. All have capacitance in the range of 0.1 – 0.47 μF. All are billed as "digital bypass capacitors." Items 1 and 2 have a miserably large ESR. Item 3 has an unusually large lead inductance. Items 4 and 5 are about as good as leaded bypass capacitors get.

Item 6 is a bypass capacitor already mounted in a DIP socket (see Figure 8.12). The axial capacitor stretches directly between pins 8 and 16 of the socket. The manufacturer of this component touts its performance, stating that the capacitor lead length is as short as possible. True enough, but bonding the capacitor directly to the power and ground planes is a better arrangement. This is proved in line 7, where the authors removed the

TABLE 8.1 CAPACITOR PERFORMANCE

Capacitor	Lead spacing (in.)	ESR (Ω)	Lead inductance (nH)	Comment
1	0.4	1.1	4	Low profile
2	0.3	0.5	6	Yellow
3	0.4	0.1	10	Fat legs
4	0.3	<0.1	7	DIP 0.3-in. type
5	0.2	<0.1	6	Square body
6	0.7	0.2	16	DIP socket
7	0.3	0.2	6	Same as item 6 but removed from socket
8	0.1	0.1	1.1	SMT 1206

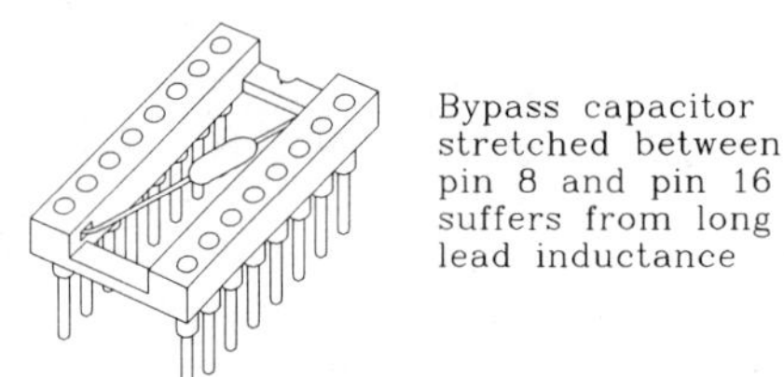

Figure 8.12 Capacitor mounted in a DIP socket.

capacitor from its DIP frame and bonded it as shown in Figure 8.13 to two strips of copper representing power and ground planes. Then, from a distance 0.7 in. away they measured its lead inductance. The measured inductance fell from 16 nH (in the socket) to 6 nH (bonded to power and ground planes).

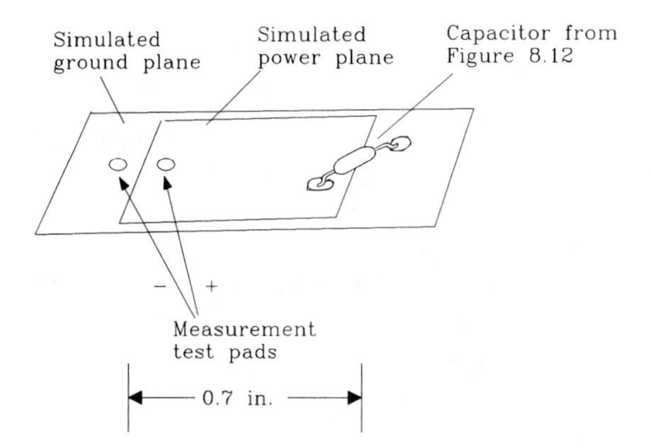

Figure 8.13 Same capacitor mounted with power and ground planes.

Item 8 is a 1206 surface-mounted capacitor.

Figure 8.14 diagrams impedance magnitude versus frequency for capacitors in Table 8.1. This chart combines information about the ESR, inductance, and capacitance of each part.

The top half shows capacitors 1, 2, and 8. All three have the same capacitance, and they are indistinguishable below 1 MHz. Around 10 MHz, the different ESR values are clearly visible. Above 100 MHz, only the lead inductance matters.

The bottom half of Figure 8.14 shows capacitors 6 and 7. This is actually the same capacitor mounted two different ways. The difference in lead inductance makes an 8-dB difference in the impedance above 10 MHz.

POINTS TO REMEMBER:

For large-valued capacitors, smaller packages have higher series inductance and ESR than larger packages.

Capacitor performance varies widely.

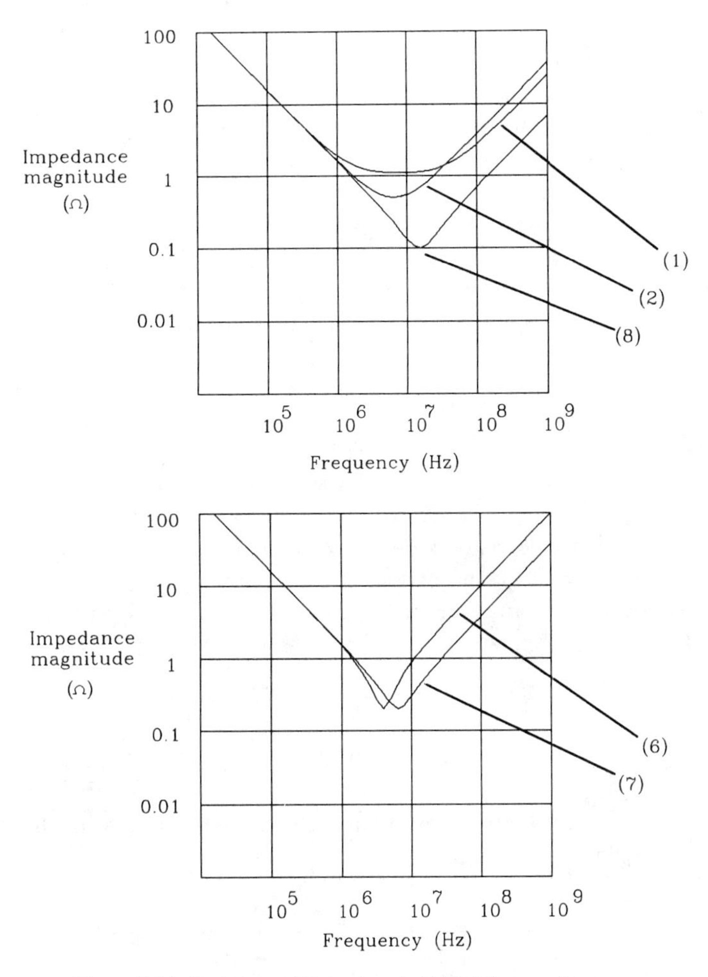

Figure 8.14 Impedance of capacitors in Table 8.1 versus frequency.

8.4.3 Surface-Mounted Capacitors

Surface-mounted capacitors solder directly to a circuit board with no intervening leads. This cuts lead inductance a lot.

Surface-mounted package sizes have names associated with their length and width. A package 0.12 in. long and 0.06 in. wide is termed a 1206 package. Other popular sizes are 1210 (0.12×0.10 in.) and 0805 (0.08×0.05 in.).

Standard 1206-sized surface-mounted bypass capacitors have excellent characteristics compared to leaded varieties. The ESR of surface-mounted capacitors may not be lower, but the lead inductance drops to the range of 1 nH.[5] Smaller packages such as the 0805 package have even less series inductance.

[5]Standard 1206 package.

When using any surface-mounted bypass capacitor, don't destroy its effectiveness by connecting it to a long, skinny via leading down to the power or ground plane. Use a larger via, or multiple vias, for connecting bypass components. Also, use as short and as fat a trace as possible leading from the via to the capacitor.

Surface-mounted capacitors can save a lot of board area when mounted on the back side of a circuit board. This requires extra manufacturing steps and costs extra. When space is precious, the space savings are well worth the cost.

When mounting components on the back side of any printed circuit board, determine whether your manufacturing shop will use the reflow- or wave-solder assembly method. If you have through-hole-mounted components on the board, they will almost certainly use wave soldering. Wave soldering imposes constraints on the placement of back-side components. With reflow soldering, components can be placed closer together.

When using the wave soldering method, insist on a dual-action wave or a vibrating wave. Anything is better than the old single-action laminar flow wave machines. The problem we need to avoid when using a solder wave is called *shadowing*. One component may disturb the wave flow, causing components behind it to receive less than their share of solder. Dual-wave machines and vibrating wave machines both tend to counteract this tendency.

With wave soldering, reasonable design rules suggest orienting components so they ride into a solder wave broadside (not skinny end first). Leave as much space between components as the width of a component body. These two rules help avoid problems with shadowing.

POINT TO REMEMBER:

Ask whether your board will be assembled with wave or reflow soldering.

8.4.4 Capacitors Mounted Under the IC Body

Two recent advances in capacitor packaging deserve mention. They are both available from Circuit Components, Inc.[6] The first is the Micro/Q Series 1000 package intended for use underneath DIP components (excellent for wire-wrap design). The second is the Micro/Q 3500SM intended for use underneath large PLCC packages.

Both packages have low lead inductance. And both packages also save board area because they fit underneath existing components.

Sketches of these packages appear in Figure 8.15. Both packages are broad and flat, for low inductance. The Micro/Q[7] 3500SM has wide mounting ears which contribute to its remarkably low lead inductance (0.3 nH on some parts). These capacitors also have low ESR (typically well below 0.1 Ω).

[6]Circuit Components, Inc., 2400 South Roosevelt Street, Tempe, Ariz.

[7]Micro/Q is a registered trademark of Circuit Components, Inc.

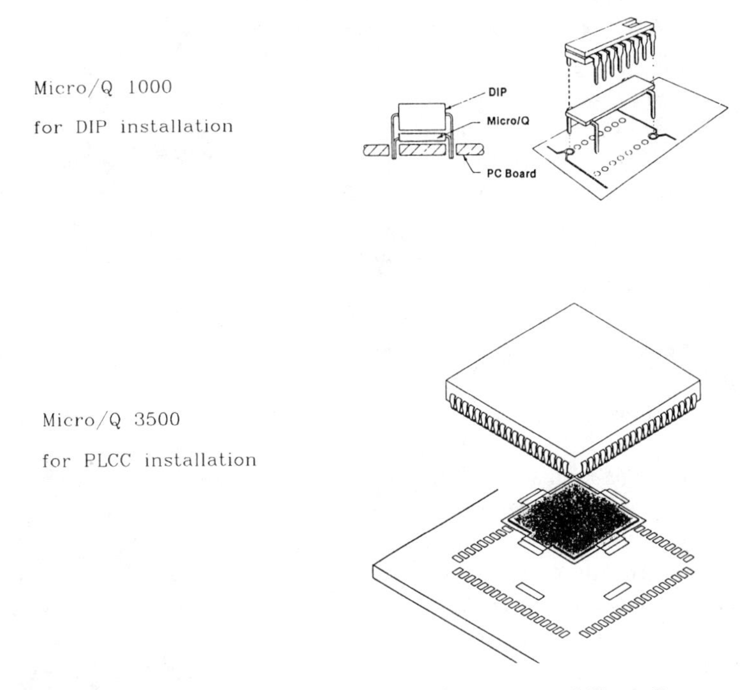

Figure 8.15 Micro/Q capacitor packages. (Drawings courtesy of Circuit Components, Inc.; formerly a division of Rogers Corporation.)

8.4.5 Three Types of Dielectric

Bypass capacitors represent only a small fraction of all capacitors manufactured. Even so, there are many types and classifications of bypass capacitors. A primary classifier of bypass capacitors is their dielectric material.

Dielectric materials used for bypass capacitors all have a relatively high dielectric constant, on the order of 1000–10,000 or more. A higher-dielectric-constant material can pack more capacitance into a smaller space than a lower-dielectric-constant material. Unfortunately, the materials with the highest dielectric constants also have the worst temperature coefficients.

Given a particular dielectric material, the volume of a capacitor is roughly proportional to its capacitance and also to its maximum voltage rating.

The next sections summarize three popular dielectric materials. For more on dielectric properties and the use of bypass capacitors, see the booklets provided by Johanson Dielectrics[8] and by Circuit Components, Inc.[9]

[8]Johanson Dielectrics, *Understanding Chip Capacitors,* Johanson Dielectrics, 2220 Screenland Drive, Burbank, Calif. 1974.

[9]Michael Scott Hyslop, *Use Power Bypassing and Busing for High Performance Circuits.* Reprints available from Circuit Components, Inc., 2400 South Roosevelt Street, Tempe, Ariz. 1990.

8.4.5.1 Aluminum Electrolytic Dielectric

Aluminum electrolytic dielectric capacitors supersede the older paper and oil-filled capacitors used in the days of electron tubes. Aluminum electrolytics are the workhorse capacitors most often used for board-level bypass. Their characteristics are similar to those of tantalum, which has an even higher dielectric constant at a slightly higher cost.

Aluminum electrolytic capacitors are manufactured from a roll of double-layer foil. First, two foil layers are insulated from each other by a dielectric film which is chemically formed on the foils before they are pressed together. Then the sandwich is rolled into a cylinder, using a thicker spacer between foils, this time to prevent shorting. Because the first chemical dielectric can be made surprisingly thin, we get a fair number of farads per cubic inch with this design.

At low voltage ratings, the thickness of the dielectric layer is only a small fraction of the overall foil sandwich thickness. Because of this effect, a capacitor rated at 3 V is not much smaller than one rated at 10 V. Aluminum electrolytics tend to have better energy storage densities at their higher voltage ratings.

One foil layer is connected to each terminal. These packages are typically cylindrical, reflecting the rolled construction. Different form factors, from short and squat to tall and skinny, are available.

Aluminum electrolytic capacitors commonly have an initial tolerance of ±20%. The least expensive varieties may have a +80 to –20% tolerance. Add to that an aging factor, which amounts to ±15% after 1000 h at their maximum rated temperature. Finally, remember to include a temperature derating factor, about –5% at 0°C. Check all these figures on your data sheet before using them. Altogether, the initial tolerance plus aging plus temperature derating can take away 40% of the capacitance you thought you were buying.

ESR is extremely sensitive to temperature. The test circuit used in Figure 8.16 to show ESR applies a 300-mV pulse sourced at 4.2 Ω to the capacitor under test. The capacitor in this case is a 33-μF 16-V aluminum electrolytic. The response is plotted for ambient temperatures of –30, 0, 25, and 60°C. At the highest temperature the step response, after

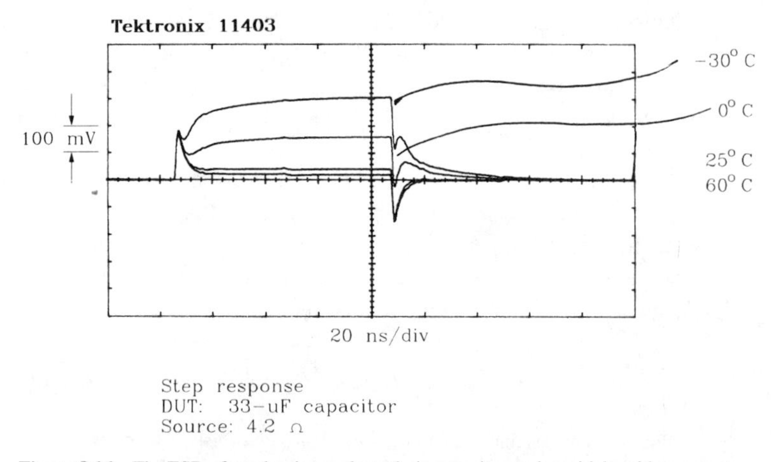

Figure 8.16 The ESR of an aluminum electrolytic capacitor varies widely with temperature.

the initial inductive kick, is 20 mV tall. The slow-ramp portion of the response looks completely flat on this time scale. Use Equation 8.33 to find the ESR at 60°C:

$$ESR_{60} = \frac{R_S X}{\Delta V - X} = \frac{(4.2)(0.020)}{0.300 - 0.020} = 0.3\ \Omega \qquad [8.35]$$

ESR on this capacitor deteriorates rapidly as the temperature drops. At 60°C the step response rises to only 20 mV, indicating an ESR of about 0.3 Ω. At 0°C the step response rises to 150 mV, or 4.2 Ω. This ESR ratio of 14:1 from 0 to 60°C is not unusual for an aluminum electrolytic capacitor. At temperatures below 0°C, the ESR on this particular capacitor renders it useless for bypassing digital signal currents. This is not an impressive capacitor.

The measured area under the inductive kick is about 720 pV-s. Use Equation 8.32 to find the lead inductance. Lead inductance, being a property of the mechanical construction, does not change with temperature.

$$L = \frac{R_S A}{\Delta V} = \frac{(4.2)(720)\ \text{pV-s}}{0.300} = 10\ \text{nH} \qquad [8.36]$$

The 33-μF capacitance is large enough that the extremely slow slope in Figure 8.16 is not visible.

Taking measurements of ESR on aluminum electrolytics, since their capacitances are so large, is easy. Instead of building a 4.2-Ω source, try using an ordinary 50-Ω pulse generator directly connected to the capacitor. The 50-Ω source shrinks the *L*/*R* decay time of the initial inductive kick, making it hard to see, but slows down the capacitive rise. The overall effect broadens the zone dominated by ESR. A step repetition rate of 100 kHz is about right.

8.4.5.2 Z5U Dielectric

Monolithic ceramic capacitors are constructed from layers of metal sandwiched between ceramic insulating layers. The entire structure is then fired, along with metal end caps, into a solid mass. The end caps connect to alternating metal layers inside the finished structure. The ceramic insulator becomes the dielectric.

You can buy these capacitors in either surface-mountable form (no leads) or in molded plastic packages with leads welded onto the end caps. In the surface-mounted package they are called *chip capacitors*.

The Z5U dielectric material has a higher dielectric constant than X7R but worse temperature and aging properties.

General specifications from Vitramon[10] list a standard tolerance range of ±20%, plus a more economical 80 to –20% range. The aging factor for Z5U is proportional to the logarithm of time since firing, about –2% per decade. Preaging the components by 100 h or so ensures no more than a 2% loss of capacitance in the first 1000 h, 2% more in the next 10,000 h, etc. Soldering a chip capacitor resets its aging clock. Finally, remem-

[10] Vitramon, Inc., is a subsidiary of Thomas & Betts Corporation.

ber temperature derating is horrible with Z5U materials. Vitramon quotes a +22 to –56% variation over the range 10–85°C. Below 10°C, Z5U is not recommended. Summing the initial tolerance, aging for 100,000 h, and a temperature of 10°C, we lose two-thirds of the advertised capacitance.

ESR ratings below 0.1 Ω are easily attainable at room temperature. An ESR ratio of 3:1 over the 10–85°C operating range is typical. Surface-mountable 1206 packages have about 1 nH of lead inductance. Leaded packages have about 5 nH of lead inductance.

Z5U capacitors in the surface-mounted 1206 size are available up to 0.33 μF at 50 working volts. Larger values are available in larger packages.

8.4.5.3 X7R Dielectric

X7R is another dielectric material used to construct monolithic capacitors. These parts are available in either surface-mountable or leaded packages. In the surface-mounted package they are called *chip capacitors*. The X7R dielectric material has a lower dielectric constant than Z5U, but better temperature and aging properties.

General specifications from Vitramon list standard tolerance ranges of ±5, 10, and 20%. The aging factor is proportional to the logarithm of the operating life, about –1% per decade (half that of Z5U). Preaging the components by 100 h or so ensures no more than a 1% loss of capacitance in the first 1000 operating h, 1% more in the next 10,000 h, etc. Soldering a chip capacitor resets its aging clock. Finally, remember temperature derating. Vitramon quotes a ±15% variation over the range –55° to 125°C. This is an excellent capacitor to use for wide temperature range applications. Between the initial tolerance (10%), aging for 100,000 h, and a temperature of ±55°C, we lose only 29% of the advertised capacitance.

ESR ratings below 0.1 Ω at room temperature are easily attainable. An ESR ratio under 2:1 is typical for the 0–70°C temperature range. Expect an ESR variation of 4:1 over the extended temperature range –55 to +125°C. Surface-mountable 1206 packages have about 1 nH of lead inductance. Leaded packages have about 5 nH of lead inductance.

X7R capacitors in the surface-mounted 1206 size are available up to 0.12 μF at 50 working volts. Larger values are available in larger packages.

As with any component, manufacturers have a difficult time producing parts at the upper end of the capacitance range. Your purchasing department will experience delays and possibly higher costs if you demand the largest capacitor that fits in a certain package. Move up to the next larger package or drop down the value to avoid this headache.

POINTS TO REMEMBER:

Higher-dielectric-constant materials pack more capacitance into a smaller space but have poor temperature coefficients and aging instability.

Aluminum electrolytics do not work well in cold applications.

8.4.6 Safety Margin for Voltage Ratings and Lifetime

Failure in capacitors is a statistical phenomenon. It accelerates at high voltages. When a manufacturer quotes a working voltage rating, it doesn't mean the capacitor will never fail if operated at that voltage. It only means it tends not to fail very often if operated at that voltage or below.

When operated below their maximum working voltage, capacitors display markedly longer useful service lives. For high-reliability engineering, discuss this subject with your manufacturer. A 50% voltage derating may significantly improve a capacitor's expected lifetime.

POINT TO REMEMBER:

Failure in capacitors is a statistical phenomenon, accelerating at high voltages.

9

Connectors

The faster we go, the harder it is to build good connectors. Evidence of this principle abounds. The cost per pin of a popular DIN connector, useful up to only a few tens of megahertz, is 100 times less than the cost of individually hand-assembled SMA hard line connectors useful up to 25 GHz. Why? What differences could possibly be so significant among connectors?

This chapter examines connector properties important to high-speed system designers. After reading this chapter, you will know what properties matter in your application and how to test a connector system.

The primary electrical factors affecting high-speed performance in connectors are

- Mutual inductance—causes crosstalk
- Series inductance—slows down signal propagation and creates electromagnetic interference (EMI)
- Parasitic capacitance—slows down signal propagation.

9.1 MUTUAL INDUCTANCE—HOW CONNECTORS CREATE CROSSTALK

The current loops in Figure 9.1 illustrate basic mutual inductive coupling at work. Current leaving gate A returns to its source through signal return path X. Because current paths X, Y and Z overlap, magnetic fields from path X induce electric noise voltages in signal paths Y and Z.

The noise induced in path Y is larger than that in path Z because it shares more overlapping area with path X. The paths needn't overlap at all to create mutual inductive noise; any two nearby current loops will interact a little bit.

Connectors also have parasitic capacitance between pins, which turns out to create less crosstalk in digital circuits than mutual inductance. Section 9.4 shows how to mea-

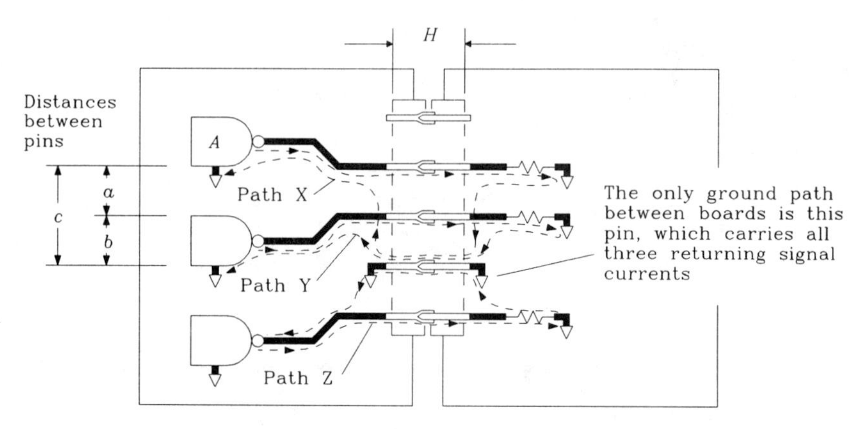

Figure 9.1 Mutual inductive coupling in a connector.

sure the overall coupling coefficient for a connector which combines the inductive and capacitive components. For now, we concentrate on the larger term, the inductance.

9.1.1 Estimating Crosstalk

We can use relationships developed in Chapter 1 to estimate the amount of signal crosstalk between any signal pins in Figure 9.1. Such an estimate requires three facts:

- The mutual inductance between two loops
- The maximum rate of change of the source signal *dI/dt*
- The impedance of the receiving network and whether it is source- or end-terminated.

Regarding the mutual inductance of two loops, we seek the worst-case crosstalk, so let's focus on just the interaction between directly overlapping loop pairs like X and Y.

The contributions to total magnetic flux in loop Y come from two places. The first contribution is from currents flowing out of gate A along its signal wire. The second contribution is from returning signal currents flowing in the ground wire. The mutual inductance formula therefore shows two terms. The second term (the ground wire term) is the bigger of the two.

$$L_{X,Y} = 5.08H \ln\left(\frac{c}{a}\right) + 5.08H \ln\left(\frac{b}{D/2}\right) \quad [9.1]$$

where a = distance signal X to signal Y, in.
b = distance signal Y to ground wire, in.
c = distance signal X to ground wire, in.
D = diameter of connector pin, in.
H = pin length in connector, in.
$L_{X,Y}$ = mutual inductance between loops, X and Y, nH

Equation 9.1 assumes a single row of pins and a relatively long connector (big H/a ratio). Even when these assumptions are not true, the forgiving nature of the logarithm function in Equation 9.1 yields an answer easily accurate to within an order of magnitude. This determines accurately enough whether or not connector crosstalk performance will be a significant issue. If connector performance might matter in your system, buy a connector and measure its performance.

Next we need the maximum dI/dt for the system in question. Use either Equation 2.41 or 2.42 to estimate the dI/dt.

The last fact involves the topology used in the noise receptor circuit. Figure 9.2 shows the choices. Case I involves a driver connected close to the connector. *Close* in this context means within the electrical length of a rising edge (see Equation 1.3). Case II covers all other configurations, including source termination.

The noise coupled into a case II configuration splits half in either direction. The noise coupled into a case I situation reflects promptly off the low-impedance driver, doubling the coupled noise on the receiving side.

Here are the formulas for the height of a noise pulse induced in loop Y due to a single step input from gate A. The pulse duration will be comparable to the input pulse rise time:

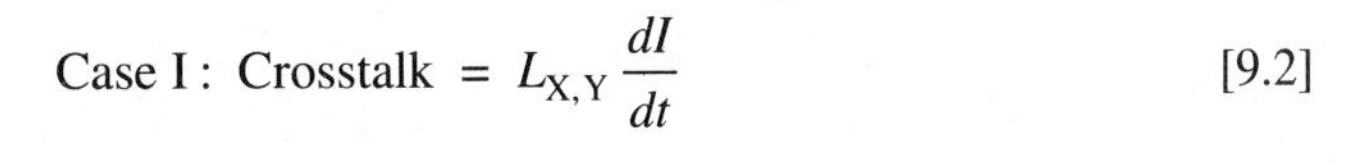

$$\text{Case I: Crosstalk} = L_{X,Y} \frac{dI}{dt} \qquad [9.2]$$

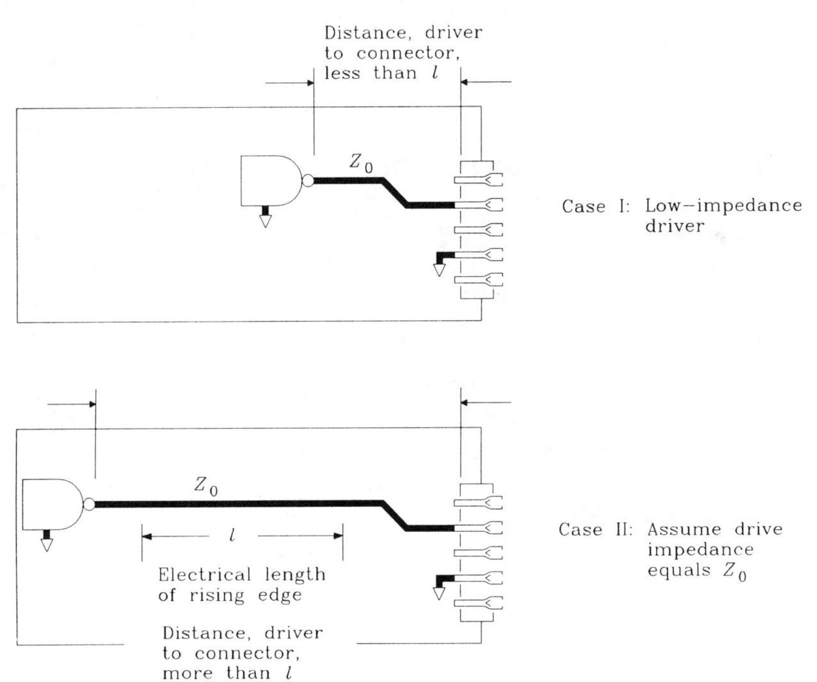

Figure 9.2 Nearby drivers have lower drive impedance.

$$\text{Case II: Crosstalk} = \frac{1}{2} L_{X,Y} \frac{dI}{dt} \qquad [9.3]$$

Slowing the rise time of the driving signal produces immediate rewards in crosstalk reduction. Reduce the driving rise time with a capacitor on the source side of the connector as in Figure 9.3. Placing the capacitor on the receiving side just increases the surge current flowing through the connector when the driver switches, making the problem worse.

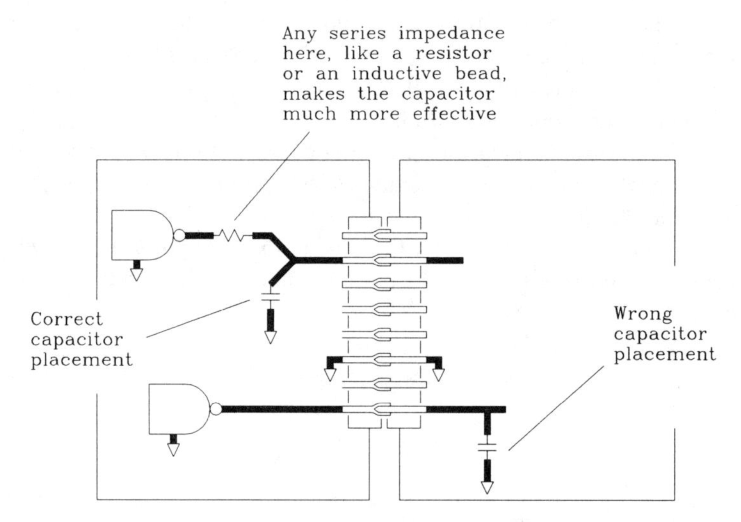

Figure 9.3 Method for slowing the rise time of signals.

9.1.2 How Grounds Alter the Return-Current Path

The following four rules of connector behavior, combined with Equation 9.1, can help estimate the behavior of various connector grounding arrangements. These rules work well when figuring adjustments to an existing system. Using them, we can predict what will happen after various proposed changes.

Rule 1. Changing the pattern of ground connections in Figure 9.1, we can decrease (or increase) the mutual inductance between particular wires. If we move the ground wire further from signal wires X and Y, increasing dimensions b and c, both terms in Equation 9.1 grow. The mutual inductance $L_{X,Y}$ goes up. Conversely, moving the ground closer to signals X and Y decreases their mutual inductance. The change in inductance is proportional to the logarithm of distance.

Rule 2. Adding extra grounds has a more direct effect. Remember in Equation 9.1 how the second term (the ground wire term) was the largest? The ground wire intimately couples loops X and Y, and so current flowing in the ground wire has a big impact on loop Y. If we could divide the ground wire current in half, the mutual inductance $L_{X,Y}$ would drop by almost a factor of 2.

Divide the ground current in half by adding a second ground wire above the X signal wire as shown in Figure 9.4. By symmetry, we know the ground current will split, half going to each ground wire. The mutual inductance $L_{X,Y}$ drops accordingly. Adding more ground wires splits the ground current even further but not nearly as much as the original split by two.

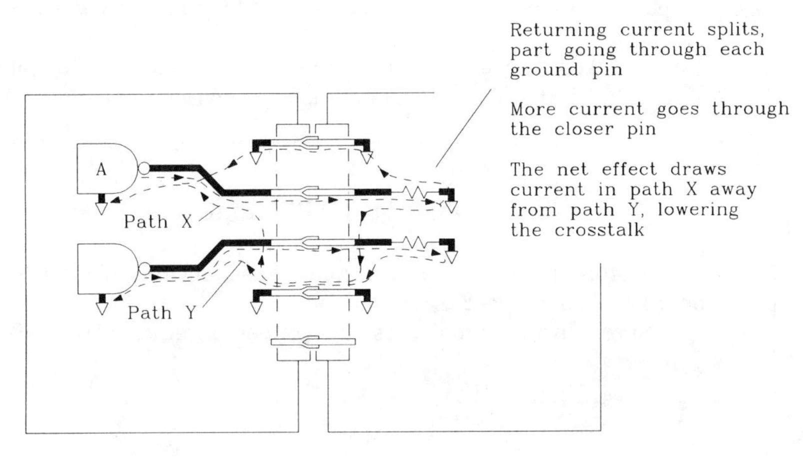

Figure 9.4 Adding a second ground divides the ground current.

Rule 3. Interposing ground wires between signals X and Y makes a much bigger difference than adding grounds outside X and Y. If we add N grounds between X and Y, spacing X and Y further apart as shown in Figure 9.5, the coupling between them falls off proportional to

$$\text{Coupling is proportional to: } \frac{1}{1+N^2} \quad [9.4]$$

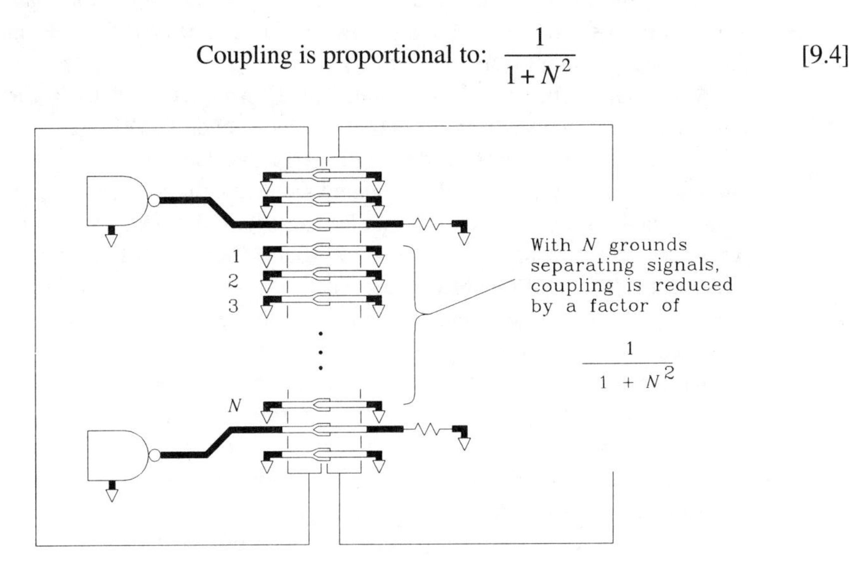

Figure 9.5 Separating signals with multiple grounds reduces coupling.

Rule 4. Noise coupled onto any given wire is contributed by each of the other wires in the connector. Simply reducing the number of signals in a connector cuts the aggregate crosstalk. Alternately, partitioning a connector into several weakly interacting signal groups by placing intervening grounds between each group accomplishes the same thing. Partitioning effectively cuts the number of wires that crosstalk strongly onto any given receptor. Crosstalk is roughly proportional to the number of wires between grounds.

Rule 5. Adding extra grounds at the end of a connector does almost nothing to reduce crosstalk. Giant ground lugs at the end of a connector also do nothing.

> POINTS TO REMEMBER:
>
> Mutual inductance, not mutual capacitance, is mostly responsible for crosstalk generated by connectors.
>
> Spreading grounds across the connector reduces crosstalk.

9.2 SERIES INDUCTANCE—HOW CONNECTORS CREATE EMI

Electromagnetic interference (EMI) emanates from signal current flowing in large loops.

The situation in Figure 9.6 illustrates a common electromagnetic interference problem. Card A sources a 64-bit bus through connector B onto mother card C. The mother card could be either a main CPU card or a passive conduit to other daughter cards. Signal return current from these 64 lines flows from mother card C back onto card A, mostly through the ground pins in connector B.

A small fraction of the returning signal current flows back to card A through a different path. It is that small fraction that causes no end of EMI grief.

High-frequency currents flowing in large loops radiate lots of electromagnetic energy and will not pass FCC- or VDE-mandated radiated emission tests. The battle in EMI design is to contain all signal flow to loops having small cross sections. For example, high-frequency current flowing out over a solid ground plane tends to return along a path directly underneath the signal conductor (see Section 5.1). A 6-in. circuit trace spaced 0.010 in. above a ground plane encompasses a total loop area of only 0.06 in.^{2} Such a loop area is often acceptable for EMI purposes. In Figure 9.6, we may ignore the loop area between signal and ground as the 64-bit bus flows across the solid ground planes on cards A and C.

Any interruption or discontinuity in the return-current path, such as a diversion through the ground pins of a connector, creates a bubble in the current loop. Whether this bubble presents enough area to violate radiation regulations depends on the total dI/dt of the signal currents flowing in the loop.

The bubble of loop area in Figure 9.6 occurs mostly inside connector B as a result of the separation of signal and ground pins. This bubble is labeled G_1. The majority of loop inductance in the 64-bit bus signal path is contributed by the inductance of loop G_1.

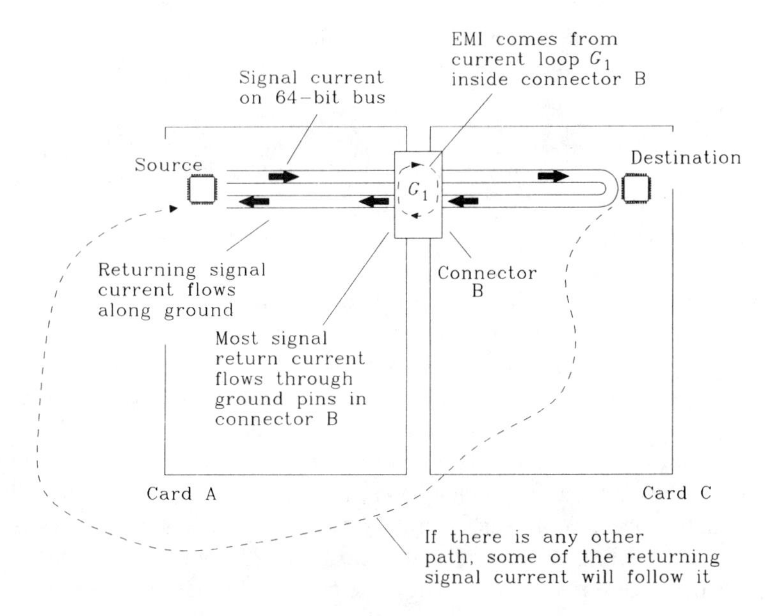

Figure 9.6 Signal return current from a 64-bit bus.

What other paths are available for returning signal current depends on the physical construction of connector B and other details of the card cage frame surrounding cards A and C. Any current returning to the source on card A by a path other than through connector B will encompass a huge loop area and radiate substantially.

For example, suppose cards A and C share two connectors as in Figure 9.7. Call the additional connector D and locate it some distance away from connector B. Some portion of the returning signal current can now flow back to A through the ground wires on connector D, illustrated in Figure 9.7 as loop G_2.

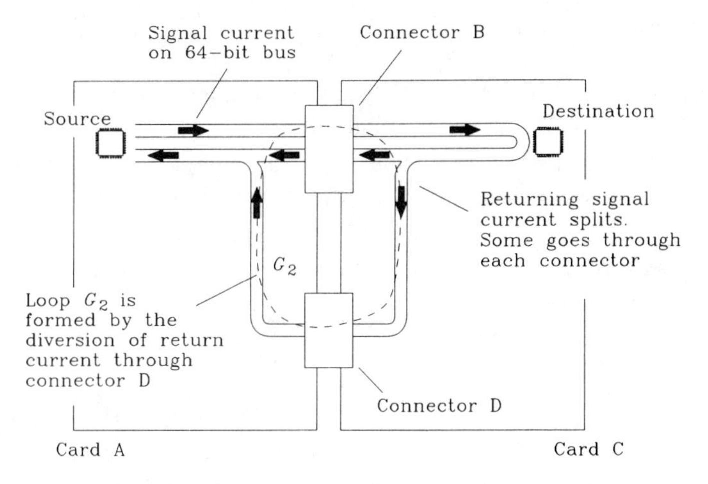

Figure 9.7 Returning current diverts through connector D.

The proportion of high-speed returning signal current that flows through connector D depends on the ratio of the inductance of loop G_1 (Figure 9.6) to the inductance of loop G_2 (Figure 9.7).

$$\text{Current through D} = (\text{return current from A})\frac{L_{G1}}{L_{G2}} \qquad [9.5]$$

At very low frequencies, the amount of current returning through D depends on a ratio of resistances. At higher frequencies it is determined by the ratio of inductances in Equation 9.5. Since EMI is a high-frequency problem, we need only concern ourselves here with the inductance ratios of the two loops.

Loop G_1 having the smaller area, has less inductance than loop G_2. Therefore, only a small portion of the returning signal current flows through path G_2. Even that small portion can be enough to violate radiation regulations. Above 30 MHz, both FCC and VDE limits are approximately 100 µV/m, as measured a distance of 3 m away from your equipment.[1] For further details on emission regulations and design techniques to combat them, see Ott,[2] Mardiguian,[3] and Keiser.[4]

Calculating precise radiated intensity levels from a digital product is a chancy business because so many factors affect the outcome. Equation 9.6 expresses a simple restriction on loop area, peak current, and rise time which should, in an open field measurement test, just meet the FCC and VDE radiated-emission limits above 30 MHz.

$$E = 1.4 \times 10^{-18} \frac{A I_p F_{\text{clock}}}{(T_{10-90})} < 10^{-4} \text{ V/m} \qquad [9.6]$$

where E = radiated electric field, V/m, at 3 meters
A = radiating loop area, in.^{2}
I_p = peak current, A
T_{10-90} = signal rise time, s
F_{clock} = clock frequency, Hz

Caveats concerning Equation 9.6:

- In finished products, radiation levels can easily vary 20 dB from the predicted level in Equation 9.6. Include a big fudge factor.
- Remember that emission tests sum[5] emissions from all wires in the system. If one wire just meets specifications, then 100 additional wires certainly will not.

[1]The FCC/VDE limit steps up to 200-µV/m at higher frequencies. For initial planning, just use the 100-µV/m number.

[2]Henry Ott, *Noise Reduction Techniques in Electronic Systems,* 2nd ed., John Wiley, New York 1988.

[3]Michel Mardiguian, *Interference Control in Computers and Microprocessor-Based Equipment,* Don White Consultants, Inc., 1984, Gainesville, Vir.

[4]Bernhard Keiser, *Principles of Electromagnetic Compatibility,* Artech House, 1987, Norwood, Mass.

[5]For random signals, the total emission is proportional to the square root of the number of signals. For correlated signals (such as clock lines), the emission can be as much as directly proportional to the number of signals.

- Build a mock system containing just a few clock sources routed through your connector system for testing before finalizing your design. This sounds wasteful, but it saves a lot of money in the long run. The cost of redesigning mechanical packaging and shielding grows dramatically towards the end of a project.

EXAMPLE 9.1: Noise Radiated from a Connector

Figure 9.8 shows a typical 16-bit bus. We will step through calculation of the inductances for paths G_1 and G_2, the radiation from path G_1, and then the radiation from path G_2.

Inductance of path G_1 (from Appendix C, for the case of a rectangular loop):

$$\begin{aligned} r &= 0.025/2 \text{ (pin radius, half of diameter, in.;} \\ &\quad \text{notice we use } h/r \text{ instead of } 2h/d) \\ w_1 &= 0.2 \text{ (separation signal to ground, in.)} \\ h &= 0.4 \text{ (connector pin length, in.)} \\ \tfrac{1}{2} &= \text{fudge factor to account for ground wires on either} \\ &\quad \text{side of signal (rule 2, Section 9.1)} \end{aligned}$$

$$L_{G1} \approx \frac{1}{2}\left\{10.16\left[w_1 \ln\left(\frac{h}{r}\right) + h \ln\left(\frac{w_1}{r}\right)\right]\right\}$$

$$= \frac{1}{2}\left\{10.16\left[0.2 \ln\left(\frac{0.4}{0.013}\right) + 0.4 \ln\left(\frac{0.2}{0.013}\right)\right]\right\}$$

$$= 9.0 \text{ nH} \qquad [9.7]$$

Inductance of path G_2 (from Appendix C, for the case of a rectangular loop):

$$\begin{aligned} r &= 0.025/2 \text{ (pin radius, half of diameter, in.)} \\ w_2 &= 6.0 \text{ (separation, signal to connector D, in.)} \\ h &= 0.4 \text{ (connector pin length, in.)} \end{aligned}$$

$$L_{G2} \approx 10.16\left[w_2 \ln\left(\frac{h}{r}\right) + h \ln\left(\frac{w_2}{r}\right)\right]$$

$$= 10.16\left[6 \ln\left(\frac{0.4}{0.013}\right) + 0.4 \ln\left(\frac{6}{0.013}\right)\right]$$

$$= 234.0 \text{ nH} \qquad [9.8]$$

Assuming each driven signal travels in a 50-Ω transmission line with a typical TTL amplitude of 3.7 V, the peak-to-peak signal current is 74 mA. The peak current is half that, or ±37 mA.

Use Equation 9.5 to find peak currents in path Y:

$$I_{G1} = 0.037 \text{ A} \qquad [9.9]$$

$$I_{G2} = 0.037 \frac{9.0 \text{ nH}}{234 \text{ nH}} = 0.0014 \text{ A} \qquad [9.10]$$

Now use Equation 9.6 to estimate the radiation from loops G_1 and G_2. We will do loop G_1 first:

$$A = 0.08 \text{ (0.4-in. pin length times 0.2-in. signal-to-ground, in.}^2\text{)}$$
$$I_{G1} = 0.037 \text{ (peak current, A)}$$
$$T_{10-90} = 5 \times 10^{-9} \text{ (signal rise time, s)}$$
$$F_{\text{clock}} = 10^8 \text{ Hz}$$

$$E_{G1} = 1.4 \times 10^{-18} \frac{(0.08)(0.037)\left(10^8\right)}{\left(5 \times 10^{-9}\right)} = 82\ \mu\text{V/m} \quad [9.11]$$

The radiation from one signal wire amounts to 82 μV. Radiation being proportional roughly to the square root of the number of wires involved,[6] the radiation from all 16 wires amounts to

$$E_{G1,\text{total}} = 82 \times 10^{-6} (16)^{1/2} = 328\ \mu\text{V/m} \quad [9.12]$$

As designed, this connector arrangement does not pass regulations. Now take a look at connector D (Figure 9.7):

$$A = 2.4 \text{ (0.4-in. pin length times 6-in. signal-to-ground, in.}^2\text{)}$$
$$I_{G2} = 0.0015 \text{ (peak current, A)}$$
$$T_{10-90} = 5 \times 10^{-9} \text{ (signal rise time, s)}$$
$$F_{\text{clock}} = 10^8 \text{ Hz}$$

$$E_{G2} = 1.4 \times 10^{-18} \frac{(2.4)(0.0014)\left(10^8\right)}{\left(5 \times 10^{-9}\right)} = 94\ \mu\text{V/m} \quad [9.13]$$

The radiation from one signal wire amounts to 94 μV. The aggregate radiation from all 16 wires amounts to

$$E_{G2,\text{total}} = 94 \times 10^{-6} (16)^{1/2} = 376\ \mu\text{V/m} \quad [9.14]$$

Loop G_2 actually radiates more than path G_1. This happens because the inductance L_{G2} grows only as fast as the logarithm of the distance between connectors B and D. Meanwhile,

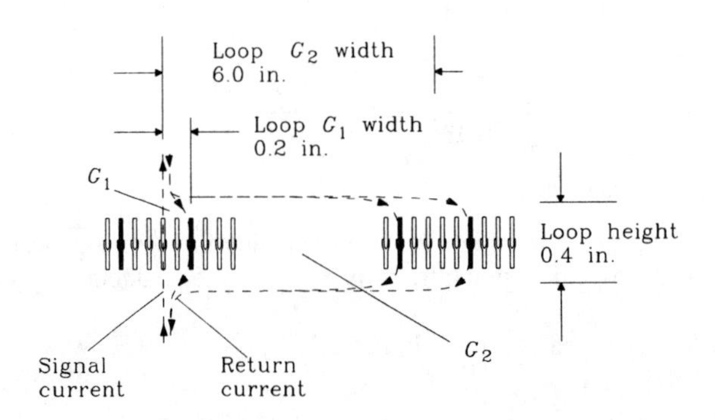

Figure 9.8 Example of returning current diversion.

[6]Assumes random, uncorrelated signals on all wires. If all signals switched together, as in a worst-case bus test, the radiation would be much worse.

the loop area of path G_2 grows directly proportional to the distance between connectors B and D. The growing inductance reduces the current flow through path G_2, but the growing loop area increases emissions even faster. Moving connectors B and D further apart will actually worsen the problem.

Here are some rules useful for reducing connector emissions:

Rule 1. Use more grounds in connector B. This brings grounds closer to each signal, lowering the effective radiating loop area in connector B.

Rule 2. Adding more grounds in connector B also lowers the inductance of connector B. This reduces the current flow in remote loops according to Equation 9.5.

Rule 3. Disrupt or remove remote return paths by placing all mother card connectors close together on card A.

Rule 4. Place a continuous ground contact all along the edge of cards A and C. This provides a very-low-impedance return path, lowering the remote loop current according to Equation 9.5.

Rule 5. Never attach I/O cables to the outer edge of card A. This creates a huge remote return-current path from mother card C, through earth ground, and back into card A through the I/O cable. Attach the cables instead to the mother card or bypass them at high frequencies to a point on the mother card near connector B.

Rule 6. Use driver gates with as slow a rise time as practical. Emission in Equation 9.6 is proportional to the inverse of rise time.

POINTS TO REMEMBER:

EMI emanates from signal current flowing in large loops.
Provide a low-inductance return-current path for every connector.
Disrupt or eliminate remote return-current paths.

9.3 PARASITIC CAPACITANCE—USING CONNECTORS ON A MULTIDROP BUS

Multidrop bus applications place more of a burden on the connector system than point-to-point applications. In a point-to-point application, transmitted signals must traverse each connector only once. In this case, a connector's series inductance usually dominates its performance.

Multidrop situations are much different. In reference to Figure 9.9, only one transmitter at a time is enabled. Other transmitters attached to the bus remain disabled, but still connected, until it is their turn to send. This particular bus is terminated at both ends to prevent reflections. The rise time of signals propagated on this bus can be a small fraction of the total bus length.

Signals transmitted from one location distort successively as they pass by each tap on the bus. The cumulative effect of the parasitic capacitance from many connectors can distort the transmitted signal more than the series inductance of the source connector. For multidrop applications we want a connector with very low parasitic capacitance, even at the expense of higher inductance.

For fast bus operation, we need to minimize the lumped capacitance to ground of each bus tap. Section 4.4.2 discusses the effect of lumped capacitance on transmission line signals and shows why less capacitance is better.

The lumped capacitance at each tap consists of three parts, only one of which is due to the connector:

(1) Pin-to-pin capacitance of the connector and its pads on the printed circuit board
(2) Capacitance of the circuit trace leading from the connector to the local drivers and receivers

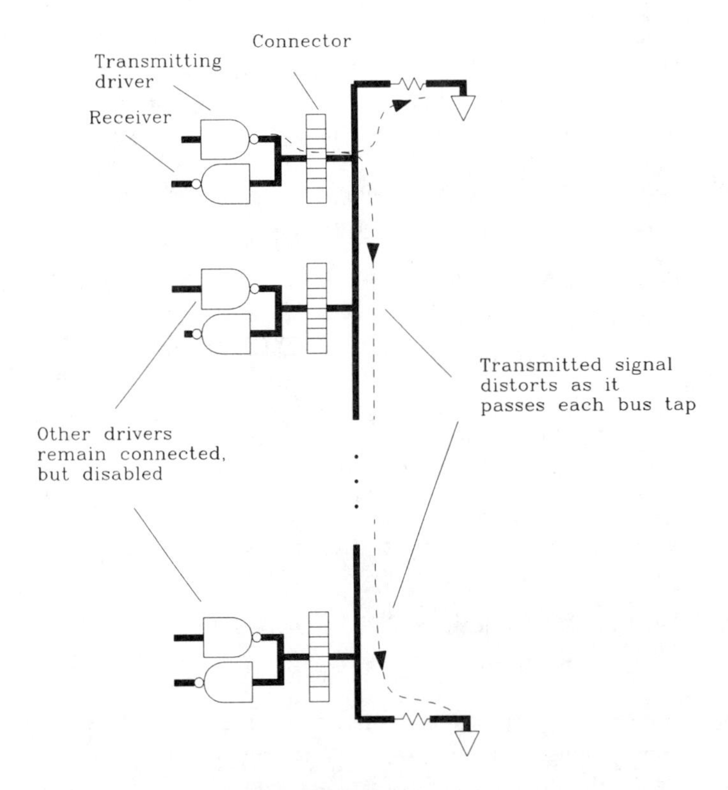

Figure 9.9 Multidrop bus.

(3) Input capacitance of the local receiver plus the output capacitance of the local driver in its disabled condition.

9.3.1 Pin-to-Pin Capacitance

This term is easy to measure. Mount the connector on a board with all pins grounded except one signal pin. Measure the capacitance to ground of the signal pin using an ordinary capacitance meter.

If you do not have a capacitance meter, use the circuit in Figure 1.6 to make the measurement.

Most connectors manufactured with 0.1-in. pin spacings have a few picofarads between any signal pin and ground. The capacitance of the pads on a printed circuit board will add perhaps 0.5 pF on either side of a board-to-board connector.

Some connectors have particularly large gaps between pins or have their pins arrayed in a staggered fashion. This helps reduce capacitance. It also leaves more space between pads on the solder side of the board, which manufacturing technicians appreciate. Staggering pins becomes increasingly important at connector pin pitches of 0.05 in. and below.

9.3.2 Circuit Trace Capacitance

Equation 9.15 calculates circuit trace capacitance per inch from a knowledge of the trace impedance and the trace propagation delay:

$$C_{\text{per inch}} = \frac{T_d}{Z_0} \qquad [9.15]$$

where T_d = trace propagation delay, ps/in. (see Section 1.3)
Z_0 = trace impedance, Ω (see Section 4.5)
C = capacitance, pF/in.

9.3.3 Capacitance of Receivers and Drivers

Many manufacturers state the capacitance of high-speed receivers on their specification sheets. If no specification exists, measure a sample using the arrangement in Figure 1.6. When making the measurement, adjust the pulse generator to produce a pulse whose voltage is centered in the active region of the receiver, and with a pulse height approximating the actual conditions in use. Turn the power to the receiver on. Values in the range of 2–10 pF are typical.

The capacitance of a three-state driver when switched to its off state is much larger. Many manufacturers do not state this capacitance, hoping you won't think about it. The driver incorporates some rather large transistors which have significant parasitic capacitance when switched off.

The only way to get a realistic view of driver capacitance is to measure it. This measurement proceeds the same way as for a receiver. Turn the transmitting gate's power on, but leave its outputs disabled. Bias the pulse generator in the gate's active region. Values as high as 80 pF are not unusual.

EXAMPLE 9.2: Trace Capacitance

An inner trace leading from a connector goes first to a driver chip and then to a receiver. Its total length is 0.75 in. What is its capacitance?

$$T_d = 180 \text{ ps/in.} \qquad \text{(FR-4 inner trace)}$$

$$Z_0 = 50\ \Omega$$

$$C_{\text{per inch}} = 180/50 = 3.6 \text{ pF} \qquad [9.16]$$

$$C_{\text{total}} = 0.75(3.6) = 2.7 \text{ pF} \qquad [9.17]$$

9.3.4 Evenly-Spaced Loads

Section 4.4.2 discusses the impact of lumped capacitance on transmission line signals. It points out that arranging the bus taps at evenly-spaced intervals has the side effect of lowering the bus impedance and interferes less with signal propagation than lumping all the capacitance together at one place.

The even-spacing model holds true in a card cage system if the card slots are evenly spaced along the mother card and if a card is always inserted in every slot. If the system must operate with some slots empty, the even-spacing model is no longer valid.

A compromise model assumes the capacitance of a connector at every slot, but with no card present. Just the effect of the connector capacitance can measurably lower the impedance of the backplane and slow it down. A pleasant side effect of low backplane impedance is that each card, when inserted, makes only a small difference in the overall transmission characteristics.

9.3.5 Very Slow Bus

If you do not need the speed, consider source-terminating a multidrop bus. In this configuration omit the resistors shown at either end of the bus in Figure 9.9. Connect each three-state source to the bus through a series damping resistor. Receivers can connect directly to the bus.[7] An advantage of this topology in card cage applications is that no terminating components are required on the backplane.

If the driver rise time is longer than the electrical length of the bus, the bus acts as a single lumped-circuit element. Then we get no deleterious reflections, and the source resistors charge the lumped capacitance of the bus in a slow but orderly manner.

If the driver rise time is comparable to the electrical length of the bus, reflections appear on the bus. We can curtail these reflections by slowing down the rise time until the bus acts as a lumped circuit. By making the source resistor of higher impedance than the bus, we get the *RC* rise-time effect described in Section 4.4.1.2. As the resistance is

[7]Some designers advocate connecting the receiver to the driver so that both are isolated from the line by the series resistance. This has the effect of lowering the lumped capacitive load present at the bus but also of slowing the receiver response.

increased, the system acquires a slow *RC* rise characteristic, becoming monotonic. The connector capacitance, along with other trace and driver or receiver capacitances, are slowly charged through the source resistor.

The source resistor advocated in this section differs from the source termination described in Section 4.3.3. In that earlier section we set the source resistance equal to the line impedance and obtained perfect transmission with no reflections. That works only on a point-to-point line. With a multidrop bus there is no value of source resistance that will prevent reflections. Whatever value of source resistance we use, signals still bounce back and forth between the ends of the bus. Our approach here uses a source resistor much larger than the characteristic impedance of the bus to slowly, monotonically leak charge onto the bus.

If you can afford to wait for the bus to stabilize between each clock, a large source resistor has the following advantages:

- Low power—quiescent drive current is zero
- Simplicity—requires no terminators on the backplane
- Low EMI—reduced current flow through the connectors.

POINTS TO REMEMBER:

Multidrop bus applications place more of a burden on the connector system than point-to-point applications.

For multidrop applications we want a connector with very low parasitic capacitance, even at the expense of higher inductance.

9.4 MEASURING COUPLING IN A CONNECTOR

The setup in Figure 9.10 measures the performance of any connector under actual operating conditions. Adjust the pulse generator to produce pulses having about the same rise time as the driver you plan to use. The scope measures actual coupled noise as it would look in the real circuit.

By connecting the source and destination cables to different pins, we can measure the noise coupling from any source to any destination. Measure the relation between noise and separation and then estimate the *total* crosstalk noise at each receiving location. The test cables can be any length and will easily reach into an existing system to make this measurement.

9.4.1 Ground and Signal Pins

Solder all ground pins to test boards 1 and 2 as planned in the real system. Leave other signal pins (except the ones under test) disconnected.

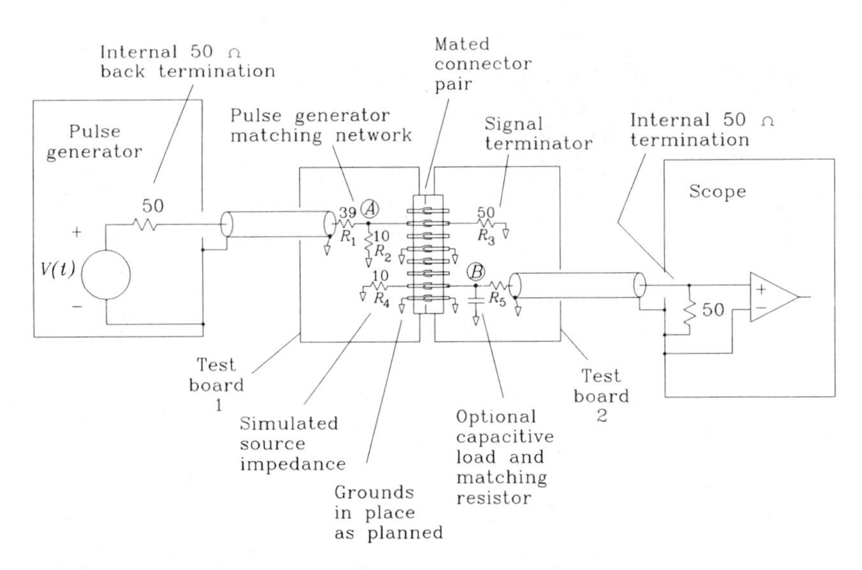

Figure 9.10 Test setup for measuring connector crosstalk.

Detaching individual ground pins temporarily from the ground plane, we can directly measure their importance. Notice that you need only detach a ground pin at one end to destroy its protective effect. Detaching one end disrupts the flow of returning signal current. Additionally detaching the other end of the same pin makes little difference. This experiment proves that most of the coupling in the connector is inductive, not capacitive. Were the pin effective as a partial Faraday shield, it would require grounding on only one side of the connector.

9.4.2 Pulse Generator and Source Impedance

Use the back termination on the pulse generator and a matched 50-Ω cable to cut reflections and ringing on the source cable. In Figure 9.10, the source cable is terminated with a matching network. The matching network provides an approximate 50-Ω termination for the incoming pulse, helping to further reduce reflections. The matching network also provides a low output impedance driving the connector, similar to the output impedance of the gate actually used. If you do not know the output impedance of the driving gate, then use 10 Ω. We aren't trying to be the National Bureau of Standards here; we just want to see if the crosstalk is going to be a problem.

Ideally, set the signal size at point *A* equal to the size of the signal you plan to use. A different test signal can be used if necessary, but remember to correct the final measured results to account for the difference.

Set the pulse generator rise time equal to the rise time of the drivers in the finished circuit.[8] If your pulse generator output is not adjustable, then use what's available, cor-

[8]Excellent pulse generators for this application are the Hewlett-Packard 8012B, which has an output rise time adjustable down to 5 ns, and the Hewlett-Packard 8082A, which is adjustable down to a rise time of 1 ns. Below 1 ns, wire in a real driver chip with the trace geometry as planned.

recting the final result assuming that noise is inversely proportional to the rise time of the driving waveform.

9.4.3 Terminating Impedance on the Transmitting Line

Terminate the transmitting line with an impedance representing the finished circuit. If the finished circuit will be resistively terminated, use a single resistor. If it will be terminated with a gate input plus a short trace, use a small capacitor. The capacitor should account for both gate input capacitance and trace capacitance.

9.4.4 Simulated Source Impedance of Receiving Line

If your driver has a known source impedance, set R_4 accordingly. Otherwise set R_4 equal to 10 Ω.

The difference in measured noise between a short circuit at R_4 and a value equal to R_3 is a factor of 2. For initial estimation, a factor of 2 is not terribly significant.

Alternately, wire a real driver in this location with its power turned on and its output enabled. Be sure to enable the output. If you leave the output disabled (three-stated) you will measure little or no noise at *B*. Try it.

9.4.5 Matching Resistor

If you plan to terminate the finished circuit resistively, put a matching resistor at position R_5 equal to

$$R_5 = Z_0 - 50 \quad [9.18]$$

where Z_0 = termination impedance for finished circuit, Ω
R_5 = matching resistor, Ω

If you plan a capacitive load, put a capacitor to ground at *B* equal to the expected gate input plus trace capacitance. Then use a 470-Ω resistor at position R_5, making a 10:1 probe.

Remember to account for attenuation in the probe when interpreting the finished measurements. The probe gain is equal to

$$G = \frac{50}{R_5 + 50} \quad [9.19]$$

where R_5 = matching resistor, Ω
G = gain of probe (multiply measured results by 1/G to get actual voltage at *B*)

The scope should use its internal 50-Ω termination. Use a 50-Ω coax leading to the scope.

POINT TO REMEMBER:

A simple test setup characterizes the crosstalk in a connector.

9.5 CONTINUITY OF GROUND UNDERNEATH A CONNECTOR

The connector shown schematically in Figure 9.11 has been misapplied and will not function properly at high speeds. The layout features a big hole in the ground plane through which all the connector pins pass. Even though there are many ground pins, the returning signal current must flow around the hole, defeating the effectiveness of interleaved grounds. We might as well have provided just a single ground pin at either end of the connector.

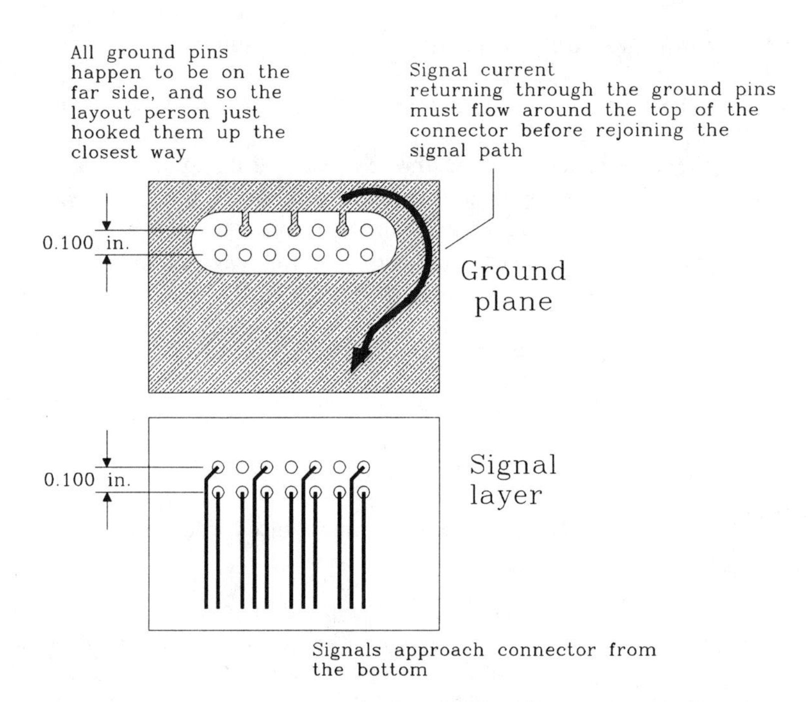

Figure 9.11 Bad connector layout.

One solution to this problem routes ground traces through the connector pin field as shown in Figure 9.12.

Miniature connectors with pins on a 0.050 pitch can use the grounding pattern shown in Figure 9.13. This approach isn't as good as the one in Figure 9.12, but at least the ground current doesn't have to flow the long way around the connector as in Figure 9.11.

Sometimes printed circuit board manufacturers can reliably produce pads on the inner layers much smaller than pads on the outer layers. Use this fact to route traces between pins on the inner layers, even when they will not fit between pins on the outer layers.

Finally, some fine-pitch connectors come with a staggered pin pattern which increases the space between pins. This feature helps a lot with ground routing.

Ground pins are tied
to a solid ground plane
which extends between
connector pins
Return current can snake
through the pin field,
staying close to the signal traces
0.100 in.
Ground
plane
0.100 in.
Signal
layer
Signals approach connector from
the bottom

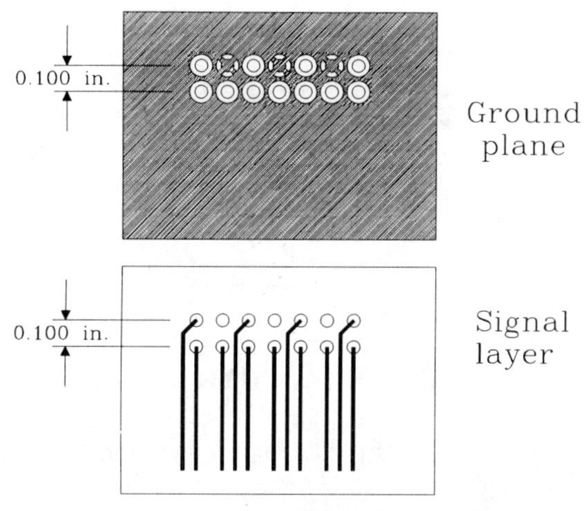

Figure 9.12 Better connector layout.

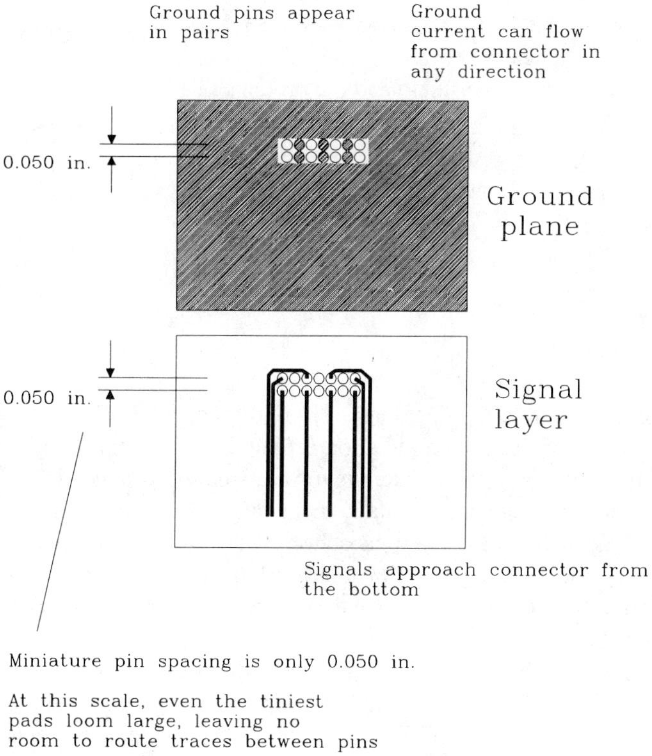

Ground pins appear
in pairs
Ground
current can flow
from connector in
any direction
0.050 in.
Ground
plane
0.050 in.
Signal
layer
Signals approach connector from
the bottom
Miniature pin spacing is only 0.050 in.
At this scale, even the tiniest
pads loom large, leaving no
room to route traces between pins

Figure 9.13 Layout for miniature connectors.

POINT TO REMEMBER:

If returning signal current from a connector must flow around a ground plane hole, it doesn't matter how many grounds we use; the performance gets no better than a connector with a single ground pin at either end.

9.6 FIXING EMI PROBLEMS WITH EXTERNAL CONNECTIONS

Exposed wiring carrying high-speed digital signals between circuit boards always fails FCC and VDE radiated emission tests. Here are three ways to reduce radiated emission below acceptable levels:

(1) Filter the signals to remove high-frequency content. This slows the rising and falling edges.
(2) Shield the cable. Shields provide a low-inductance return path for signal current, which tends to prevent it from returning along more remote paths. Proper connection between the shield and the product chassis is critical.
(3) Place a common mode choke on the cable. This increases the inductance of remote current paths, lowering their current flow. This helps with either shielded or unshielded cables.

When designing a new system, don't forget to check Section 9.8, on differential mode signaling.

9.6.1 Filtering

If you can tolerate the degradation in rise time, filter all outgoing digital signals before they exit the chassis. The radiation efficiency of current loops increases dramatically at high frequencies. Typical filters involve a small impedance in series with each logic driver. This series impedance usually feeds into a shunt capacitance to ground.

It is vital that the shunt capacitor connect to a quiet ground. Connecting the capacitor to a local digital ground in a large system is not adequate because the local digital ground usually carries more than enough noise to exceed FCC/VDE emission regulations if carried outside the chassis.

To circumvent this problem, some equipment manufacturers mount a small circuit board near their connector bulkhead which translates from internal connectors to external ones and also includes the filtering components. The ground connection on this board ties directly to the chassis, not to the digital ground on any circuit card.

Connector manufacturers have picked up on this issue and now offer "D"-size connectors with built-in filtering. These connectors have the advantage of small size and make ideal electric connection with the chassis.

9.6.2 Shielding

The most common approach taken by digital engineers is shielding. Shielding has the advantage of being an entirely mechanical approach, requiring little analog circuit design. The physics of shielding action is illustrated in Figure 9.14. By providing a continuous metal shield around the inner conductors, returning signal currents can distribute evenly around the outgoing signal wires. The effective net radiating current loop between signal and ground paths is extremely small. A perfectly conducting, symmetric shield radiates nothing.

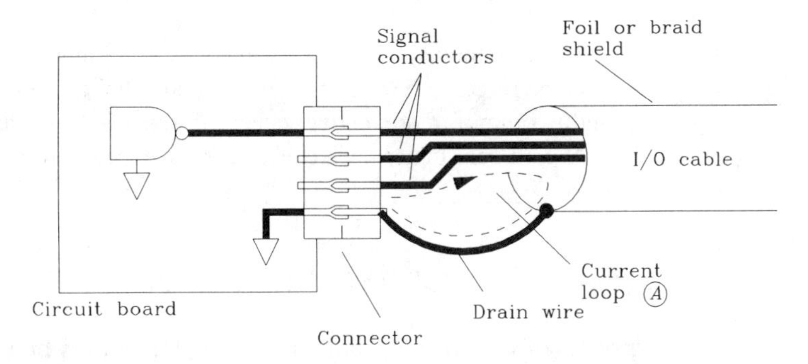

Figure 9.14 EMI radiation from a drain wire loop on a shielded cable.

Since a perfectly shielded cable does not radiate, if two cables are both individually shielded, the crosstalk between them is practically nonexistent. Conversely, for wires inside a single shielded cable, the crosstalk is high because all wires use a common ground path.

At the ends of the shield, a low-inductance path is needed for current to flow from the shield into the local chassis ground. Some manufacturers provide a drain wire to help with grounding. The drain wire intimately contacts the shield. It is made of the same gauge wire as the inner conductors, but with no insulation. *Warning: Current loop A in Figure 9.14 is often large enough to exceed FCC/VDE limits!* Drain wires work well in analog applications at low frequencies. Drain wires make poor ground connections for high-speed digital signals.

To prevent radiation from a drain-wire current loop, insist on a connector with a metal exterior shell. The foil or braid shield should clamp directly inside the metal shell, leaving no wiring exposed. The metal shell should mate using a wide, flat, low-inductance contact to the metal exterior of the product chassis.

With a plastic chassis, there is no good place to attach the shield. A metal shell connector helps little in this situation. Use method (1) or (3) for noise reduction in plastic systems.

9.6.3 Common Mode Choke

This scheme is almost never designed into a system but is applied as a patch to systems which just don't meet FCC/VDE regulations any other way.

External to the product chassis, but near the I/O connector, wind the I/O cable several turns through a large magnetic core. Since normal signal currents go out through the signal conductors in a cable and return along ground wires in the same cable, the net current through the core is practically zero. It has no effect on currents that return along the cable.

For currents that otherwise would return through some different path, the core has a big effect. These currents pass in only one direction through the core and thus encounter the full inductance of the windings. If this inductance exceeds the natural inductance of the remotely returning current loops, the core will reduce the current flowing in these remote loops.

A variety of common mode chokes are available for all kinds of cable shapes. Even flat ribbon cables can be fitted with special common mode choke cores.

Before fitting a common mode choke, check its specified impedance near the frequency of interest. Not all core materials are effective at very high frequencies.

POINTS TO REMEMBER:

Exposed wiring carrying high-speed digital signals between circuit boards always fails FCC and VDE radiated emission tests.

If you can tolerate the degradation in rise time, filter all outgoing digital signals before they exit the chassis.

A common mode choke reduces current flowing in remote return loops.

The drain wire loop marked (A) in Figure 9.14 is often large enough to radiate more than FCC/VDE limits.

9.7 SPECIAL CONNECTORS FOR HIGH-SPEED APPLICATIONS

AMP and Augat have both created special connectors for high-speed point-to-point applications. These connectors incorporate ground structures internal to the connector. The ground structures serve two functions. First, they provide low-impedance signal return paths for low crosstalk. Second, they increase the parasitic capacitance to ground of each pin, balancing it with the pin's series inductance. This balance minimizes signal distortion in point-to-point transmission line applications.

Teradyne has created a connector with special features for use with multidrop bus applications. It has a very-low-impedance signal return path which reduces radiated emission without increasing parasitic capacitance. The low parasitic capacitance is a good feature for use with a multidrop bus.

9.7.1 AMP Z-Pack Point-to-Point Connector

Cross-section and perspective views of the AMP[9] Z-pack stripline connector appear in Figure 9.15.[10] This AMP connector includes four rows of signal pins.

Thin metal plates fit between each column of four pins. The metal plates serve as low-inductance return-current carriers. Grounding pins for the plates project on either side of the connector. The flat plates inside the connector let the return current spread out, lowering the connector's series inductance.

Crosstalk in this connector is very low between columns of pins. AMP reports the connector is usable with rise times as fast as 250 ps. The crosstalk with a 500-ps rising edge, as measured by the setup in Figure 9.10, is less than 3%. Expect a propagation delay of about 150 ps going through the connector.

The flat plates increase the signal-to-ground capacitance at every pin. This capacitance balances the natural series inductance of the connector. The effective $(L/C)^{1/2}$ ratio

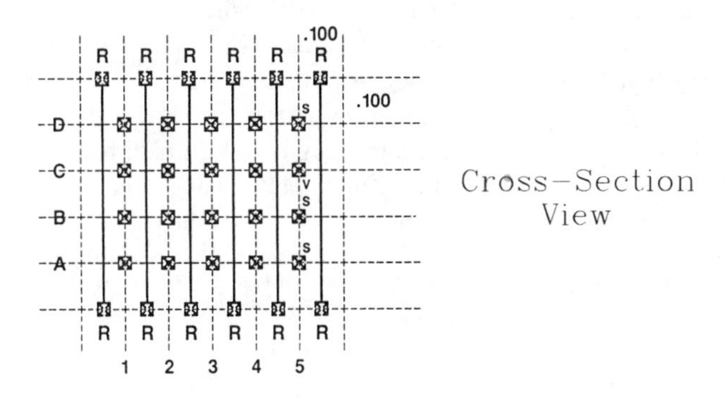

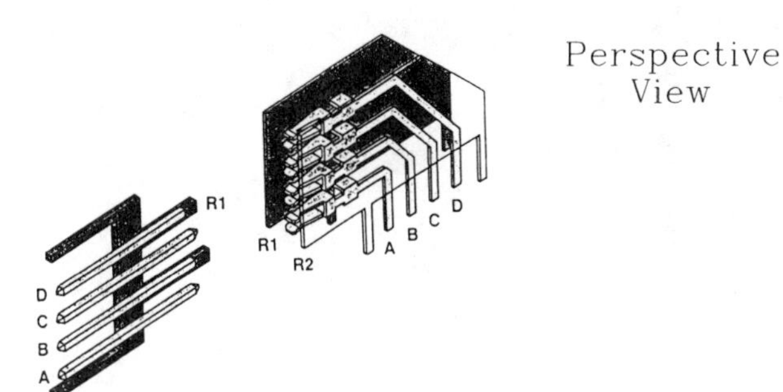

Figure 9.15 AMP Z-Pack connector for high-speed point-to-point applications. (Drawings courtesy of AMP Incorporated.)

[9]AMP is a registered trademark of AMP, Inc.

[10]Reprinted with permission of AMP, Inc., from M. Sucheski and D. Glover, "A High Density, High-Speed, Board-to-Board Stripline Connector," AMP Order No. 82509, AMP Inc., 1990, Harrisburg, Penn.

varies from 40 to 56 Ω, depending on the pad layout and the pin row under test. This connector is suitable for use in series with 50-Ω transmission lines.

9.7.2 Augat Point-to-Point Connector

Augat makes the *electronically invisible interconnect* (EII) connector. This unique connector actually incorporates a tiny ribbon that routes signals from one circuit card to another along a microstrip flex circuit. The flex circuit can be custom-manufactured to any impedance specifications. This connector is very small, possessing a total delay of 115 ps.

Augat reports the connector is usable with rise times as low as 35 ps. At a rising edge time of 900 ps, the microstrip design produces a 2% crosstalk.

The standard microstrip impedance of each pin lies between 45 and 55 Ω. This connector is suitable for use in series with 50-Ω transmission lines.

9.7.3 Teradyne Multidrop Bus Connector

The Teradyne backplane connector system provides four rows of signal pins. It also provides two additional rows of ground pins. The ground pins appear in rows on either side of the signal pin rows, as in the AMP Z-Pack, but there the similarity ends.

Teradyne installs low-impedance ground plates horizontally, as shown in Figure 9.16, not vertically. These plates provide a low-inductance signal return path but do not isolate between columns. Crosstalk in this connector is thus worse than in the AMP Z-Pack. The corresponding benefit is that the plates, being removed from the signal pins, do

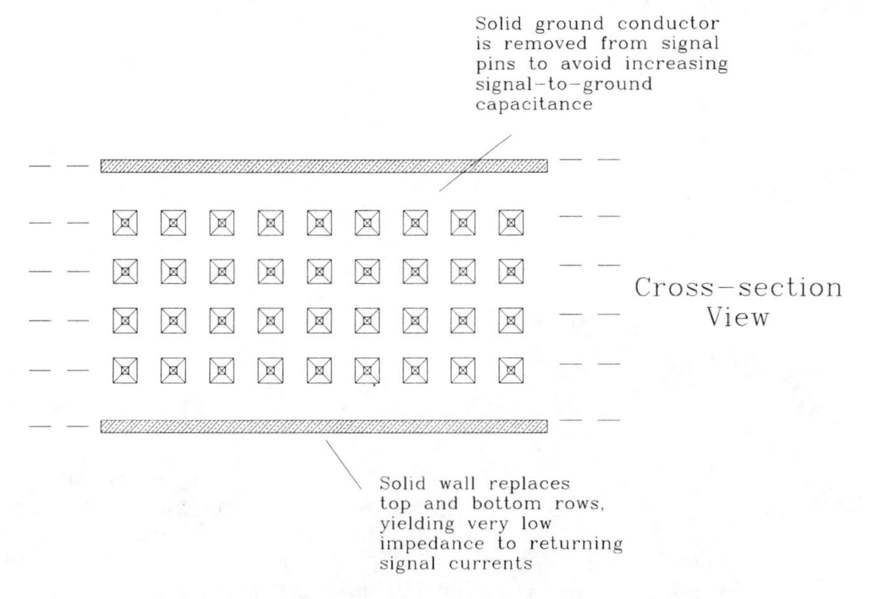

Figure 9.16 Teradyne high-speed connector for multidrop applications.

not add capacitance. The signal-to-ground capacitance is rather low, making this connector a good choice for multidrop bus applications.

> **POINT TO REMEMBER:**
>
> Crosstalk and EMI control require extraordinary connectors at high speeds.

9.8 DIFFERENTIAL SIGNALING THROUGH A CONNECTOR

Differential signaling attacks the problem of signal return current not by providing a low-impedance path for it but by eliminating it.

The theory of differential signaling is simple. Instead of transmitting one signal, transmit two. Send the signal you want, *plus a second signal equal to the negative of the first*. The return current from the first signal is positive. The return current from the second signal is negative. Together, they cancel (see Figure 9.17).

At the receiving location, compare the two signals to determine their logic polarity. The comparison requires no local reference voltage. Ground voltage shifts between the transmitter and receiver affect each line equally and so have no effect on the difference between the two lines. Differential reception is unaffected by ground voltage shifts between the transmitter and receiver.

The only signal current returning from a differential signal pair is that due to any imbalance between the two transmitted signals. If the differential signals are not exactly opposite, their currents do not exactly cancel. This imbalance in current is called *common mode current*. In a well-designed differential driver the common mode current is

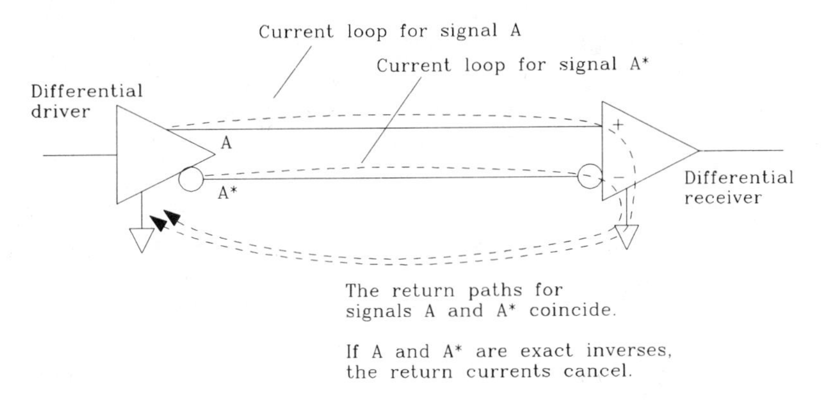

Figure 9.17 Differential signaling eliminates returning signal current.

a factor of 100 less than the primary current. Low common mode current benefits EMI performance.

When sending differential signals through a connector, keep them on adjacent pins. That way their returning signal current paths will overlap and cancel. Also, keep the traces routed close together as they move across the printed circuit board. Here is a formula for determining the amount of imbalance current, as a fraction of the signal current, that occurs when one trace is slightly longer than the other:

$$\text{Imbalance fraction} = \frac{T_p X}{T_{10-90}} \qquad [9.20]$$

where T_p = propagation delay of media, ps/in.
X = length difference between traces, in.
T_{10-90} = rise time of signals, ps

Don't worry about crosstalk between the two signals in a differential pair. They can be routed a little closer together than ordinary traces. Because the crosstalk from one trace correlates with its differential mate, there exists little observable interference. If the two wires are extremely close, the apparent impedance of each trace goes down. Compensate for this effect by using smaller terminating resistors or narrower traces than normal.

EXAMPLE 9.3: Differential Imbalance

Two signals of a differential pair route near each other in FR-4 material. The rise time is 500 ps. One signal diverts 0.3 in. to go around an inconvenient hole in the board. What is the resulting imbalance percentage?

$$T_p = 180 \text{ ps/in.} \qquad \text{(FR-4)}$$
$$X = 0.3 \text{ in.}$$
$$T_{10-90} = 500 \text{ ps}$$

$$\text{Imbalance fraction} = \frac{(180)(0.3)}{500} = 0.108 \qquad [9.21]$$

We have just taken a differential output that was probably balanced to within 1% and increased the imbalance to over 10%.

POINTS TO REMEMBER:

Differential reception is unaffected by ground voltage shifts between the transmitter and receiver.

Common mode current from a well-designed differential driver is a factor of 100 less than the primary current.

9.9 POWER-HANDLING FEATURES OF CONNECTORS

Many connectors intended for backplane applications now include staggered pin heights. Features such as dual, triple, or even four-way pin height selections are common. The variable pin height feature helps sequence soft-start power and reset operations when a card plugs into a live backplane.

Typically, ground connections touch first on the longest pins. Then power connections make contact on the next longest pins. Power sometimes sequences through two pin heights, first applying soft-start power and then full power (see Section 8.2.3). Finally, the card contacts the shortest pins which carry data signals. One of the data pins usually starts a timer which holds the card in reset for a fixed time interval. This reset feature provides enough time for the card to seat fully in its socket. The whole sequencing operation takes less than 0.1 s as the card is pressed into its slot.

When using variable pin lengths, put duplicate sets of the long pins at either end of the connector. That way, if the card tilts one way or the other as it goes into the card slot, it will still hit the long pins first.

POINT TO REMEMBER:

Variable-height pins help sequence soft-start power and reset operations as a card is inserted into a live backplane.

10

Ribbon Cables

The term *ribbon cable* refers to any cable having multiple conductors bound together in a flat, wide strip. The ribbon cable concept is simple, but its implementations vary.

The original 3-M ribbon cable pictured in Figure 10.1 embeds multiple conductors in a thick, extruded gray, plastic insulating medium. Later developments, like rainbow cable, resemble individual round wires bonded together. Finally, some ribbon cables support their wires on the surface of a tough plastic insulating strip. Each dielectric configuration has different high-frequency properties.

Whatever the dielectric configuration, ribbon cable wires always run parallel to each other at precisely controlled separations. A uniform separation facilitates *mass termination connectors* which simultaneously crimp onto every wire in the ribbon during one swift operation. This simultaneous crimping action is an inexpensive means of connecting to a multiwire cable. Ribbon cables proliferate today because, in conjunction with mass termination, they are very cheap.

Fortunately for us, the uniform separation between ribbon cable wires creates a useful side effect: Ribbon cables make excellent transmission lines.

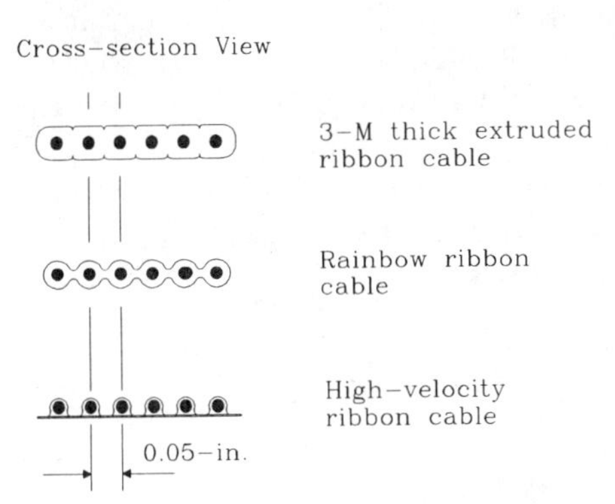

Figure 10.1 Types of ribbon cables.

10.1 RIBBON CABLE SIGNAL PROPAGATION

The rise time of any ribbon cable varies with the square of its length:

$$T_{10-90} = \frac{(L)^2}{K} \qquad [10.1]$$

where T_{10-90} = 10–90% rise time, ns
K = constant dependent on cable, ft^2- GHz
L = length, ft

The square of length is a very fast-moving function. If we cut a cable's length in half, we decrease its rise time fourfold. Cutting the length to one-tenth shortens the rise time by a factor of 100.

Does this relationship apply only to ribbon cable or to all cables? The answer is that it applies to all coax, twisted-pair, and ribbon cables. The following explanation shows why.

Regardless of the cable type, its frequency response is determined solely by the cable's inductance, capacitance, and resistance per unit length. At digital frequencies, the shape of the response curve is dominated by the skin effect (see Section 4.2.3). All cables, whether coaxial, twisted, or ribbon, share the same basic frequency-response shape described in Equation 10.2.[1] The only significant differences for digital systems lie in the constant K which appears in both Equations 10.1 and 10.2.

$$|H(f)| = e^{-0.546\left[\frac{(\text{length}^2)(f)}{K}\right]^{1/2}} \qquad [10.2]$$

where $|H(f)|$ = magnitude of frequency response at f
f = frequency, GHz
K = constant dependent on cable, ft^2- GHz
length = length, ft

Granted, RG-59U coaxial cable has a lower resistance per foot than AWG 30 ribbon cable and therefore a higher value of K. So we get less attenuation per foot on RG-59U coax than on AWG 30 ribbon cable. At any particular frequency we get a different attenuation on the two cables but, drawn on log-log paper, the shapes of the frequency response curves for the two cable types are identical.

What follows from this shape invariance? Looking at Equation 10.2, if we change either K or the length, the response shifts. If we change K but then compensate by changing the length so that the overall ratio L^2/K remains fixed, the frequency response stays

[1]Some variance from this formula exists if the dielectric properties of the insulating medium change. Other variances exist below the skin-effect frequency, approximately 100 kHz for reasonably sized cable. Finally, at the lowest frequencies a cable reverts to *RC* mode, which introduces unusual phase distortion. *RC* mode problems usually occur on very long cables which must be operated below a few kilohertz (the original trans-Atlantic telephone cables operated in this mode).

the same. This is the key to understanding cable attenuation. We get the same frequency response using a long piece of coax or a short piece of ribbon cable.

10.1.1 Ribbon Cable Frequency Response

It is surprising how well ribbon cable works at short distances. Of course, the response depends on how we connect the grounds. Here we assume a G-S-G connection pattern, as depicted in Figure 10.2. This pattern gives us a characteristic impedance between 80 and 100 Ω, depending on the dielectric.

In reference to Figure 10.3, over a span of 10 ft we expect an attenuation of less than 3.3 dB up to about 500 MHz. The next section shows how this attenuation figure gives us a rise time of 1 ns.

The effective bandwidth varies with the inverse square of distance. For connections shorter than 10 ft, ribbon cable performance is terrific.

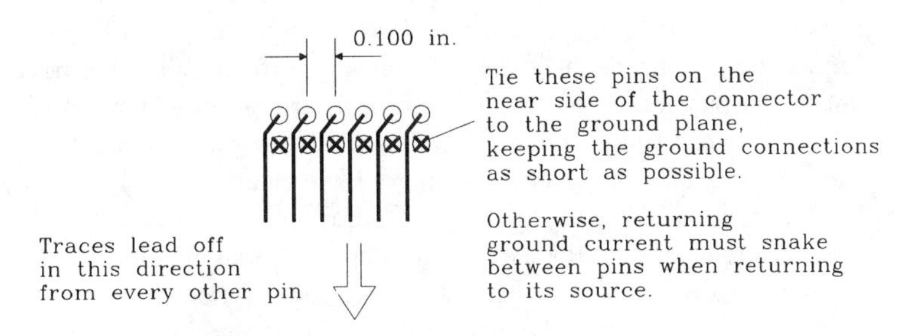

Figure 10.2 Ground-signal-ground configuration for a dual-row ribbon cable connector.

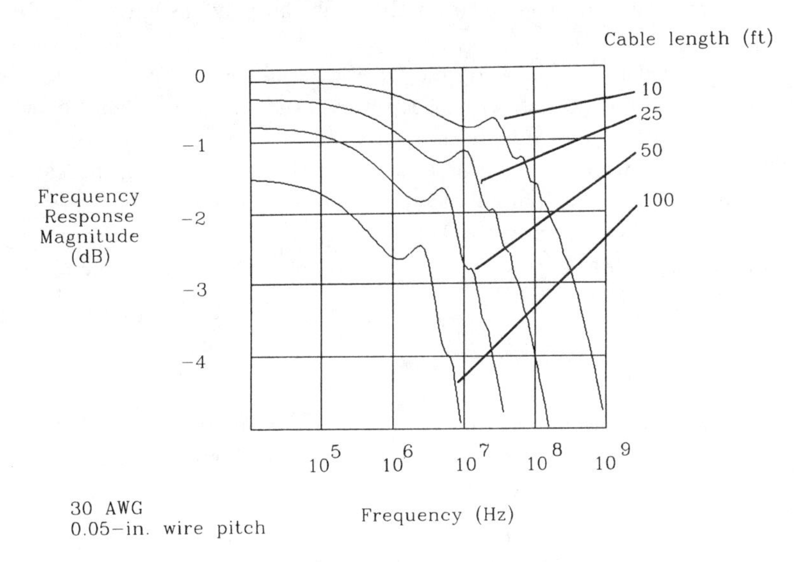

Figure 10.3 Frequency response of a ribbon cable.

At distances greater than 10 ft, the ribbon cable deteriorates markedly. At 100 ft, the 3.3-dB attenuation point occurs at 5 MHz, yielding a rise time of 100 ns.

The response plots in Figure 10.3 show the cable frequency response at lengths of 10, 25, 50, and 100 ft. The shape of the response curve is the same in all cases, but shifted. These plots were computed using MathCad,[2] assuming the cable was terminated with a resistor approximating its characteristic impedance. Because all practical transmission lines have a complex (not real-valued) impedance, the resistor is not an exact termination match. We therefore get little reflections which show up in the frequency response plot as bumps in the frequency response from 3 to 30 MHz. These bumps are only 0.25 dB tall and so have little effect on the measured step response.

Some theoreticians assume a perfect termination when drawing response curves, which therefore have no bumps. Here we use a resistor, the most common termination for practical digital circuits.

Another drawback of a purely resistive termination, besides the bumps, is that the cable resistance introduces attenuation at DC. At great distances, your logic signals may never rise to full height. Witness the 100-ft response curve in Figure 10.3, which shows an attenuation of 1.5 dB at low frequencies. The final value will never rise above 84%. This is a serious source of voltage margin deterioration which motivates the use of line receivers with carefully centered switching thresholds. Such receivers are less sensitive to voltage margin deterioration than ordinary logic gates.

The cable's dielectric configuration impacts performance in two ways. It controls the signal propagation velocity, and it controls the attenuation. Propagation velocity, in ft/ns, is inversely proportional to the square root of electric permittivity. Cables having dielectric material completely surrounding the wires exhibit a higher effective permittivity and thus a lower overall speed. Cables supporting their wires on a thin, flat plastic sheet carry most of their electric fields in the air and thus have low permittivity and high speed.

Attenuation depends on the ratio of series resistance to cable impedance. At high frequencies, the skin effect causes series resistance to rise with the square root of frequency. This makes attenuation also go up with the square root of frequency. The dielectric configuration influences attenuation by changing the cable's characteristic impedance. Cables having dielectric material completely surrounding the wires exhibit a higher effective permittivity, causing a lower impedance and more attenuation. Cables built on a thin, flat plastic sheet have high speed, less attenuation, and a faster rise time.

10.1.2 Ribbon Cable Rise Time

We may evaluate the rise time of a cable by taking the inverse Fourier transform of its frequency response. This calculation appears in Figure 10.4 for the four ribbon cable lengths used in the previous section.

[2]MathCad is a registered trademark of MathSoft, Inc.

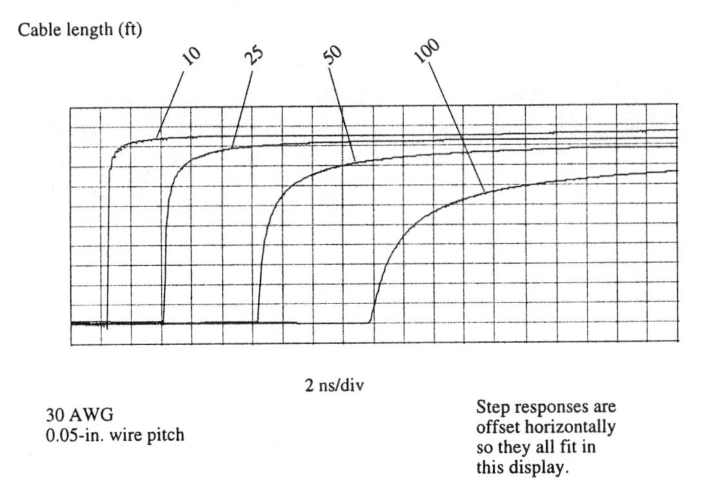

Figure 10.4 Step response of a ribbon cable.

As the cable length grows, the rise time stretches even further. The rise time is proportional to the square of the length. At great lengths, the rising signal never reaches full strength. This effect is due to the attenuation at DC caused by the cable resistance.

These cable waveforms are not like the ordinary gaussian rising waveforms that come from logic gates or complex systems. The cable waveform is the result of its unique $1/\sqrt{f}$ frequency response. The middle section of each pulse rises quickly but is preceded and followed by long, slow-moving tails. This behavior is common to all conductive cable types (but not to fiber optics).

At very long lengths, when the cable reverts to an *RC* mode of operation, the response becomes even more asymmetric. The precursor shrinks, but the final, slowly rising tail gets worse. These long tails introduce significant intersymbol interference in long-distance transmission systems. For normal digital applications, you should keep the system clock much slower than the 10–90% rise time of the cable to avoid overlap between pulses.

Sometimes we are given a cable specification and asked to predict its rise time. If we have a complete frequency response curve for the cable, that is easily accomplished. First find the value of *K* that makes Equation 10.2 best fit the frequency response of your cable. Given just one value of attenuation at a specific length and frequency, we can invert Equation 10.2 to find *K*:

$$K = \frac{L_0^2 F_0 (22.5)}{A_0^2} \quad [10.3]$$

where K = constant dependent on cable, ft^2-GHz
L_0 = cable length at which cable is specified, ft
F_0 = frequency at which cable is specified, GHz
A_0 = attenuation, dB

Catalogs usually list several values for attenuation at specific frequencies and lengths. Try the calculation of K at various specification points. It usually wanders around somewhat at different frequency values. This is due to imperfections in the dielectric which cause the permittivity to change slightly as a function of frequency. Specification points near the intended working frequency give the best answers.

Knowing K, we can then use Equation 10.1 to solve for the rise time:

$$T_{10-90} = \frac{(L)^2}{K} \qquad [10.4]$$

where T_{10-90} = 10–90% rise time, ns
K = constant dependent on cable, ft^2-GHz
L = length, ft

10.1.3 Measuring Rise Time

Measure cable propagation under these conditions:

(1) Terminate the cable at its far end with a resistor. The source impedance must be low compared to the characteristic impedance of the cable. If the source impedance is not low, then at least make sure it is purely resistive. Alternately, use the driver you actually plan to put in your circuit.

(2) The termination resistance value is equal to $(L/C)^{1/2}$, where L and C are the inductance and capacitance per foot of cable. This is the best resistive termination. It gives a small, but not zero, reflection.

(3) The input must be a step function with a rise time much shorter than the cable rise time. Your oscilloscope should also be faster than the cable. If the oscilloscope and pulse generator (or driver) are not much faster than the cable, then subtract the effect of the scope and pulse rise times from your measurement. First connect the scope directly to the pulse generator (or driver) and measure the drive rise time T_{drive}. Then install the cable and measure the rise time again.

$$t_{cable} = \left[(t_{measured})^2 - (t_{drive})^2 \right]^{1/2} \qquad [10.5]$$

where t_{cable} = actual rise time of cable, s
$t_{measured}$ = measured rise time of cable, s
t_{drive} = measured rise time of scope and pulse generator (or driver), s

(4) Sense the output using a probe which does not load the line. Most commercial 10:1 scope probes are inadequate for this purpose. Build a shop-made 10:1 probe or buy a special low-capacitance, active high-frequency probe for this measurement. Common 10-pF probes have an impedance magnitude of $-j31\ \Omega$ at 500 MHz. Such a probe will interact heavily with your measurements.

POINTS TO REMEMBER:

The rise time of any ribbon cable varies with the square of its length.

All cables, whether they be coaxial, twisted, or ribbon, share the same basic frequency response. Their frequency response, in decibels, is proportional to the inverse square root of frequency.

A ribbon cable's dielectric configuration impacts both signal velocity and attenuation.

10.2 RIBBON CABLE CROSSTALK

Crosstalk in ribbon cables varies with the placement of grounds among the signal conductors. Any level of crosstalk attenuation is achievable given enough grounds. How many grounds are enough?

10.2.1 Basic Calculation of Crosstalk

Crosstalk in ribbon cables results from both inductive and capacitive coupling. As described in Section 5.7, the capacitive and inductive crosstalk components are approximately equal in intensity. Their near equality causes a large reverse coupling coefficient but almost no forward coupling.

Since forward coupling is a small difference between two large coupling mechanisms, it is almost impossible to calculate analytically. Direct measurement is the best approach. When measuring forward crosstalk, realize that fields surrounding a ribbon cable extend out into the space surrounding the cable. The results vary if other conductive or magnetically permeable materials are present near the surface of the ribbon. Support the ribbon a few inches above any conducting or dielectric objects when making this measurement.

Reverse coupling is quite large and is easy to calculate for simple geometries. Since the inductive and capacitive components very nearly balance, you can compute only the inductive coupling and then double your result. The calculation of inductive reverse coupling proceeds in three parts. First, model the magnetic field patterns emanating from the signal wires. Next, integrate to find the total flux captured between the receiving wires. Finally, convert the change in flux per unit time to a voltage.

The simplest geometry for computing crosstalk is the four-wire example shown in Figure 10.5.

Wire B in Figure 10.5 carries the transmitted signal. We will assume all the signal current flowing out along B returns to its source along ground wire A. The currents on

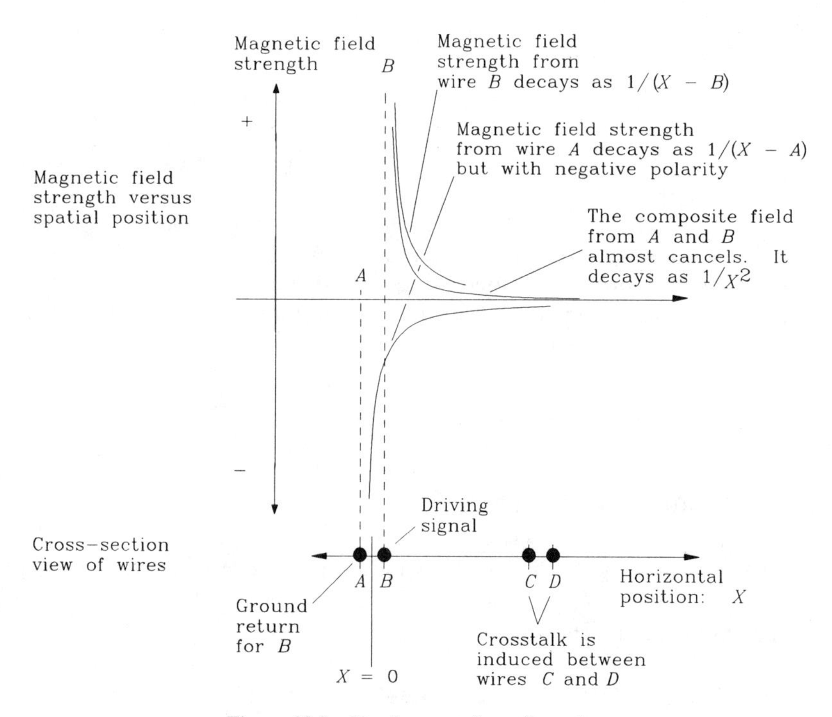

Figure 10.5 Simple crosstalk configuration.

wires *A* and *B* are therefore equal and opposite. We are interested in calculating crosstalk induced between wires *C* and *D*.

Let's use our intuition to see if we can deduce some properties of the final answer. We have the following facts at our disposal:

- Magnetic field intensity, as a function of distance from each wire, varies as $1/R$.
- Magnetic fields from the signal wire and its return-current path partially cancel. The overall magnetic field profile is thus proportional $1/R^2$, which is the derivative of $1/R$. The integrated flux over any small area remote from *A* and *B* will vary as $1/R^2$.
- The partial cancellation of fields from signal and return wires is proportional to the separation between them, called Δ_1.
- The total flux captured between the two receiving wires is proportional to the separation between them, called Δ_2.

From these four facts we deduce the following about coupling from wire pair (A, B) to wire pair (C, D):

$$\text{Coupling is roughly proportional to } \frac{\Delta_1\Delta_2}{R^2} \qquad [10.6]$$

The constant of proportionality turns out to be a function of the cable characteristic impedance and cable delay. Equation 10.7 lists the overall reverse coupling coefficient between two wire pairs:

$$K_r \approx 2538\ (\text{delay}) \frac{1}{Z_0} \frac{\Delta_1 \Delta_2}{X^2} \qquad [10.7]$$

where K_r = reverse coupling coefficient
delay = delay of line, ps/in.
Z_0 = characteristic impedance of line, Ω
X = separation between wire pairs, in.
Δ_1 = separation between wires in pair 1, in.
Δ_2 = separation between wires in pair 2, in.

An astute reader may recognize Equation 10.7 as half the ratio of the mutual inductance between pairs to the self-inductance of the signal loop. You can find the reverse coupling coefficient by using a low-frequency inductance meter to measure the mutual inductance and self-inductance and then taking half their ratio.

In a cable having many ground wires, the coupling ratio in Equation 10.7 is diminished at least by one-half and perhaps to one-fourth of its stated size.

10.2.2 Effect of Multiple Grounds

The previous section assumed all return current for signal wire *B* came back along ground wire *A*. Practical ribbon cable implementations have multiple ground wires. Returning signal current always splits among all the ground paths, according to the inductance of each path. More return current flows in the low-inductance paths which lie near the signal wire, and less in outlying paths.

Imagine there are *N* ground wires in a cable, indexed from 1 to *N*. Let X_0 equal the distance between the driving signal wire and its closest ground. If the driving signal wire lies evenly between two grounds, then X_0 equals the separation between the signal and either ground. Returning signal current in any ground wire *n* is determined roughly by

$$I_X \approx \frac{K_1}{1 + \left(X_n^2 / X_0^2 \right)} \qquad [10.8]$$

where X_0 = separation between signal and first ground, in.
X_n = separation between signal and ground *n*, in.
I_n = return current in ground *n*, A
K_I = constant selected so sum of all return currents equals signal current

Crosstalk on any signal wire is dominated by the amount of ground current on nearby wires. In the *ground-signal-ground*, or *G-S-G*, configuration commonly used with ribbon cables, the grounds alternate with signals. Every signal wire lies between two ground wires. Crosstalk in a ground-signal-ground cable, as a function of physical separation X from the driving signal wire, closely follows the ground current distribution:

$$V_r \approx \frac{K_2}{1 + \left(X^2 / X_0^2 \right)} \qquad [10.9]$$

where X_0 = separation between driving signal and first ground, in.
X = separation between driving signal and signal under test, in.
V_r = reverse coupling coefficient
K_2 = constant determined by cable construction

In Equation 10.9 the coefficient K_2 ranges from about one-tenth to one-fourth. This corresponds to reverse crosstalk between nearest neighbors in the range 2–5%.

For cables with fewer grounds, crosstalk is bigger but still falls off as $1/X^2$. It is also proportional to the separation Δ_1 between the transmitting wire and its nearest return wire, and also the separation Δ_2 between the receiving wire and its nearest ground.

In a sparsely grounded cable, doubling the number of grounds, which halves both Δ_1 and Δ_2, cuts the crosstalk between remote wires by one-fourth. The crosstalk between adjacent wires changes little unless we interpose grounds between them.

10.2.3 Effect of Twists

Twisted cables have unique advantages when used properly. Ideally, one arranges to have each signal wire and its nearest return path twisted tightly together. This holds the wires in close physical contact, reducing the separation Δ_1 between them. Every signal wire in a twisted cable should have its own individually twisted ground return wire.

As a signal propagates along twisted pair X, the magnetic field emanating from the pair flips polarity every time the wires twist over each other. Remember, the magnetic fields from the two wires have opposite polarity and nearly cancel. The field polarity at some distance from the pair is determined by which wire is closest. As the wires roll over each other, one or the other draws nearer and the magnetic field polarity reverses.

The result is that crosstalk from a twisted pair onto a straight, parallel wire pair is practically zero. The net crosstalk from the alternating plus and minus coupling cancels!

The net crosstalk between two adjacent twisted-pair wires is also zero, provided the wires are twisted in the same direction. This effect assumes the wires are positioned uniformly and twisted at a constant rate.

Practical twisted-pair cables incorporate a different twist rate on every wire pair. This tends to cancel out some pickup that results from slight asymmetries in the twisting process. If the twist rates on two pairs synchronize, then a slight wobble in the twisting machine may introduce a persistent pickup effect. High-quality twisted-pair cables use either different overall twist rates or random variations in the twist rate for each wire pair. Do not expect much cancellation due to twisting unless the length of a rising edge extends over many twists. The number of twists required to ensure cancellation depends not on the twist rate but on the minimum *difference in twist rates* among the wires in the cable.

Given two twisted pairs in a single cable, find the length at which pair A executes one more twist than pair B. Call this the *precession length* of the two pairs. The total coupling over exactly one precession length is zero. Coupling at a fraction of a precession length is nonzero. At nanosecond rise times, we need precession lengths on the order of 1 in. to achieve the complete benefit of twisting. This requires lots of total twists per inch to guarantee a precession length among all pairs of 1 in. or less.

Fortunately, twisting rarely ever hurts. It's better to try it than not.

Another advantage of twisting a pair of communication wires is reduced electromagnetic emissions. Most of the return current for each signal flows in its twisted ground wire, whose alternations cancel the radiated field pattern.

When used in conjunction with differential transmission, twisted cabling really shines. Differential signaling contains practically all the return current in the tiny loop area between the signal+ and signal– wires. This reduces emissions 20–30 dB below the emission levels of single-ended transmission. Crosstalk between adjacent pairs of a twisted-pair cable, when used differentially, is very low.

Be careful when using twisted-pair cable to properly assign pins on the connectors. A casual error here can result in twisting together pairs of signal wires instead of each signal with its ground. The results are hilarious, but your manager won't be pleased.

Twisted-pair cable is available in a flat ribbon configuration. This cable, popularly known as Twist 'N' Flat[3], consists of many individually twisted pairs bonded into a flat ribbon structure. Every few feet, the twists flatten out into an ordinary flat ribbon section with a predictable pin-out which accommodates mass termination connectors. Because the flat sections couple like ordinary flat ribbon cable, we do not get the full benefit of twisting, but some benefit results. The advantage is that we can cheaply attach connectors.

10.2.4 Measuring Crosstalk

Figure 10.6 illustrates the far-end (forward) and near-end (reverse) crosstalk on a typical sample of ribbon cable. This cable is 8 ft long, with a 0.05-in. wire spacing. The wires are AWG 30, having a diameter of 0.01 in.

The driving waveform at the beginning of the cable appears at the top of Figure 10.6. This waveform was recorded at three different rise times: 5, 10, and 20 ns. The various responses were recorded using a Tektronix 11403 digital oscilloscope and superimposed for this figure.

The cable has 10 wires configured with every other wire grounded at both ends. This is a ground-signal-ground configuration. Signaling occurs on even-numbered wires. The driving signal traverses wire 6, with crosstalk displayed from wire 8. Both ends of wire 8, as well as the far end of wire 6, terminate in the characteristic impedance of the cable, 100 Ω.

The second section of Figure 10.6 shows the near-end (reverse) crosstalk on wire 8. The vertical scale of the near-end noise waveform is blown up 25 times larger than the

[3]Twist 'N' Flat is a registered trademark of the Amphenol Corporation.

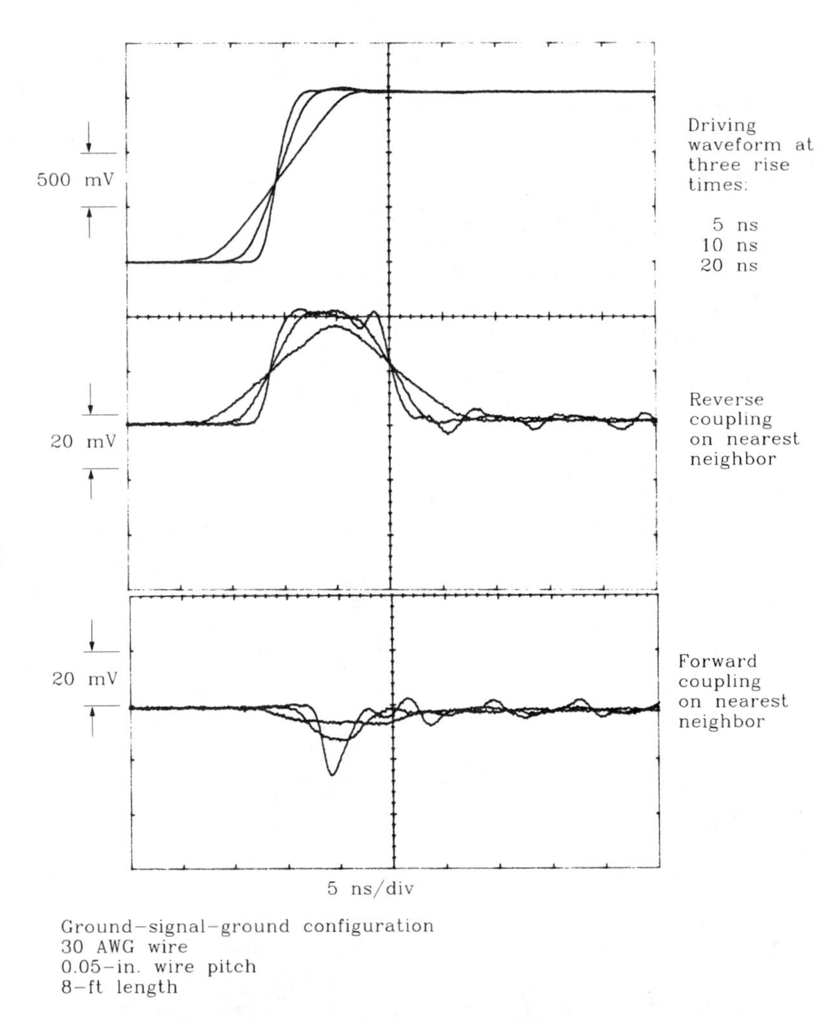

Figure 10.6 Forward and reverse coupling on a ribbon cable.

driving signal so that we can see the crosstalk. The ratio of crosstalk to signal is about 2.5%. Values in the 2–5% range are common for G-S-G-wired cables.

The near-end crosstalk pulse begins immediately along with the driving waveform, with rising and falling edges matching those of the driving waveform. The crosstalk forms an elongated pulse. This pulse length equals 22 ns, twice the one-way delay time of the cable (see Section 5.7). The one-way delay time must therefore be 11 ns. The cable delay per inch on this 8-ft cable is

$$D = \frac{11 \text{ ns}}{96 \text{ in.}} = 114 \text{ ps/in.} \qquad [10.10]$$

The effective relative permittivity of the cable must be

$$\varepsilon_r = \left(\frac{D}{C}\right)^2 = \left(\frac{114}{85}\right)^2 = 1.8 \qquad [10.11]$$

If we slowed the driving rise time further, the near-end crosstalk would not have time to climb to its full amplitude. For driving rise times shorter than 20 ns, the near-end crosstalk has a constant amplitude. Near-end crosstalk between individual wires is bounded at a maximum amplitude of 2–5% on G-S-G cables.

Crosstalk onto each signal wire aggregates from every wire in the cable. Assume both wires on either side of wire 8 generate the same crosstalk. Then add more crosstalk from remote wires according to the $1/(1 + n^2)$ crosstalk rule. We get an aggregate near-end crosstalk in the range 8–20%.

The third section of Figure 10.6 shows far-end crosstalk. The scale of the far-end crosstalk waveform is also blown up 25 times larger than the driving signal. The far-end crosstalk is a short blip, not an elongated pulse like the near-end crosstalk. Reaching a maximum amplitude of 1.6%, the far-end crosstalk for this configuration causes less trouble than near-end crosstalk. Also, because the far-end crosstalk decays quickly, we may clock these data lines safely after only a short delay. Near-end crosstalk persists for two full cable delays. When using low-impedance drivers, the near-end crosstalk converts to far-end crosstalk as discussed in Section 5.7.4, fouling up the quick decay property of far-end crosstalk. The quick decay property works only with source-terminated lines.

Far-end crosstalk follows the derivative of the driving signal and so grows without bound when we shorten the driving rise time. A rise time of 1 ns in Figure 10.6 would generate an 8% crosstalk on this cable. A rise time of 100 ps would probably not generate an 80% crosstalk because crosstalk that large interacts heavily with the driving signal. Our crosstalk calculations assumed small crosstalk, saving us the trouble of computing these interactions. In any case, we suspect that a rise time of 100 ps would generate too much crosstalk. If we separated the signals by two or three grounds instead of one, we could expect the far-end crosstalk to drop by a big factor.

Far-end crosstalk accumulates as it propagates down the cable. Far-end crosstalk on a shorter cable would be much less, and on a longer cable much more.

Near-end crosstalk stays the same amplitude regardless of cable length but elongates in time as the cable is stretched.

10.2.5 Stacking Ribbon Cables

Crosstalk increases markedly when wires come close together. This applies to ribbon cables as well as to other cable types.

Two ribbon cables stacked tightly on top of each other can easily display more crosstalk between cables than among the wires in one cable. When routing ribbon cables, always use cable spacers to keep the cables separated.

When folding a cable to fit inside a round shield, you will observe a similar increase in crosstalk.

POINTS TO REMEMBER:

Any level of crosstalk attenuation is achievable given enough grounds.

Crosstalk falls off as $1/X^2$. It is also proportional to the separation Δ_1 between each transmitting wire and its nearest return wire, and the separation Δ_2 between any receiving wire and its nearest ground.

In the ground-signal-ground, or G-S-G, configuration the reverse crosstalk coefficient between nearest neighbors is about 2–5%.

In a twisted cable, if the rising edge spreads out over N precessions of the twist cycle, we can expect coupling to be about $1/N$ the amount in an ordinary parallel wire cable.

Far-end crosstalk accumulates as it propagates down a cable.

Near-end crosstalk stays the same amplitude regardless of cable length but elongates in time as the cable is lengthened.

10.3 RIBBON CABLE CONNECTORS

Mass termination connectors attach simultaneously to all wires in a ribbon cable in one quick stroke. When pressed onto flat ribbon cable, pins in the mass termination connector penetrate the cable insulation, make contact with the inner conductors, and form a permanent gas-tight seal on every wire. These connectors, also called *insulation displacement connectors*, are good for one and only one insertion. Do not remove an insulation displacement connector for reuse. The pins are permanently strained during the insertion process and will not properly seat during a second insertion.

Mass termination connectors crimp to a ribbon cable on one side and have other connecting terminals on their other side. The other connecting terminals may be either female or male pins designed for removable interconnection with other connector types. Some mass termination connectors have solder pins intended for permanent direct mounting on printed circuit boards.

Whatever the mechanical scheme, ribbon cable connectors always introduce parasitic inductance and capacitance. As with any other connector, the performance of your digital signaling loop degrades because of these parasitic effects.

10.3.1 Connector Inductance

Equation 10.12 estimates the self-inductance of a loop formed by a single signal pin and one ground pin.

$$L = 10.16x \ln\left(\frac{H}{r}\right) \qquad [10.12]$$

where L = inductance, nH
H = pin separation, in.
x = connector pin length, in.
r = connector pin radius, in.

Using typical values of $r = 0.0125$, $x = 0.4$, and $H = 0.1$, we arrive at an inductance of 8 nH for a pair of pins. In a G-S-G configuration like Figure 10.2, halve this approximation to account for the presence of multiple nearby grounds.

An inductance of L henries, in series with a transmission line of Z_0 ohms, introduces a rise-time degradation of

$$T_{10-90} = 2.2\frac{L}{2Z_0} \qquad [10.13]$$

Our 8-nH single-pin inductance in series with a 100-Ω line generates a rise time of 100 ps. In a G-S-G configuration, the degradation would be less.

10.3.2 Connector Capacitance

Equation 10.14 estimates the parasitic capacitance between a single signal pin and one ground pin.

$$C = 0.7065\frac{x}{\ln(H/r)} \qquad [10.14]$$

where C = capacitance, pF
H = pin separation, in.
x = connector pin length, in.
r = connector pin radius, in.

Using typical values of $r = 0.0125$, $x = 0.4$, and $H = 0.1$, we arrive at a capacitance of 0.136 pF between each pair of pins. In a G-S-G configuration like Figure 10.2, more than double the approximation to account for the presence of multiple nearby grounds.

A capacitance of C farads, shunted across a transmission line of Z_0 ohms, introduces a rise-time degradation of (see Equation 4.76)

$$T_{10-90} = 2.2C\frac{Z_0}{2} \qquad [10.15]$$

Our 0.136 pF single-pin capacitance, when shunting a 100-Ω line, generates a rise time of 15 ps. In a G-S-G configuration, the degradation would be greater.

10.3.3 Staggering Connections to Reduce Parasitic Effects

When working at subnanosecond speeds, parasitic connector effects play a big role. Anything that reduces their impact is useful.

AMP makes a mass termination ribbon cable connector which works particularly well for high-speed circuitry. Their connector part number is AMP 1-111037-1. This connector provides insulation displacement pins on one side for pressing onto a ribbon cable. The other side provides solder pins for permanent direct attachment to a printed circuit board. The absence of any detaching feature shortens the pins, reducing overall parasitic effects.

The AMP 1-111037-1 connector also staggers its pin configuration, as shown in Figure 10.7. The staggered effect raises the series inductance while lowering shunt capacitance. Staggering the pins is a good tradeoff for use as a multidrop bus connector.

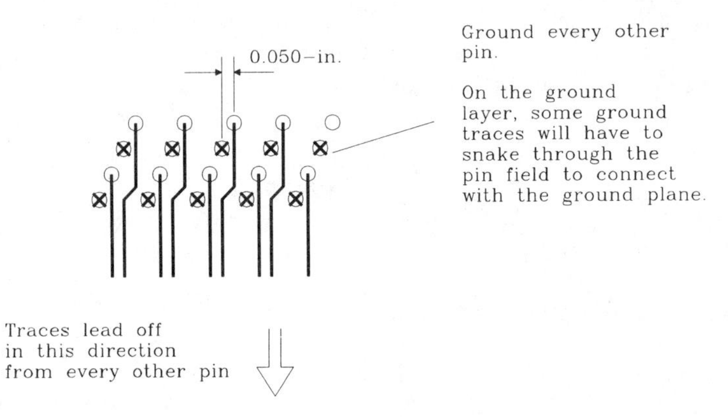

Figure 10.7 Staggered ribbon cable connector pins reduce parasitic capacitance.

> POINTS TO REMEMBER:
>
> Ribbon cables proliferate today because, in conjunction with mass termination, they are very cheap.
>
> Whatever the connector scheme, it always introduces parasitic inductance and capacitance.

10.4 RIBBON CABLE EMI

Ribbon cables suffer from severe EMI problems when routed between cabinets. In response to those problems, manufacturers of these cables have created several types of shielded ribbon cable.

Shields provide a very-low-inductance return path for signal currents. Since the returning current follows the path of least inductance, a shield tends to concentrate returning currents in the shield. This reduces current flow in other more remote paths.

10.4.1 Flat Foil Wrap

A long strip of foil folded or spiraled around a ribbon cable makes a continuous solid shield. With a spiral wrap, be sure that each joint between overlapping spiral layers makes firm electric contact with the previous layer. Otherwise, returning current has to spiral around the signal wires to get back to its source.

Grounding a flat wrap to a product chassis is an awkward procedure. Connectors are not available to handle both the foil shield and signal connections in an integrated fashion. To avoid the drain wire effect discussed in Section 9.7, you must first run the shield inside your cabinet. Then slit the shield along both its edges and peel it away from the ribbon on both sides (top and bottom). Both the top and bottom shield sections must connect electrically to the chassis at the point of entrance.

10.4.2 Flat Shield on One Side

A flat copper braid, bonded to one side of a flat ribbon cable, offers several advantages. First, the close proximity of the copper braid acts like a ground plane, reducing crosstalk between individual wires in the cable. The bonded copper braid is more uniform than a foil wrap and offers better transmission line characteristics. Second, the copper braid provides a low-inductance path for returning signal currents.

The braid usually sports a drain wire for electric connection to ground. The drain wire is a weak spot in this system. If you find a way to connect the braid directly to your ground plane, the braid will be much more effective.

Some manufacturers of flexible circuits are now able to produce two-sided flex cables with a ground on one side and signal traces on the other. These flex cables can have plated-through vias, bringing all the signals and grounds onto the same side of the cable for soldering or other attachment.

10.4.3 Folded (Round) Shielded Cables

Manufacturers can scrunch or fold a ribbon cable into a round shape that fits inside an ordinary round shield and jacket. In this form, the shielded ribbon cable looks just like a regular shielded multiwire cable.

These cables enjoy both the advantages of mass termination connectors and the advantages of shielding. The shielding, however, is only as good as the connection between the shield and a good ground.

POINTS TO REMEMBER:

Shields provide a very-low-inductance return path for signal currents.

With a spiral wrap, be sure that each joint between overlapping spiral layers makes firm electric contact with the previous layer.

A drain wire is a weak spot in any shield.

11

Clock Distribution

Clock signals toggle faster than any other signals in a digital system. For every data transition some clock must transition twice, completing a full cycle. Not only are clocks the fastest signals, they are also the most heavily loaded. Clocks connect to every flip-flop in a system, while individual data wires fan out to only a few devices each.

Because they are so fast and heavily loaded, clock signals deserve special attention. This chapter examines clock drivers, special clock routing rules, and peculiar circuits used to improve the distribution of clock signals.

11.1 TIMING MARGIN

The circuit in Figure 11.1 is a 2-bit *ring counter*, also called a *switch-tail counter*. When clocked at low speeds, the bit pattern at Q_1 repeats forever (...00110011...).

As we raise the clock frequency in Figure 11.1, the circuit emits the same pattern until at some high frequency the circuit fails. The circuit fails because of a lack of setup

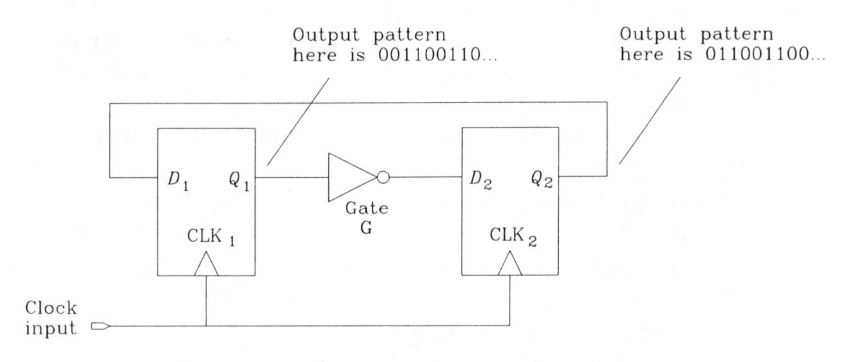

Figure 11.1 A 2-bit ring counter.

time for flip-flop 2. At the failure frequency, each transition at Q_1 emerges from gate G too late to meet the setup time requirement of D_2. Figure 11.2 diagrams this failure mode. When clocked at or beyond the failure frequency, the circuit no longer produces an 0011 output sequence. This type of failure is called a *timing margin failure*.

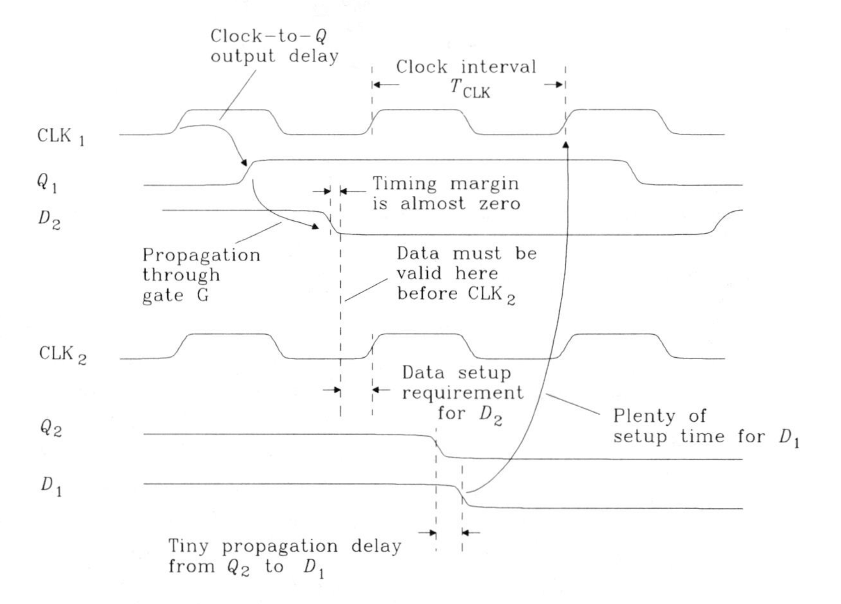

Figure 11.2 Timing analysis of a 2-bit ring counter.

The *timing margin* is defined in this circuit as the amount of time remaining between

(1) The time when signals actually emerge from gate G and
(2) The time when signals at D_2 must be valid to meet the setup requirement of flip-flop 2.

The timing margin measures the slack, or excess time, remaining in each clock cycle. A system with a big timing margin on every circuit can usually run at a higher clock speed without error.

As the clock speed in Figure 11.1 approaches its failure frequency, the timing margin drops to zero. Never operate a circuit near its failure frequency. Reduce the maximum operating speed for any circuit somewhat below the failure frequency, leaving a small positive timing margin under all operating conditions. A positive timing margin protects your circuit against signal crosstalk which may slightly perturb the edge transition times, general miscalculations that often occur when counting logic delays, and later minor changes in the board design or layout.

Many designers aim for a positive timing margin equal to about one gate delay. When working with slow logic families, this rule of thumb allots more timing margin than when working with fast logic families. This keeps the timing margin fixed as a percentage of delay over a wide range of designs. You will have to decide how much excess timing margin is acceptable.

The timing margin depends on both the delay of logic paths and the clock interval. Either too long a delay or too short a clock interval can cause a timing margin failure. As explained in the next section, differential delays between the clock signals CLK_1 and CLK_2 can also cause a timing margin failure.

POINTS TO REMEMBER:

Timing margin measures the slack, or excess time, remaining in each clock cycle.

Timing margin protects your circuit against signal crosstalk, miscalculation of logic delays, and later minor changes in the layout.

11.2 CLOCK SKEW

Let's take a closer look at timing margins. Figure 11.3 dissects our ring counter circuit, showing the components of timing margin analysis.

We seek the worst-case timing margin. Figure 11.3 calculates the latest possible time of arrival for pulses emerging from gate G, comparing that to the earliest possible arrival time required by the setup conditions of flip-flop 2.

The latest possible arrival time for a pulse coming through gate G is

$$T_{slow} = T_{C1,max} + T_{FF,max} + T_{G,max} \qquad [11.1]$$

where T_{slow} = slowest arrival time for pulse from gate G, s
$T_{C1,max}$ = maximum delay of path C_1, s
$T_{FF,max}$ = maximum delay, clock to Q, of flip-flop 1, s
$T_{G,max}$ = maximum delay of gate G, including circuit trace delay, s

In Equation 11.1 we use maximum delay times for all elements. We also assume that the clock pulse of interest occurs at time zero; no absolute time reference appears in Equation 11.1.

The pulse from G gets clocked into flip-flop 2 on the next clock pulse. This clock occurs at time T_{CLK} and propagates through path C_2 to input CLK_2. The earliest possible arrival for the next clock at CLK_2 is $T_{CLK} + T_{C2,min}$. Flip-flop 2 requires a valid input at least T_{setup} seconds before this CLK_2. The arrival time required by flip-flop 2 is

$$T_{required} = T_{CLK} + T_{C2,min} - T_{setup} \qquad [11.2]$$

where T_{required} = elapsed time by which data from G must arrive, ns
T_{CLK} = interval between clocks, s
$T_{C2,\text{min}}$ = minimum delay of path C_2, s
T_{setup} = worst-case setup time required by flip-flop 2, s
data at D_1 must arrive at least T_{setup} seconds before CLK_2

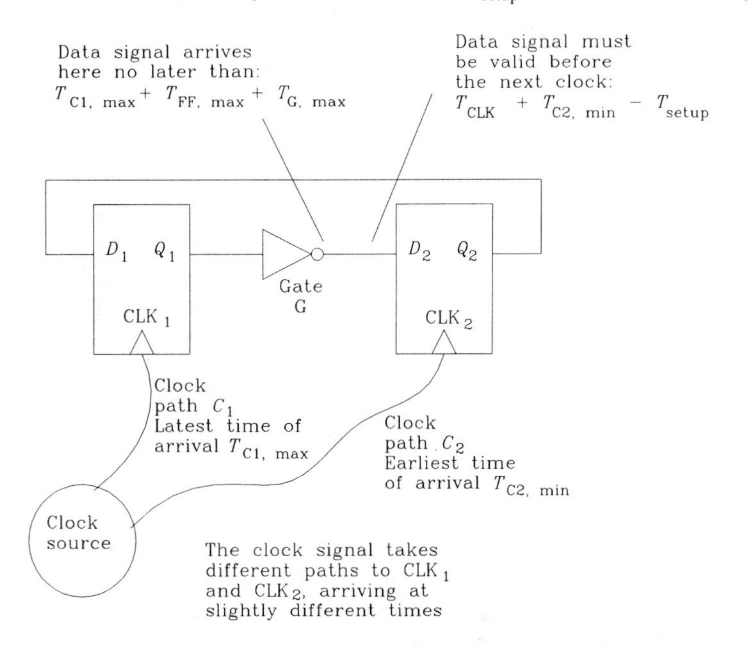

Figure 11.3 Timing analysis showing clock skew.

Equation 11.2 uses the minimum delay time for path C_2, which moves the required data arrival time to the early side. This would be the worst condition.

Data from G must arrive before T_{required} to properly set flip-flop 2. In mathematical terms, we require

$$T_{\text{slow}} < T_{\text{required}} \qquad [11.3]$$

This constraint may be expanded using Equations 11.1 and 11.2:

$$T_{\text{CLK}} > T_{FF,\text{max}} + T_{\text{G},\text{max}} + T_{\text{setup}} + \left(T_{C1,\text{max}} - T_{C2,\text{min}}\right) \qquad [11.4]$$

In words, the clock interval must exceed the flip-flop delay, the gate G delay, and the setup time. These three terms make perfect sense because all three events must occur in sequence each cycle. The last term takes more explaining. It involves the *difference in clock arrival times* at nodes CLK_1 and CLK_2. This difference is called *clock skew*. If the clock arrives late at flip-flop 1, then output Q_1 also occurs late, and our timing margin deteriorates. If delay C_2 is unusually small, flip-flop 2 gets clocked earlier, and data must be valid earlier to meet the setup time. This also deteriorates our timing margin. In either case we must increase the clock interval, slowing down system performance, to fix the problem. Clock skew always affects timing margins.

What if clock 2 is late instead of early? That gives a *greater* timing margin. Some designers make use of this effect in sequential processing nesting, carefully positioning the clock at each stage for maximum performance. Such an approach does not work if there is any feedback from later stages back to the beginning. In Figure 11.3 retarding the clock on flip-flop 2 adds timing margin to D_2, but takes it away from D_1. Most designers just try to minimize clock skew.

In Equation 11.4 only the *difference* in clock propagation delays will matter. The absolute amount of clock delay, as long as it is balanced between the two paths, does not matter.

In the practical world, a clock interval is usually crystal-controlled, and so there is very little uncertainty in the parameter T_{CLK}. If the clock is not crystal-controlled, its nominal frequency must be lowered slightly to guarantee a clock interval greater than T_{CLK}.

EXAMPLE 11.1: System Timing Budget

Here is a system-level timing budget, in picoseconds, for a system constructed with 10E type ECL. This budget includes four categories of delay:

(1) Flip-flop propagation 10E131	700	
+ setup time	150	
	850	850
(2) Logic delay between flip-flops		
10E171 MUX	850	
+ circuit trace 4 in.	740	
× three sections	×3	
	4770	4770
(3) Clock skew		
Max-min per gate 10E111	50	
+ circuit trace 2-in. skew	370	
	420	420
(4) Timing margin		
15%	1065	1065
Clock interval (=135 MHz)		7105

The system operates well below the flip-flop maximum toggle frequency of 1100 MHz. The biggest budget goes for the three stages of logic planned between each latch.

The clock distribution system consists of one clock source driving two 10E111 clock fan-out gates. The maximum skew between gates, plus an allowance for different trace lengths, accounts for 10% of the timing budget. Greater emphasis on such a good clock distribution system will provide little reward in this system.

Equation 11.4 tells us that clock skew has as much of an impact on overall operating speed as any other propagation delay. Common sense tells us there are many fewer clock signals than data signals on a typical circuit board. To get a big increase in timing margin for only a little work, look at your clock lines.

Some manufacturers have produced driver chips configured for driving multiple clock lines. The internal construction of these chips reduces the clock skew among their various outputs. The Motorola MC10E111 consists of one input and nine differential 50-Ω ECL outputs. All nine MC10E111 outputs switch within 50 ps.

POINT TO REMEMBER:

Clock skew has as much of an impact on overall operating speed as any other propagation delay.

11.3 USING LOW-IMPEDANCE DRIVERS

The brute force method for low skew has two parts:

(1) Locate all clock inputs close together.
(2) Drive them from the same source.

If a system has many clock inputs that cannot be physically colocated, the simple brute force method fails.

In that case, try the spider distribution network. This network, drawn in Figure 11.4, distributes clocks from a single source to N remote destinations. Reflections are damped by resistive terminations R at the end of each spider leg. The drive circuit experiences a total load of R/N.

Using a transmission line impedance of 75 Ω, a network of three spider legs presents a 25-Ω composite load to its driver. Some commercial chips drive loads that low, but not many.

To service more spider legs, we need a more powerful clock driver. Two or more driver outputs connected in parallel make a convenient and simple high-powered driver. Always draw the paralleled outputs from a common integrated circuit. Outputs from the same chip have only a small skew between them and are thus unlikely to burn each other out when connected in parallel.

A discrete, low-impedance amplifier can drive many spider legs. The ECL power driver in Figure 11.5 uses a common emitter gain stage with transformer output coupling. The transformer converts the high-impedance, high-voltage output to a low-impedance, high-current output. The transformer also performs a DC-level shift.

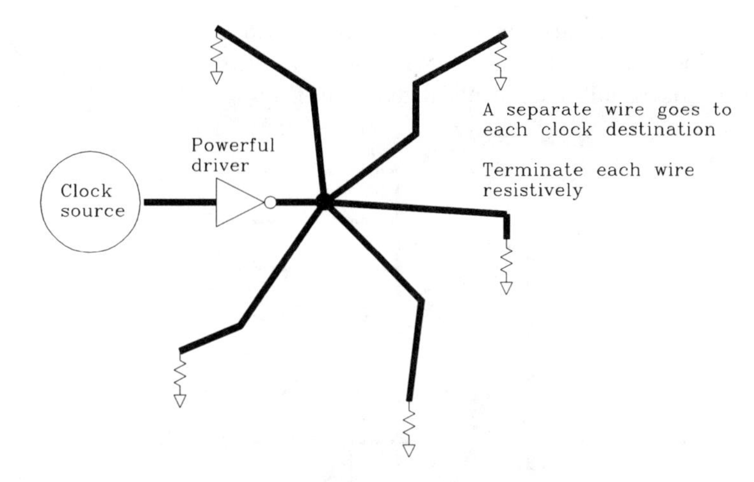

Figure 11.4 Spider-leg clock distribution network.

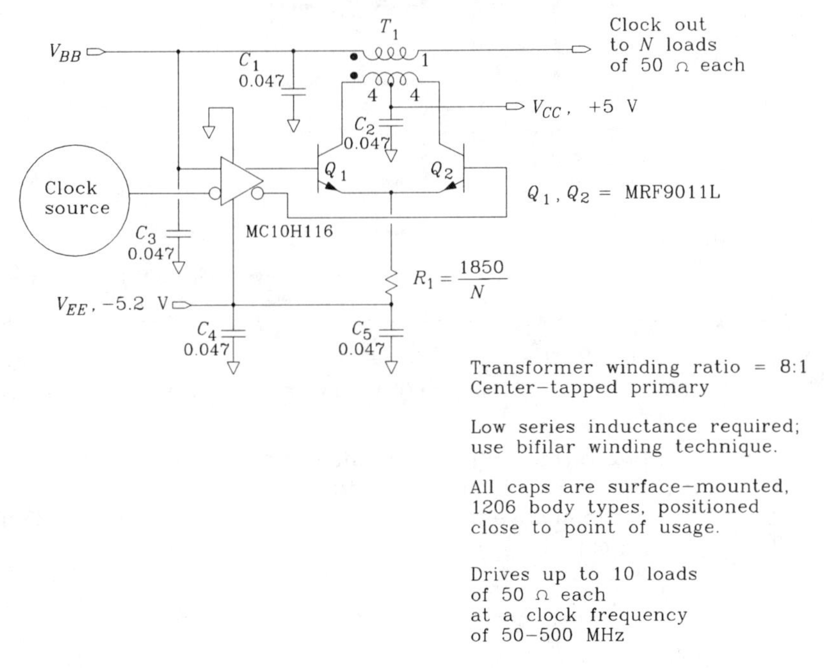

Figure 11.5 ECL clock amplifier.

A similar circuit for TTL would require higher bias voltages (+12, –5) and larger transistors. The total drive power required for TTL clock signals is 25 times that of ECL circuits, owing to the larger TTL voltage swing (4 V versus 800 mV).

When using an ECL power driver circuit, remember to size resistor R_1 according to the required output drive current.

The clock distribution tree in Figure 11.6 trades quantity for power. This scheme distributes clocks through a tree network to their final destinations. Balancing the tree with equal numbers of identical gate types helps reduce clock skew.

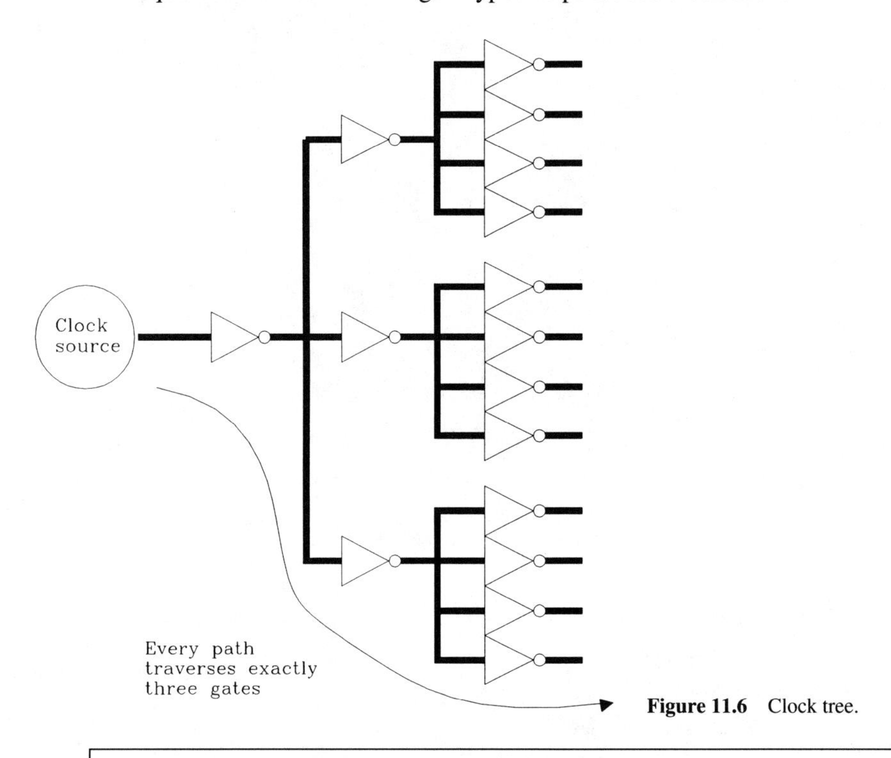

Figure 11.6 Clock tree.

> **POINTS TO REMEMBER:**
>
> Two or more driver outputs connected in parallel make a convenient and simple high-powered driver.
>
> The total drive power required for TTL clock signals is 25 times that of ECL circuits.

11.4 USING LOW-IMPEDANCE CLOCK DISTRIBUTION LINES

The clock line in Figure 11.7 services many clock inputs. As the clock signal passes each input, its rise time lengthens and a small reflected pulse propagates backward along the line. The reflected pulses interfere with reception.

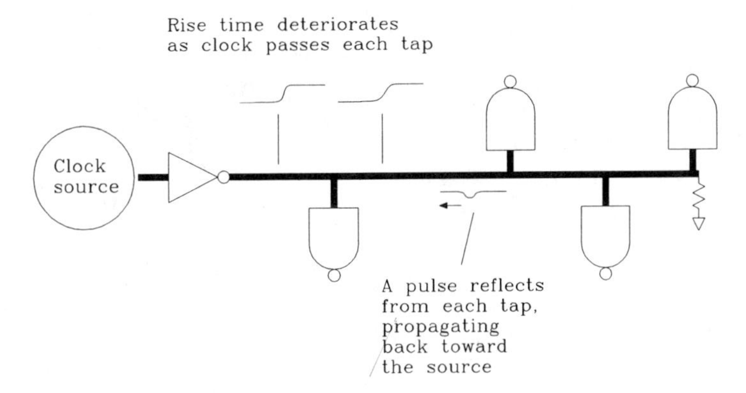

Figure 11.7 Clock driver serving many loads.

The reflected pulses follow the derivative of the input signal (Section 4.4.2.1). Each reflected pulse height is proportional to $-C(Z_0/2)$. The three means of reducing reflected pulse height are

(1) Slow the rise time of the driver. This lessens its derivative, shrinking the reflected pulse.
(2) Lower the capacitance of each tap.
(3) Lower the characteristic impedance of the clock distribution line, (Z_0).

Item (1) points out how logic too fast for your application works against you. Use a driver just fast enough to meet your clock skew budget.

Know the input capacitance of your clock receiver. On a multidrop bus two other factors affect item (2): the parasitic capacitance of the connector, and the capacitance of the printed circuit traces leading to the receiver.

Item (3) depends on the physical geometry of the clock traces (Section 4.5). Clock driver specifications must accommodate the impedance of the clock distribution lines. A doubled driver chip feeding a 20-Ω clock line is 2.5 times less sensitive to the capacitance of clock taps than a 50-Ω line. In stripline form, a 20-Ω clock line has a W/B ratio of 2:1 (Figure 4.34). In microstrip form a 20-Ω clock line has a W/H ratio of 7:1 (Figure 4.32).

In multidrop bus applications devices frequently are inserted and removed from service. Mutlidrop clock distribution systems must therefore tolerate varying loads. Lowering the distribution impedance helps prevent varying loads from affecting clock skew.

When planning a low-impedance line, use the characteristic impedance formulas in Appendix C—not the simple formulas given in Chapter 4. The simple formulas do not work well for predicting the characteristic impedance of lines at low impedances.

POINT TO REMEMBER:

A 20-Ω clock line is 2.5 times less sensitive to the capacitance of clock taps than a 50-Ω line.

11.5 SOURCE TERMINATION OF MULTIPLE CLOCK LINES

On the basis of Figure 11.8, some engineers attempt to drive multiple source-terminated lines from a single driver. This figure shows that the input impedance of a source-terminated line is twice that of an end-terminated line. Not only that, the drive current requirement drops to zero after $2T$ seconds, lowering the average power drain. These facts tempt us to assume that a single gate can drive multiple source-terminated lines.

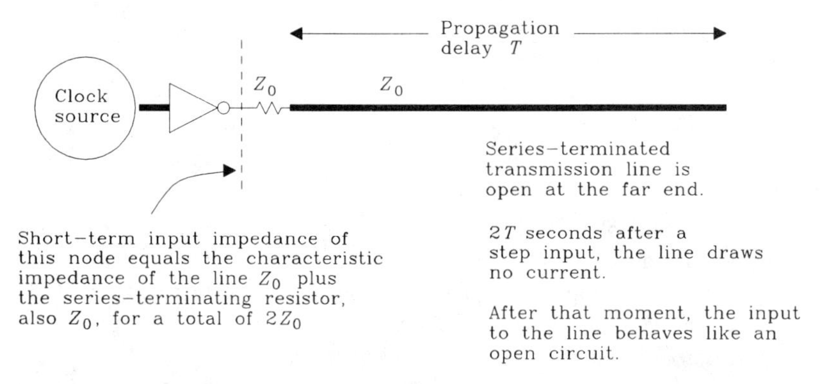

Figure 11.8 Short-term impedance of a series-terminated line.

A careful examination of initial conditions (Section 6.2.4) reveals that the *peak* drive current requirements of source- and end-terminated lines are the same. At high speeds, where every edge must be perfect, our circuits demand peak drive capability, in addition to average.

Nevertheless, some driver circuits have enough theoretical current output capacity for two source-terminated lines. Is it possible to drive two or more source-terminated lines from such a driver? Yes, but only in the limited configuration diagramed in Figure 11.9.

The trick to understanding this figure is to realize that the lines are coupled together into a jointly resonant structure. We cannot properly analyze just one line without seeing what happens to all the lines. The coupling happens because of the finite output impedance of the driver.

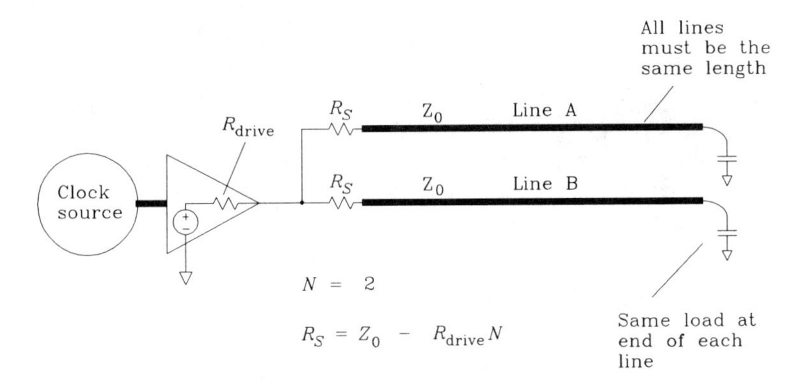

Figure 11.9 Single clock driver feeding two source-terminated lines.

If the driver output impedance were zero (it never is), there would be no cross-coupling between lines and we could simply use a separate series-terminating resistor of value $R = Z_0$ on each line. Unfortunately, the reality of finite driver impedance forces us to contemplate joint resonance. The paragraphs below show how to jointly analyze the system.

Skipping ahead to the answer: *Multiple source termination with nonzero driver impedance works only if the lines are equally long and the loads at each end are balanced.* The source-termination resistors must equal

$$R_S = Z_0 - R_{drive}N \quad [11.5]$$

where R_S = source termination resistor, Ω
Z_0 = driven line impedance, Ω
R_{drive} = effective output resistance of driver, Ω
N = number of driven lines

When driving one line ($N = 1$), Equation 11.5 matches the characteristic impedance Z_0 to the total source impedance ($R_S + R_{drive}$). This is a normal source termination. When driving multiple lines, Equation 11.5 prescribes smaller source-terminating resistors. With N too large, Equation 11.5 goes negative, implying that no practical solution exists.

Let's analyze the lines in Figure 11.9 one at a time to see what happens. Adding up the responses, we will have our joint analysis.

In Figure 11.9, a pulse travels down line A toward the load. This pulse reflects off the far end of line A, returning to the driver. In the usual application of source termination, the source termination matches the characteristic impedance of the line, and so there is no reflection at the driver. In Figure 11.9 the effective source impedance is not matched; it is lower than the characteristic impedance of the line. The returning pulse on line A will bounce off the driver, producing a negative reflection (Equation 4.54). So far, the negative reflection looks like a problem.

Another effect occurs at the same time. As current from the returning pulse on line A surges into the driver chip and through R_{drive}, it generates a voltage at the driver output pin. This voltage couples into line B. The polarity of the crosstalk pulse coupled onto line B is positive.

So far, we know the consequences of a pulse on line A include a negative reflection on line A and a positive crosstalk pulse on line B.

Now imagine pulses reflecting off the far ends of lines A and B at the same time. If these pulses return to the driver at the same moment, they will each induce on themselves some negative reflections and on each other some positive crosstalk. If we choose the resistor values carefully, we can get the negative reflections and positive crosstalk to cancel exactly. The result is a perfectly damped system.

The conditions under which perfect cancellation may be achieved are very restrictive, namely:

(1) The lines must be equally long (this guarantees the reflected pulses will arrive at the same time).
(2) The loads must be balanced (this guarantees the reflected pulses will have the same shape).
(3) The resistors must be calculated according to Equation 11.5.

Equation 11.5 sets the source-terminating resistance so that line A experiences a negative reflection pulse exactly compensated for by the positive crosstalk pulse from line B. Equation 11.5 works with any number of lines, as long as they are equal in length and identically loaded.

Perfect balance rarely occurs in practice. If the lines are not perfectly balanced, the reflections and crosstalk from each line will not cancel. Incomplete cancellation makes the system ring.

POINT TO REMEMBER:

A single driver can service two or more source-terminated lines under restricted circumstances.

11.6 CONTROLLING CROSSTALK ON CLOCK LINES

Chapter 5 makes clear the relation of crosstalk to line separation. Over a solid ground plane, doubling line separation divides crosstalk by 4. Clocks being delicate signals, we favor extra crosstalk protection for them. Gaining extra crosstalk protection involves two aspects: the physical means of providing more crosstalk protection, and the logistical means of getting the correct physical result.

The physical means of providing extra protection are simple: Leave extra gaps around traces or put clock traces on separate layers encapsulated between ground planes.[1]

The logistical means of providing extra protection are more complex. One must first complete the error-prone task of identifying each clock trace by either hand-drawn marks on a schematic or a list of schematic net names. The special routing requirements must then be communicated to your layout person. The layout person then either accommodates your requests or ignores them. Remember that you and the layout people rarely share the same boss. No offense is meant to layout professionals, but realistically, a layout person generally has more than enough to do without accommodating a long list of intricate special requests.

Because written instructions for routing clock traces on a separate, protected layer are compact and easily understood, many engineers use this approach. Wasting a routing layer is, in their estimation, worth the cost if it accomplishes their goal.

A more elegant approach sets up different trace separation specifications by net class. Nets classified as clock nets stay farther from other traces, creating less crosstalk. Each year, more automatic routing packages incorporate this feature, but few digital designers make use of it.

If your layout package does not support different net routing classes, it will certainly support different trace widths. Designating all the clock nets as fat traces will force other traces away from the clock nets during routing. After routing, globally change the clock nets to a narrower width. A major disadvantage of this approach is that fat clock traces won't fit between the pins of an integrated circuit.

To enforce spacing requirements, some designers insert guard traces during layout only to remove them at the last minute. Their temporary guard traces force other traces away from high-speed lines during routing, thus reducing crosstalk problems.

POINT TO REMEMBER:

The physical means of providing extra crosstalk protection are simple; the logistical means are complex.

11.7 DELAY ADJUSTMENTS

The clock skew term in Equation 11.4 comprises a difference between two propagation delays. Precise balance between the two clock propagation paths produces low clock skew.

[1]You may encapsulate between either power or ground planes, provided there is a very low impedance between them.

Sometimes a small positive (or negative) skew may be desirable. A retarded (or advanced) clock usually improves timing margins for one part of the circuit but worsens them elsewhere. Use purposeful clock skew only when you have a good timing model of the whole circuit.

Because we are sometimes interested in a purposeful, nonzero skew, engineers often cast the clock distribution problem in terms of reducing the uncertainty in clock arrival time rather than simply attaining everywhere a low skew.

Clock adjustments can achieve either low skew or purposeful skew, according to their design. The same principles apply in both cases. Adjustments to clock delays are sometimes called adjustments to the *clock phase*. This term reminds us that the clock is a repetitive waveform, roughly sinusoidal.

11.7.1 Fixed Delays

The simplest form of clock adjustment is a *fixed delay*. A fixed delay provides a predetermined amount of clock delay that does not change after assembly.

A fixed delay compensates for nominal delays elsewhere in the circuit, bringing the nominal value of clock skew to its desired value. Because its delay is frozen at the time of design, a fixed delay cannot cancel variations in board fabrication or active component delay, neither of which is known until the circuit is tested.

Fixed delays are built from three basic building blocks: transmission lines, logic gates, and passive lumped circuits. Each has its advantages (Table 11.1). Delay lines are good for short delays and are very accurate. Gate delays use less board area than delay lines but are considerably less accurate. A lumped-circuit delay element covers the widest range of possible delays. Its delay variation depends mainly on the quality of its analog components.

TABLE 11.1 FIXED DELAY ELEMENTS

	Practical amount of delay (ns)	Variation in delay (%)
Delay line	0.1–5	10
Gate delay	0.1–20	300
Lumped-circuit delay	0.1–1000	5–20

Delay lines directly printed on a printed circuit board exact a terrible size penalty. Figure 4.28 illustrates a typical printed delay line fabricated on the outer layer of a 0.010 in.-thick FR-4 substrate. Using inner-layer traces with a 0.025-in. trace pitch, each nanosecond of delay consumes about 0.135 in.^2 of board area. A 7-ns delay using these dimensions will occupy 1 in.^2. That's a lot of area.

When using a printed circuit trace as a transmission line delay, keep in mind the variation in relative permittivity of the trace with temperature. For FR-4 material, this variation results in a 10% change in propagation speed over a temperature range of 0–70°C.

Some commercial delay lines surround a transmission line with magnetically permeable material. The permeable material radically increases the delay per inch, shrinking the delay line. These delay lines are available with or without buffering in a DIP or surface-mounted format.

A spare gate makes an effective delay element. The problem with using a gate for a delay element is that while all manufacturers specify maximum propagation delay, few talk about minimum gate delay. The total variation in gate delay is so large that it sometimes hinders clock skew rather than helping. Unfortunately, inside a gate array or custom chip, there may be no choice but to use a gate for a delay element.

The lumped-circuit delay in Figure 11.10 produces clean, repeatable delays when used with CMOS gates. The slow rise time of the *RC* circuit retards the propagation of pulses from the first gate to the second gate.

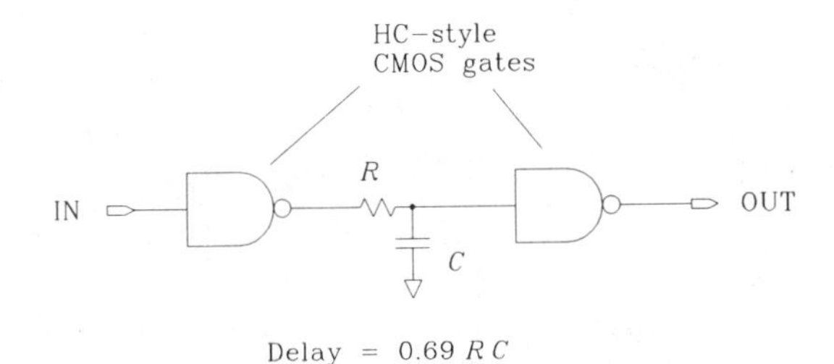

Figure 11.10 Lumped-circuit delay element.

The accuracy and stability of this circuit depends in part on the accuracy of the analog component values *R* and *C*. Accuracy and stability also depend on the value of the parasitic input capacitance of the second gate.

A second issue surrounding the use of lumped delay circuits concerns uncertainty in the switching thresholds. The circuit in Figure 11.10 will switch when the input to the second gate climbs beyond its switching threshold. If the threshold is uncertain, the switching time is uncertain.

The exact switching moment is altered by asymmetry in the switching threshold. This problem plagues logic families like TTL and HCT. The switching threshold of both TTL and HCT families lies nearer ground than V_{CC}.[2] With the switching threshold offset toward ground, the *RC* delay circuit delays less on rising edges than on falling ones. An ideal circuit delays rising or falling edges equally.

A differential receiver with its negative input terminal connected midway between HI and LO logic levels has a symmetric threshold. These receivers are available in the TTL and HCT logic families. Such a circuit preceded by a one-pole *RC* filter produces nominally equal rising and falling edge delays.

With TTL gates, the circuit in Figure 11.10 works poorly. CMOS gates work better in this application because they require no net DC input current. When the input to the second gate reaches its asymptote, there is no current flowing into that CMOS gate input. The voltage drop across *R* is therefore zero. This circuit, when used with CMOS gates, results in no net loss of voltage margin.

[2]The switching thresholds of HCMOS and ECL logic are centered about their logic swings.

When used with TTL circuitry, the net TTL input current required in the LO input state must trickle through *R*. To maintain our voltage margins we need a resistor *R* of less than 100 Ω. Alternately, try an inductive bead or a wound inductor as the series loss element in Figure 11.10. An inductor has no loss at DC and so passes the TTL input current without introducing a voltage drop.

With regular gates, don't try shifting the clock by more than 12% of a clock cycle in a single stage. Cascade several stages of delay, buffered by gates, to build up more delay. When an *RC* circuit delays a square wave by 12%, the *RC* response does not have time to decay fully between pulse edges. The output waveform slurs between 10 and 90% of its nominal value. More delay slurs the waveform to an even smaller size.

Commercially available fixed-delay circuits all use some combination of transmission line, logic gate, and lumped-circuit delay elements.

Whatever form of fixed delay you choose, incorporate its uncertainty in delay into your timing-margin calculations.

11.7.2 Adjustable Delays

An adjustable delay can compensate for actual delays in a circuit, as well as nominal delays. Technicians must make adjustments after assembly as part of the final test process. Each adjustment, if properly set, reduces the uncertainty in clock skew caused by variations in board fabrication and active component delay.

Do not assume your manufacturing staff will understand the meaning of the adjustments provided. Write a test procedure for each adjustment showing how to measure the clock delay at that point and indicating the limits of proper adjustment.

The three basic delay building blocks are the transmission line, the logic gate, and the passive lumped circuit. All three building blocks come in adjustable versions.

A delay line adjusts in quantized steps. The layout in Figure 11.11 illustrates a typical adjustable delay. The transmission line in Figure 11.11 has five adjustment taps.

A more flexible arrangement appears in Figure 11.12, which produces 16 different delays with only 8 jumpers. The jumper sizes in Figure 11.12 are tuned to one, two, four,

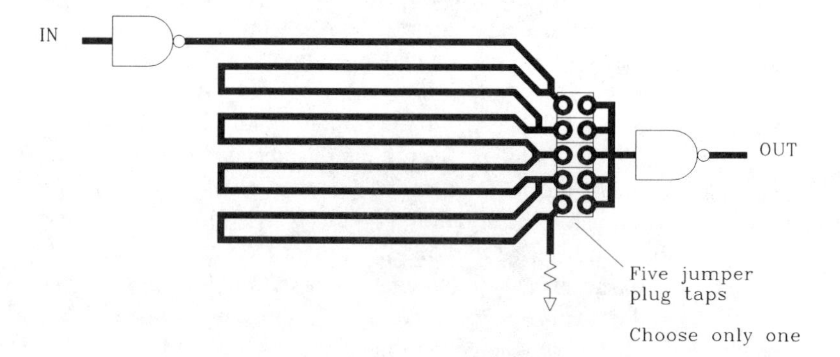

Figure 11.11 Adjustable transmission line delay.

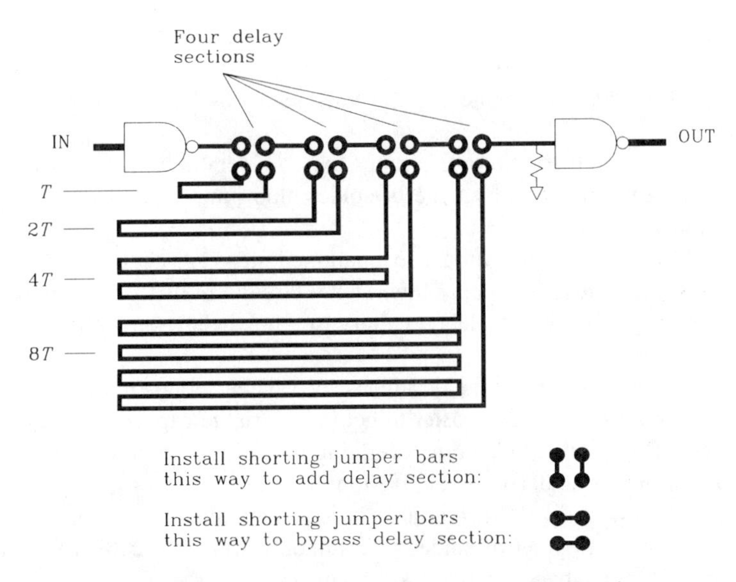

Figure 11.12 Adjustable transmission line delay with 16 settings.

and eight times a basic delay T. The switches can select any combination of delay sections. Although the circuit in Figure 11.12 is technically more powerful, its complexity works against us. Simpler adjustments cause fewer adjustment mistakes.

A *shorting jumper bar* makes good adjustable taps at low frequencies. These tiny, removable plugs fit onto a pair of 0.025-in.-square posts separated by 0.100-in. (Figure 11.13). Some people call shorting jumper bars *software jumper plugs* because of their prevalence as option jumpers on personal computer add-on cards. Above 100 MHz you will notice the inductance of shorting jumper bars. Their inductance varies according to how far down the posts the installer pressed the plug.

If the inductance of jumper plugs is not acceptable, try *solder blob jumpers* (Figure 11.13). A solder blob jumper consists of two 0.50-in.-square pads separated by a 0.006-in.

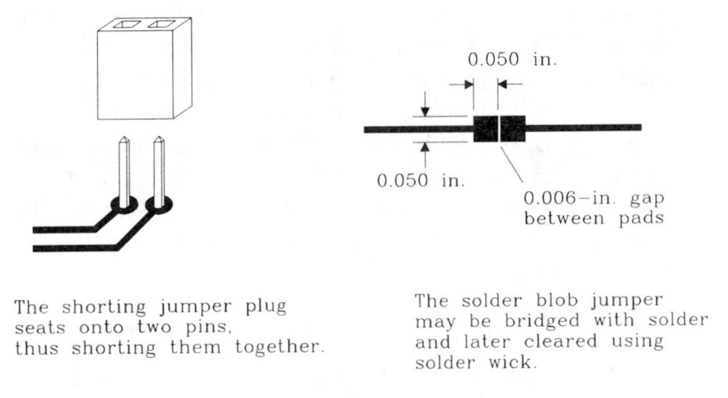

Figure 11.13 High-frequency circuit switches.

space. Always put this structure on the component side of a circuit board. The 0.006-in. gap is wide enough to prevent solder bridging during assembly, yet narrow enough to be easily bridged by a technician. A bridged solder blob clears quickly and cleanly with solder wick.

Compared to shorting jumper bars, solder blob switches take very little circuit board area. Another advantage of solder blob jumpers is that they don't fall off or get moved after assembly.

Gate delays also adjust in quantized steps. A chain of gates tapped at discrete points makes a usable delay line. Delay circuits built from gates suffer from the basic inaccuracy of each gate delay. Otherwise, they behave much like a tapped transmission line.

A lumped-circuit delay adjusts by varying either R or C. Continuously variable resistors are cheaper and easier to get than variable capacitors. With either type, provide some mechanism for clamping or gluing the adjustment after setting it. Adjustable components are particularly susceptible to vibration.

A new style of step-variable passive components incorporates several component values plus a tiny set of solder blob jumpers all on a 1206 surface-mount body. These parts allow quantized tuning of RC delay circuits.

11.7.3 Automatically Programmable Delays

An ideal delay circuit would be continuously variable, would be stable over a wide temperature range, and would adjust itself in production. Sound impossible? Read on.

First let's see how to make a continuously programmable delay. Two approaches show promise in this arena. The oldest involves a *varactor diode*. The varactor diode is a diode whose parasitic capacitance varies as a function of applied reverse bias. Normally a hindrance to design, parasitic capacitance in the varactor is its primary selling point.

The circuit in Figure 11.14 shows varactors in use as variable-delay elements. This circuit differs from the circuit in Figure 3.23 in three respects. First, it is programmed digitally. That feature could also be added to Figure 3.23. Second, Figure 11.14 uses LC delay elements, which provide a wider range of delay adjustment without attenuating the signal. Third, the lumped circuit in Figure 11.14 cascades two passive delay sections without buffering. The impedance of the second stage is three times larger than the first stage. This impedance scaling prevents the second stage from loading the first, which would distort the combined response. The cascade connection reduces the number of buffers in the critical path from three to two. Since the delay of each buffer varies with temperature and power supply voltage, we want as few buffers as possible.

A second programmable delay approach uses a chain of gates. If all the gates are internal to one gallium arsenide integrated circuit, the delay between them is on the

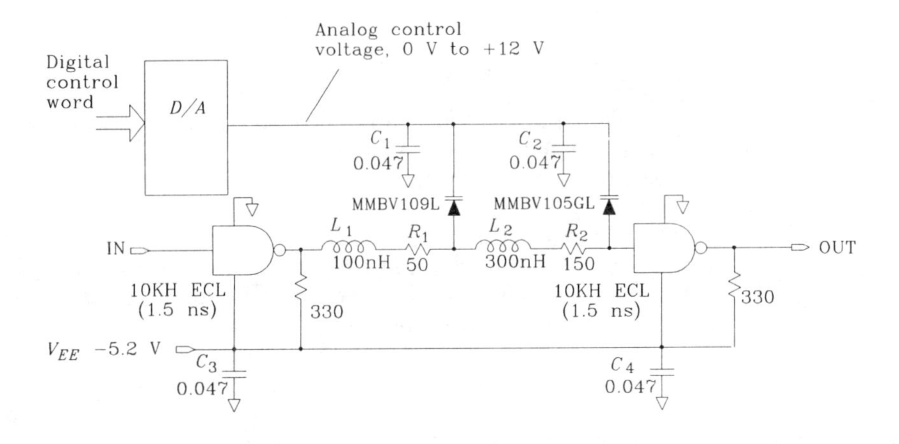

Varactors:

MMBV105GL, 4–15 pF
MMBV109L, 8–40 pF

Delay:

From lumped circuit	1–6 ns
From gates	3 ns
Total delay	4–9 ns

Maximum working frequency: 50 MHz

Scaling for other frequencies:
L_1, L_2 and varactor capacitances are scaled proportionally to desired delay. Maximum working frequency is inversely proportional to delay.

Figure 11.14 Programmable delay element using a varactor diode.

order of only 100 ps. Such a chain can be very long inside one integrated circuit. A tapped version of the gate chain, with a giant multiplexer to select between taps, forms a useful digitally programmable delay. The design of the multiplexer must prevent glitches which might occur when switching from one tap to the next.

With either the varactor or the "chain-of-gates" circuit, we can store a table listing proper adjustment settings as a function of temperature. That's how the system achieves temperature stability.

The last issue regards how a delay circuit might automatically tune itself. As a clock skews out of adjustment in either direction, every system shows a marked increase in its error rate. Detecting that increase and then centering the clock between error-prone zones is one way to achieve automatic adjustment.

A less direct form of automatic adjustment involves sensing the switching times of data signals on a bus. The clock can be automatically adjusted to match specific transition times in the data waveform. This method resembles clock recovery systems used in serial data transmission.

POINTS TO REMEMBER:

Delay elements are built from three basic building blocks: transmission lines, logic gates, and passive lumped circuits.

A fixed delay cannot cancel variations in board fabrication or active component delay.

An adjustable delay compensates for actual delays, not just nominal delays, elsewhere in the circuit.

Whatever form of delay you choose, incorporate its uncertainty in delay into your timing margin calculations.

11.8 DIFFERENTIAL DISTRIBUTION

Differential clock signals survive tougher noise environments than single-ended clock signals. This happens for two reasons: signal size and differential balance. Since the total voltage swing between the two wires of a differential pair is twice that of a single-ended signal, a differential pair can tolerate twice the interference. Even better, if noise equally affects the two parts of a differential clock line, it cancels completely in the differential receiver, producing no net timing jitter. Noise that affects both sides of a differential line equally is called *common mode noise*. Differential lines tolerate enormous amounts of common mode noise.

Crosstalk problems are particularly acute in TTL systems that use an ECL clock distribution backbone. The ECL distribution backbone distributes clock with low skew, which is an advantage. The disadvantage of ECL clock distribution involves the low amplitude of ECL signals. The larger-amplitude TTL signals easily generate enough crosstalk to interfere with nearby ECL clock receivers. Differential ECL signaling helps overcome TTL crosstalk problems.

Differential signaling helps only if the interfering noise affects both signals equally. It does not help with the kind of crosstalk induced between circuit traces which run too close together. That crosstalk usually affects one line far more strongly than the other, creating a truly differential noise signal.

Differential signaling helps a lot when communicating between two circuit boards whose ground planes carry different noise voltages. Differences in the ground voltages cancel out in the differential receiver. Differential ECL signaling handily overcomes TTL ground noise between daughter cards on a large backplane.

POINT TO REMEMBER:

Crosstalk, as long as it affects the two parts of a differential clock line equally, induces no timing jitter.

11.9 CLOCK SIGNAL DUTY CYCLE

The ideal duty cycle for a clock signal is 50%. The falling edge of an ideal clock signal precisely bisects successive rising edges. This feature permits use of the inverted clock as an intermediate timing waveform.

The average DC value of an ideal clock lies halfway between the HI and LO states. This property permits the design of simple feedback mechanisms which keep the duty cycle fixed at 50%, as we shall see.

The reason clocks become unbalanced, drifting away from 50% duty cycle, is that clock repeaters have an asymmetric response to rising and falling waveforms. Careful measurements reveal that the propagation delay for any gate differs for rising and falling edges. A pulse propagating through an asymmetric gate is either shortened or lengthened by this difference in propagation delay. This effect is called *pulse width compression*, *pulse width expansion*, or *pulse width distortion*.

When we cascade a long series of identical gates, the pulse width distortion in each stage adds. Suppose the input pulse is positive-going and the delay of rising edges exceeds the delay of falling edges. Positive pulses will emerge shorter from each gate than from the gate before them. Somewhere along the chain, the positive pulses simply disappear.

A clock signal traversing this same chain of gates looks like a train of pulses. If, at each stage, a positive-going pulse shrinks, the duty cycle of the clock will drop as we progress along the chain. At some point along the chain, the clock fails to elicit any response, and subsequent stages fall silent.

Two clever tricks have saved generations of engineers from the misfortune of losing a clock signal as a result of asymmetric propagation delays. The first trick inverts the clock signal at every stage. This alternately converts rising edges to falling ones, and vice versa, as signals progress down the chain. The overall effect cancels pulse width compression in adjacent stages. Clock signals propagate much farther, and with a better duty cycle, in an inverting repeater chain than in a noninverting one.

The second trick requires some analog circuitry. The circuit shown in Figure 11.15 works only with logic families having symmetric switching thresholds. For nonsymmetric thresholds, find an analog design engineer to change the circuit into a feedback system suitable for your logic family.

This circuit knows that when a clock's duty cycle changes, so does its average DC value. By measuring the average DC value, we can infer the duty cycle.

The same circuit also uses a well-known relation between switching threshold and duty cycle. We know that, because of the finite rise and fall times of a clock signal, adjusting the input switching threshold on a clock repeater will change the output duty cycle. The last principle used in Figure 11.15 is feedback.

The circuit in Figure 11.15 measures the average DC value of the clock output, storing the result on capacitor C_2. The voltage on C_2 in turn adjusts the input switching threshold to achieve an output duty cycle closer to 50%.

Figure 11.15 implements a relatively low-gain feedback circuit. At 300 MHz, where this gate naturally induces a pulse width compression of 200 ps, the corrected circuit induces a pulse width compression of only one-fourth that amount.

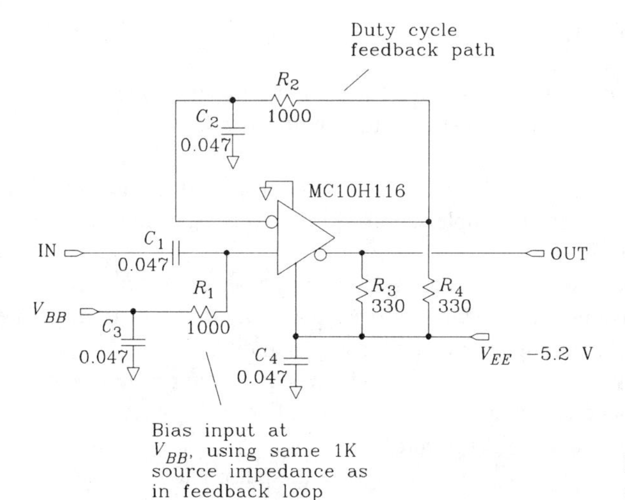

Figure 11.15 Feedback circuit to help maintain a 50% duty cycle.

An integrating feedback circuit coupled between the clock output and the input switching threshold adjustment does a better job of regulating output duty cycle but at the cost of more parts.

Adjustments to the input-switching threshold control where the gate will switch on each pulse. Lowering the threshold will advance the switching of rising edges and retard falling ones. The amount of switching shift depends on the rise time of the input signal. A slow signal shifts more than a fast one.

The adjustments made by this circuit are tiny, but in a chain of repeaters they can dramatically improve performance.

POINTS TO REMEMBER:

A clock's duty cycle changes as it progresses along a chain of repeaters.

Clock signals propagate much farther, and with a better duty cycle, in an inverting repeater chain than in a noninverting one.

11.10 CANCELING PARASITIC CAPACITANCE OF A CLOCK REPEATER

When a new device connects to a multidrop bus, the parasitic capacitance of its clock receiver shifts the received clock phase of all devices on the line. Clocks received both downstream and upstream of the new device are affected.

The amount of shift induced is proportional to the total parasitic capacitance of the new clock receiver. If you can reduce this capacitance by changing the layout, specifying a better receiver, or using another connector, then do so. When faced with using the components at hand, try the circuit in Figure 11.16.

The inductor in Figure 11.16 presents a negative reactance at the clock frequency that partially cancels the parasitic capacitance of the clock receiver circuit. RF engineers call this a matching network. The inductor-cancellation trick works only at one frequency, the fundamental. The third and higher harmonics present in the clock waveform get no relief from this technique. When canceling a parasitic capacitance, use a clock driver with slow rise and fall times. The resulting clock has less harmonic content (it looks more sinusoidal), and the neutralizing effect works better.

The two resistors are optional. In a fixed installation, where the clock receiver is never disconnected from the line, the resistors add nothing to the circuit. In a *hot plugging* environment, where cards are plugged into a bus with the power turned on and the clock running, the resistors provide a vital service. They help charge capacitor C_1 before it connects to the clock bus.

When the card is powered off, capacitor C_1 is discharged to 0 V. When the circuit is operating, capacitor C_1 is charged to the midpoint between HI and LO logic levels.

In the absence of R_1 and R_2, when the card first connects to its clock bus, the surge of current required to charge capacitor C_1 will seriously distort clock signals on the bus. This effect can be circumvented by a prepower arrangement. A properly designed hot plugging card receives power connections before touching the clock bus. Once power comes on, resistors R_1 and R_2 precharge capacitor C_1 to the middle voltage, where it stays until contacting the clock bus. This design feature prevents any sudden current surges from affecting the clock bus.

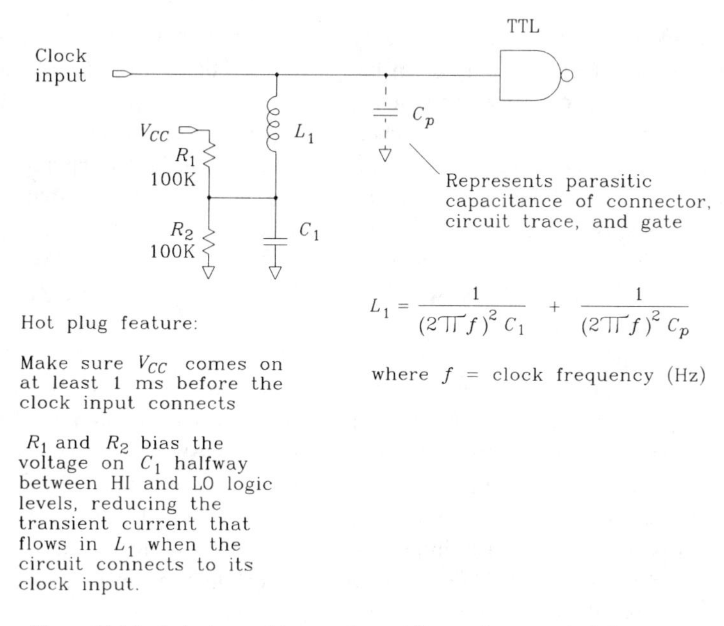

Figure 11.16 Inductor used to cancel parasitic capacitance at clock frequency.

Help shorten the precharge time on C_1 by keeping its capacitance small. The minimum value for C_1 is about 100 times C_p. Compute values for L_1 and C_1:

$$C_1 = 100C_p \qquad [11.6]$$

$$L_1 = \frac{1}{(2\pi f)^2 C_1} + \frac{1}{(2\pi f)^2 C_p} \qquad [11.7]$$

The amount of time the precharge circuit takes bringing C_1 to within 1% of its final value is

$$\text{Precharge time} = 4.6\frac{R_1 R_2}{R_1 + R_2} C_1 \qquad [11.8]$$

> POINT TO REMEMBER:
>
> An inductor can partially cancel the parasitic capacitance of a clock receiver.

11.11 DECOUPLING CLOCK RECEIVERS FROM THE CLOCK BUS

In some situations, clock taps on a clock distribution bus may seriously distort the passing clock waveform. This often happens when there are a lot of taps, when the clock receivers have too much parasitic capacitance, or when operating at high speeds.

One way to reduce the impact of each tap, at the expense of requiring more voltage gain in each clock receiver, is to build a 3:1 attenuator at the input to each clock gate. Try inserting an impedance in series with each gate input that is twice the expected input impedance of the gate at the clock frequency. The attenuating network may include a resistor in parallel with a capacitor.

For CMOS circuits, which draw little DC bias current, the attenuating network alone is sufficient. TTL gates may require a DC biasing network in addition to the attenuating components.

The advantage of a 3:1 attenuating network is that it triples the apparent input impedance of the receiver. The disadvantage is that the signal received by the gate is smaller. Fortunately, most gates have a lot of excess voltage gain.

Common differential receiver circuits, having plenty of gain and a very precise input-switching threshold, work well as attenuated clock receivers. When using ordinary gates (which have a very imprecise switching threshold) as attenuated clock receivers, better biasing is needed. Try a DC bias network that senses its own output duty cycle and then adjusts the input-switching threshold to maintain 50%.

POINT TO REMEMBER:

An attenuating network can increase the effective input impedance of a clock receiver.

12 Clock Oscillators

At one time, computer architects vigorously debated the advantages of Hartley versus Colpitts oscillators. When there wasn't very much computing hardware around, oscillators seemed very interesting.

Nowadays, the typical computer architect buys oscillators in a can. Standard industry practice has changed from *designing* oscillators to *specifying* oscillators. The subject of this chapter is how to properly specify and use oscillators and crystals.

12.1 USING CANNED CLOCK OSCILLATORS

As depicted in Figure 12.1, the canned oscillator gets its name from the hermetically sealed metal case surrounding its inner components. Usually constructed as a thick-film hybrid circuit on a tiny substrate, canned clock oscillators are ubiquitous in modern digital designs. Some newer versions now come in low-cost plastic housings.

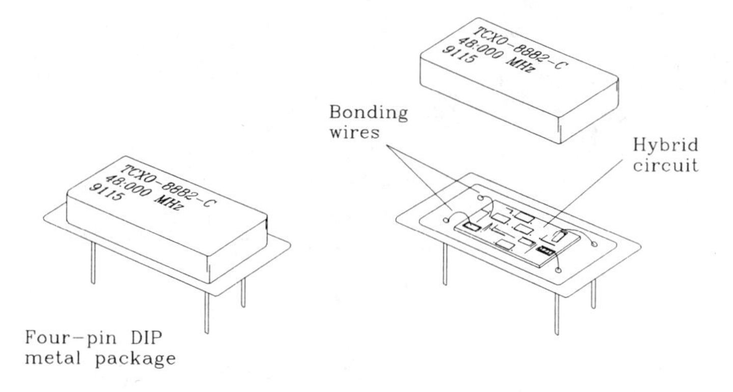

Figure 12.1 Typical canned oscillator.

Of the many oscillator circuits available, *piezoelectric (quartz) crystal oscillators* appear most often in high-quality digital equipment. System designers value their highly accurate output frequency.

Piezoelectric clocks are so accurate, compared to the delay of digital circuits, that we sometimes ignore variations in clock frequency. In reference to Equation 11.4, variations of 0.01% in T_{CLK} hardly require any adjustment in other delay parameters. The practice of ignoring crystal clock variations works well for simple state machines that use only a single clock.

More complex digital architectures severely constrain clock performance. For example, try transferring data between two separate digital machines running on independent clocks. If we hook the machines together using an asynchronous *first-in-first-out (FIFO)* buffer, the FIFO will grow (or shrink) at a rate proportional to the difference in clock speeds between the two machines. Any design involving the frequency difference between two clocks may require especially accurate or stable oscillators.

Military and other high-reliability applications impose their own peculiar requirements. Special military requirements involve the oscillator's response to vibration, shock, humidity, and high temperature. Some programs require accelerated aging, screening, or other postproduction tests. These tests rule out most commercial oscillator vendors and raise production costs. If you don't need special screening, don't specify a part that has it. Look for a cheaper commercial version.

Table 12.1 summarizes the major data sheet parameters for piezoelectric crystal oscillators, showing which parameters are particularly important for various special applications. The special application categories are communications, military, and surface mount. A discussion of each parameter follows.

12.1.1 Frequency Specifications

- Frequency
- Stability
- Aging
- Voltage sensitivity

The frequency parameter quotes the *nominal operating frequency*, or *center frequency*, assuming conditions of room temperature, normal operating voltage, and no aging. Canned oscillators have nominal operating frequencies of anywhere from 10 kHz to 300 MHz. The fundamental operating frequency of the crystal inside these oscillators ranges only as high as 40 MHz. Vendors synthesize higher-frequency clocks by filtering and enhancing harmonics of the crystal's fundamental operating frequency. The nominal operating frequency is always specified in hertz (or kilohertz or megahertz).

The actual operating frequency may drift either above or below the center frequency. The data sheet lumps all variations in operating frequency together under the *stability* specification. The units of stability are percent (for poorly performing parts) or parts per million (ppm) (for better parts). One-hundred ppm equals 0.01%. Stability sometimes appears after the frequency specification: 50.00 MHz ±100 ppm.

TABLE 12.1 CANNED OSCILLATOR FEATURES*

Parameter	Units	CX	MIL	SMT
Operating frequency				
Frequency	Hz	x		
Stability	±ppm	x		
Aging	±ppm	x		
Voltage sensitivity	ppm/V	x		
Allowed operating conditions				
Temperature	°C		x	
Input voltage	V			
Shock	G's, s		x	
Vibration	G's, Hz, or G_{RMS}			
Humidity	% relative humidity		x	
Electrical				
Output type	TTL, CMOS, ECL			
Maximum load	N, pF			
Duty cycle	% HI or LO		x	
Rise/fall times	ns or ps			
Input current	mA			
Mechanical configuration				
Package footprint	DIP, $\frac{1}{2}$ DIP, or SMT			x
Construction	Metal or plastic			x
Manufacturing Issues				
Solderability	°C, s			x
Cleaning	Permissible fluid types			x
Package leak rate	Atm cc/s		x	
Reliability				
Functional screening	% screening		x	
Aging	°C, h	x	x	
Bells and whistles				
Differential output	Yes/no			
Enable	Yes/no			
VCO	ppm/V	x		
Tuning	ppm	x		

*CX, Communications; MIL, military; SMT, surface mount.

At one time, vendors customarily indicated the precision and stability of an oscillator by the number of zeros trailing the frequency marked on its case. For example, an oscillator labeled 4.00000 MHz was considered better than one rated at 4.00 MHz. This is no longer true. The number of trailing zeros now has no bearing on stability and should be ignored.

The *stability* specification consolidates variations due to temperature, manufacturing processes, operating voltage, and aging. It quotes a single number showing the worst-case expected drift over all allowed combinations of these four parameters. Of the four, temperature variation induces the largest drift. To combat temperature drift, piezoelectric digital oscillators come in (at least) three grades of increasing performance: noncompensated, temperature-compensated, and oven-controlled.

The output frequency of a *noncompensated oscillator* varies according to the natural resonant frequency of its crystal. A *temperature-compensating oscillator*, also called a *TXCO*, contains circuitry that combats temperature drift. TXCO parts are naturally more expensive. The most exotic oscillators heat their crystals in a temperature-controlled oven to maintain a precise operating temperature.[1] These *oven-controlled oscillators* offer the best stability over a wide temperature range. Figure 12.2 shows what stability we can expect from these three grades of product over various operating temperature ranges.

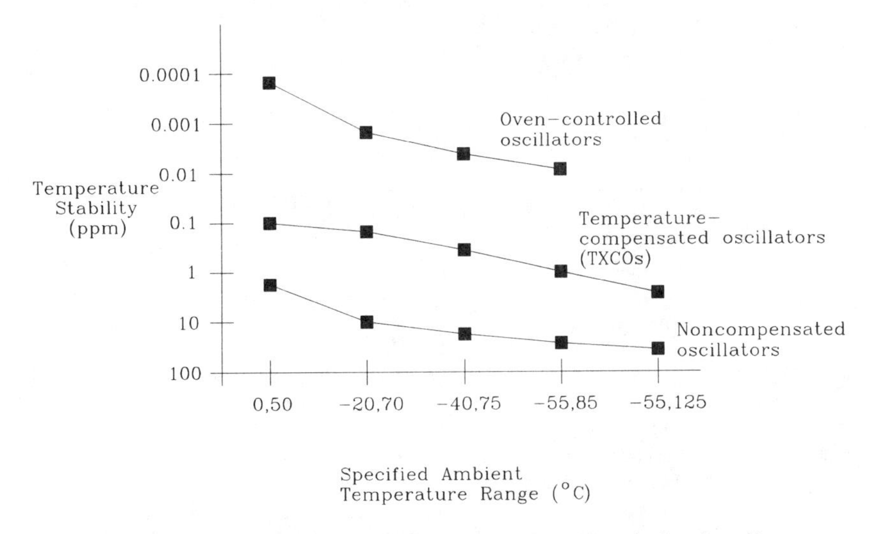

Figure 12.2 Frequency stability of three oscillator types. (Generic data from Vectron Laboratories.)

Aging, on the best data sheets, is split out separately from other aspects of stability. This is appropriate because aging causes the frequency of any crystal to drift by a few ppm each year. After 50 years, we get a lot of drift. Commercial manufacturers sometimes assume a useful product life of only a few years, and so out of a 100-ppm stability specification they just ignore aging. Engineers on the Viking space probe project look at aging differently. Aging is specified in ppm/year. New crystals age a little faster than old ones, so expect specifications like: "5 ppm in the first year, 3 ppm every year thereafter." A really good crystal, in an expensive package, may attain an aging rate as low as 1 ppm/year.

Every oscillator exhibits variations in frequency as a function of operating voltage. The data sheet lumps these variations in with the overall stability specification. Sometimes *voltage sensitivity* is called out separately. If so, it is listed in units of ppm/volt. When a system's voltage tolerance differs from the oscillator's specified operating range, we can use the ppm/volt number to calculate the expected frequency variation over the expected operating voltage range.

[1]Some designs enshrine their crystal inside two nested ovens for even greater temperature stability. The inner oven and the temperature control circuitry for the inner oven are both contained inside the first oven. Such an oscillator is called a *double-oven* oscillator.

12.1.2 Allowed Operating Conditions

- Temperature
- Input voltage
- Shock
- Vibration
- Humidity

Temperature ranges for electronic parts are always quoted in degrees Celsius, like this: 0–70°C. Always stay within the specified operating temperature range for a crystal oscillator. If you must go outside that range, buy a crystal designed to operate over a wider range. The following background material about crystals will help explain why.

Frequency variations result from every crystal's natural sensitivity to temperature. Quartz crystals, like all materials, react to temperature stress. The anisotropic structure in quartz crystals reacts to this stress by bending, flexing, or otherwise slightly changing shape as the ambient temperature changes. Any variation in shape affects the operating frequency. For any crystal we may plot a curve of its operating frequency as a function of temperature. This curve is fixed for each particular crystal and will not change with time.

Figure 12.3 plots curves of frequency versus temperature for several different crystals; look at the striking differences. For operation over the extended temperature range –50 to 100°C, curve *D* is best. It varies no more than 25 ppm from –50 to 100°C. Over the limited range 0–50°C, curve *A* wins. It varies less than 5 ppm over that range but would vary almost 100 ppm over the extended –50 to 100°C range. No one curve is best over all temperature ranges.

The wonderful thing about the curves in Figure 12.3 is that they are very repeatable. Each particular curve results from cutting the quartz crystal at a precisely known angle. Since all manufacturers use the same material (quartz) and all know about the cut-

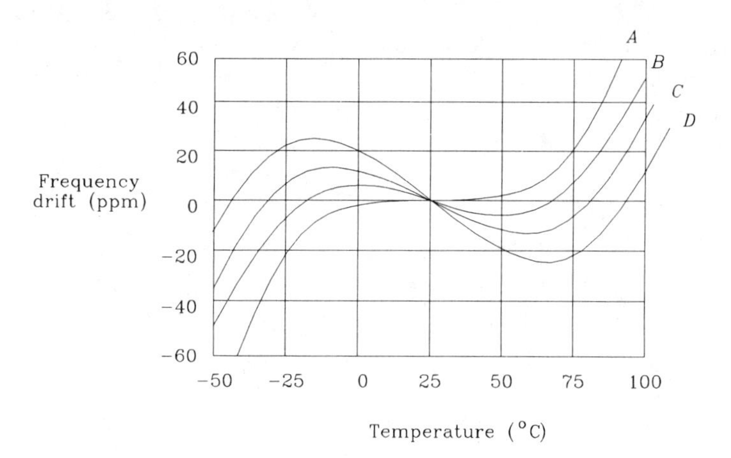

Figure 12.3 Temperature drift of quartz crystals.

ting angles needed to produce various temperature curves, we can expect similar specifications for temperature drift across all manufacturers.

Note that temperature drift in crystals is *not* linear with temperature. Do not assume that by testing an oscillator at high (hot) and low (cold) temperatures that you have characterized its temperature drift.

Input voltage (or power supply voltage) is like the V_{CC} specification for an integrated circuit. It appears either as a range (4.5–5.5 V) or as a percentage deviation (+5 $V \pm$ 10%). A few of the TXCO or VCO models may require dual-voltage inputs, but most just use a single supply.

Shock, in oscillator terminology, refers to mechanical, not electric, shock. Oscillators must endure testing by pneumatic machines that intentionally slam the oscillator into a fixed target with calibrated ferocity. A shock specification measures the sudden deceleration applied when the oscillator slams into its target and the duration of that deceleration. Deceleration is calibrated in units of *G*'s, where 1 *G* equals the acceleration of Earth's gravity. The deceleration lasts only a few milliseconds. A good shock test applies the shock in both polarities along all three geometric axes.

Vibration, similar to shock, violently shakes the oscillator. For this test, operators bolt the oscillator to a moving plate, called a shake table, which applies calibrated vibrations. As with shock testing, vibration tests span all three geometric axes. Sinusoidal vibration at a fixed frequency and amplitude resembles repetitive shock testing. It exercises the product repetitively but does not check for mechanical resonances at other frequencies. A swept frequency test covers a range of frequencies at a fixed vibrational amplitude. This is a better test because it checks for mechanical resonances. The random vibration test applies white noise waveforms to the shake table motors, keeping the RMS vibrational amplitude at a constant value. In all cases, the vibration amplitude is specified in *G*'s.

Shock and vibration tests are appropriate for military and aerospace products, portable products, and any other products likely to encounter mechanical misuse.

Relative humidity is the measure of atmospheric wetness. At 100% relative humidity, water begins to condense out of the air. Hermetically sealed or potted packages work easily in conditions of 100% relative humidity. All equipment passes this test.

12.1.3 Electrical

- Output type
- Maximum loading
- Duty cycle
- Rise and fall times
- Input current

Most digital oscillators have TTL, CMOS, or ECL outputs. When using ECL outputs, ask if they are 10K- or 100K-compatible, whichever matches with your circuitry. The 10K and 100K standards specify different tracking of the HI and LO logic levels with temperature. Connecting 10K to 100K logic, or vice versa, deteriorates the voltage margins at extreme temperatures.

In a poorly buffered oscillator, loading the output beyond its specified maximum load can shift the operating frequency. In a well-buffered oscillator, heavy loading just reduces the output amplitude. Loading specifications list either a fan-out or (better) a maximum load capacitance.

The ideal duty cycle is 50%. Practical oscillators, when they specify this parameter, may show 40–60% or 50 ± 10%. At high frequencies it becomes increasingly difficult to guarantee a good duty cycle. If a system will use both clock edges, make sure to get the duty cycle specification in writing.

Rise and fall times of 10–90% are given in nanoseconds. Beware of the occasional manufacturer who quotes a 20–80% rise time.

Input current, measured in milliamps, is a function of frequency. At high frequencies, an oscillator wastes a lot of energy charging and discharging its output load capacitance. Low-power applications demand low operating frequencies and light loading.

12.1.4 Mechanical Configuration

- DIP
- Half DIP
- Surface mount

Quite a bit of circuitry lies under the lid in these parts. Their package must accommodate the quartz crystal, its associated amplifiers, and the hybrid circuit board used to tie everything together. In the future, oscillators will not be shrinking as quickly as other digital components.

A few manufacturers are repackaging their oscillators in smaller cans. From the popular 0.300-in. 14-pin DIP package, the industry is slowly moving toward half-DIP and surface-mount packages.

12.1.5 Manufacturing Issues

- Solderability
- Cleaning
- Package leak rate

Most through-hole packages can be wave-soldered with no difficulties. Surface mount reflow-type soldering poses more of a problem because the parts must withstand greater heat for longer periods. Ask the oscillator's manufacturer to provide data about how long its parts can withstand high-temperature *IR* reflow or vapor phase reflow-soldering processes.

Circuit board assembly shops clean boards and components several times during the assembly process. Both hermetically-sealed and plastic-molded packages usually survive cleaning. It's their labeling that suffers. Make sure the oscillator is marked with an ink that withstands the cleaning fluids in use at your assembly shop. Otherwise, the labels will disappear.

The package *leak rate* measures the quality of an oscillator's hermetic seal. Assuming the package was initially filled with helium, the leak rate says how fast that gas would leak out of the package *into an evacuated chamber*. The cryptic phrase "10^{-8} atm cc/sec" means the package was probably tested using military standard MIL-STD-883, Method 1014. The units of leakage are atmospheres times cubic centimeters per second (atm-cc/s). This number expresses the total amount (pressure times volume) of helium released per second from the package under the test conditions.

12.1.6 Reliability

- Functional screening
- Aging at accelerated temperature

Functional screening means the vendor tests parts before shipping to see if they function. A 1% screening test means only 1% of the parts have been spot-checked. Spot-checking catches devastating shifts in the manufacturing process but not individually bad parts.

Aging is a proactive reliability measure. Most parts, if they are going to fail soon, will do so when stressed. If we take a batch of parts, stress them, and then weed out the ones that failed, we might hope that the remaining parts won't fail any time soon. One big drawback to this approach is that by stressing a batch of parts, we may bring some just up to the threshold of failure. Which effect wins? Does stressing make the remaining parts better, or worse?

It turns out that stress testing actually does a lot more good than harm. Manufacturers (at the urging of the military) have latched onto this idea, and we see a lot of stress-testing specifications for oscillator parts.

Aging specifications show what hurdles each part must pass before shipping. Typical tests start with an initial visual inspection and functional screening, which weeds out obviously dead parts. Next comes a long period of stressful operation under high temperatures. High-temperature operation accelerates aging effects and accelerates failures in silicon electronics. Then a few cycles of rapid cooling and heating are in order, to open weak solder joints. The aging process ends with a few blows from the shock tester and follows with another functional test.

If the device passes this battery of tests, it probably has nothing wrong with it. It will also cost a small fortune, compared to untested devices. You must decide what the failure rate is for untested components and what the cost is to you of their failure in the field. Use that information to determine what level of prescreening is appropriate.

These same prescreening tests apply to any semiconductor device.

12.1.7 Bells and Whistles

- Differential output
- Enable
- Voltage-controlled oscillator
- Tuning.

Differential outputs, when routed to a differential clock input, help overcome noisy environments. When routed to two different clock buffers, they increase the fan-out of the oscillator. If you will be using the two outputs independently, ask what the skew is between them.

An *enable* pin turns the clock on and off. This usually just disables the output driver as opposed to stopping the oscillation. Designers interested in micropower circuitry will want to actually stop oscillation. If oscillation does come to a halt, when it restarts we must wait for oscillations to build back up to a steady state. During the start-up period the output may display partial switching, a poor duty cycle, or the wrong frequency. The buildup period for a crystal oscillator will be several tens of thousands of cycles.

The operating frequency of a *voltage-controlled oscillator* (*VCO*) is electronically adjustable. Inputs to the voltage control pin cause corresponding changes in the oscillation frequency. Voltage-controlled oscillators help synchronize clocks to external phenomena like incoming serial data, television signals, or other computers. Do not assume the curve of frequency versus control voltage is linear. It usually is not.

With temperature-compensated (TXCO) or oven-controlled oscillators, we can trim out initial manufacturing variations with a tiny adjustment, usually a variable capacitor. This adjustment, if periodically updated, also compensates for aging. Noncompensated oscillators drift so much just due to temperature that it is hardly worth including a tuning adjustment in them.

POINTS TO REMEMBER:

Any design involving the frequency difference between two clocks may require especially accurate or stable oscillators.

If you don't need special screening, don't specify a part that has it.

To combat temperature drift, piezoelectric digital oscillators come in (at least) three grades of increasing performance: noncompensated, temperature-compensated, and oven-controlled.

Since all manufacturers use the same material (quartz) and all know about the cutting angles needed to produce various temperature functions, we can expect similar specifications for temperature drift across all manufacturers.

12.2 CLOCK JITTER

Every clock oscillator contains a very fine high-frequency amplifier. This amplifier detects tiny voltages in its resonant circuit and builds them up to useful logic levels. The same amplifier also picks up tiny noise voltages, amplifies them, and presents them at its output terminals along with the clock. The amplifier does not distinguish clock signals from noise; it simply amplifies whatever voltage appears at its input terminals. Someone, either the oscillator manufacturer or user, must ensure that no appreciable noise enters the amplifier.

In a modern, highly buffered oscillator, any amplified noise appears at the output in the form of clock jitter. The term *clock jitter* refers to any deviations of a clock's output transitions from their ideal positions.

Jitter results from four superimposed noise sources. First, noise emanates from the crystal itself. Like any resistive device, the crystal puts out thermal noise due to the random movement of electrons within it.[2] Second, any mechanical vibration or perturbation of the crystal causes noise. The third noise source is amplifier self-noise. The amplifier's contribution is often larger than the thermal and mechanical noise from the crystal. The last and potentially most troublesome noise comes from the power supply. Any coupling of the power terminals into the amplifier's sensitive input sends power supply noise roaring through the amplifier, causing massive amounts of jitter. An oscillator that couples power supply noise into its output is said to have poor *power supply immunity*. Many oscillators do.

Clock jitter from random sources is bad enough. Clock jitter induced by power supply noise, which fluctuates intermittently in a data-dependent fashion, is even worse. At least random jitter happens all the time. We have means of measuring, characterizing, and protecting ourselves against random jitter. Intermittent jitter is much harder to track down.

12.2.1 When Does Clock Jitter Matter?

Jitter is often not included on an oscillator's data sheet. For communications applications, it should be.

Any time we transfer data between two digital domains which are under the control of independent clocks, clock jitter matters. For example, suppose we have two machines, A and B, which both independently synchronize their clocks to a common reference (Figure 12.4). Let the common reference frequency be 8 kHz.[3] The clock frequency in each section is 154.4 MHz, 20,000 times the reference frequency. Data proceeds from section A, through the FIFO, into section B. Theoretically, once the FIFO gets started, it stays filled at a constant level because the input and output rates are the same. In practice, the two clocks are hardly the same. The common timing reference signal comes along only once every 20,000 clocks, leaving plenty of time for the two clocks to diverge between reference edges. In practice, jitter between the two clocks makes the FIFO gyrate wildly.

[2]Oscillator circuits using *LC* tanks or other resonant members display similar electrical and mechanical noise effects.

[3]A common telecommunications reference clock frequency.

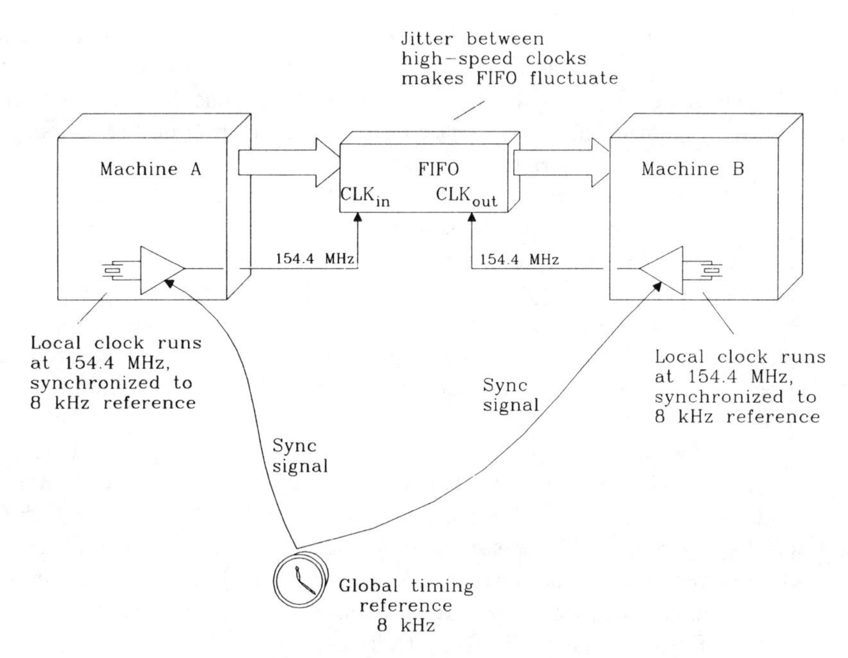

Figure 12.4 Exchanging data between synchronized digital machines.

The greater the ratio of clock to reference frequency, the worse this effect becomes. Enough jitter will cause the FIFO to either overflow or run empty. The maximum deviation in the FIFO corresponds to the maximum phase difference between the two clocks.

12.2.2 Measuring Clock Jitter

There are at least three approaches to measuring clock jitter. The three methods are spectral analysis, direct phase measurement, and differential phase measurement. The easiest measurement technique for digital engineers (given the equipment typically available to them) is differential phase measurement. Because you are likely to find references to all three in the confused literature on the subject of clock jitter, we will delve briefly into all three methods.

Spectral analysis is easy to do, given enough equipment. Just connect your jittery clock to a high-quality spectrum analyzer. The spectrum of a perfect clock consists of infinitely thin spectral peaks at harmonics of the fundamental frequency. Close examination of a jittery clock spectrum reveals a tiny amount of spreading around the fundamental frequency and around each harmonic. This spreading relates to clock jitter. Simply put, when a clock spends part of its time at frequency F, we see a peak there corresponding to what percentage of the time it lingered at that frequency. When a clock's phase jitters, so does its instantaneous frequency, causing spreading of its spectrum about the fundamental. Spectral analysis is very popular with communications engineers.

The problem with spectral analysis is that it does not directly address the issue of phase error. The spectrum tells us what frequencies the clock visited but not how long it stayed. A clock that lingers too long away from its center frequency accumulates a big phase error. A clock that deviates back and forth quickly about its center frequency may visit the same frequency for the same proportion of time, but stay so briefly each visit that it accumulates almost no phase error. From the spectrum we cannot determine the maximum phase deviation from ideal.

If you have access to an ideal clock, you can directly compare it to your jittery clock with a phase detector. The output of this *direct phase measurement* shows just what we want to know: how much the clock jitters. The obvious difficulty with this approach is getting an ideal clock. Try filtering the jittery clock through a phase-locked loop to create a smooth clock having the same average frequency. The phase error output from the phase-locked loop will be the jitter signal you seek. Of course, if we are measuring jitter from a high-quality frequency source, it may not be easy to build a phase-locked loop with less jitter.

Differential phase measurement compares a jittery clock not to an ideal clock but to a delayed version of itself. At a large enough delay, the delayed waveform is uncorrelated with the original, and we get the effect of comparing two similar, but different, jittery clocks. The resulting differential jitter is twice the actual jitter. The advantage of using a delayed version of the original clock is that it naturally has the correct average frequency. A differential jitter measurement requires an oscilloscope with a delayed time base sweep feature. First set your oscilloscope to trigger on the clock waveform. Then, using the delayed time base sweep, take a close look at the clock some hundreds, thousands, or ten-thousands of clock cycles later. Jitter shows up as a blur in the displayed waveform.

Before assuming the blur comes from jitter on the clock, take a look at a stable clock source using the same setup. If it looks clean, we can then assume your scope time base is accurate enough to perform this measurement.

While adjusting the delay interval, you may notice that the jitter gets worse or better. This is normal. Clock jitter normally is worse in some frequency bands, which leads to maxima in the expected differential jitter at certain time delays. Beyond some maximum time delay, the jitter becomes completely uncorrelated and there is no longer any change in jitter with increasing delay.

If the jitter amounts to more than half a clock period, successive edges blur together, becoming very difficult to see. In that case, divide the clock by 2, 4, or more using a counter circuit before displaying it. The division doesn't change the worst-case jitter on individual clock edges, but it does lengthen the space between nominal clock transitions so that we can see where the jitter goes.

Jitter measurements on precise crystal clocks require an extremely stable time base and can take a long time to perform. Jitter measurements performed on noncrystal oscillators used in serial data transmission are much easier to do, owing to the much greater jitter.

12.2.3 Measuring Power Supply Immunity

Since power supply noise is one of the biggest causes of clock jitter, we need a way to directly measure its effects. Let's inject some power supply noise into an oscillator and see what happens. Using the circuit in Figure 12.5, we can adjust both the frequency and amplitude of injected power supply noise.

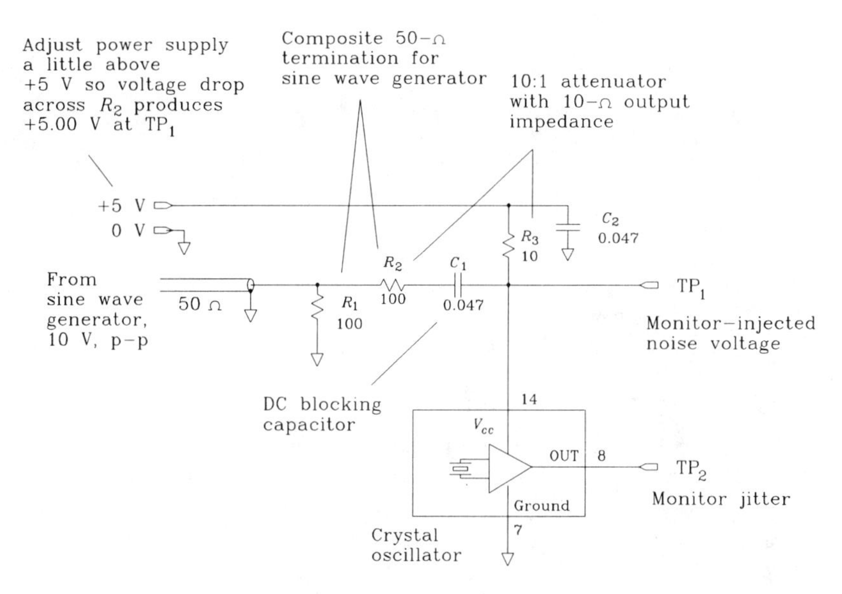

Figure 12.5 Power supply immunity test.

We will fix the injected noise amplitude at a known voltage and then make a plot of jitter versus injected noise frequency. Use the differential jitter measurement technique in Section 12.2.2 to estimate jitter. Prepare a graph showing injected noise frequencies ranging logarithmically from 10 kHz to 100 MHz or more.

Begin with the noise amplitude set at 0.5 V. For each frequency point, first trim the injected noise amplitude to 0.5 V and then set the differential time base delay to

$$\Delta T = \frac{0.5}{F} \quad [12.1]$$

where ΔT = differential time base delay, s
F = injected noise frequency, Hz

The particular time delay in Equation 12.1 will always display the worst jitter. Scan the jitter versus time delay waveform on your scope to verify that this is true. Thereafter, just record for each frequency F the differential jitter measured at delay ΔT.

Estimating the jitter at each frequency from the blur on your oscilloscope is tricky. Take photographs if possible, so that you can see the results lined up together.

If you have a point accumulation feature (the Tektronix 11403 oscilloscope does), you can run the jitter display for a fixed period of time and then analyze it to find the actual jitter variance. One simple analysis technique first counts how many points lie between the HI and LO logic levels (see Figure 12.6). The Tektronix 11403 can count these points for us. These are the points where we caught the clock edge in transition. Then look to see what horizontal displacement encompasses 80% of these points. The jitter variance is 21% of this displacement.

Sometimes we find a frequency range where the oscillator becomes very sensitive to power supply noise. This effect usually results from insufficient power supply filtering

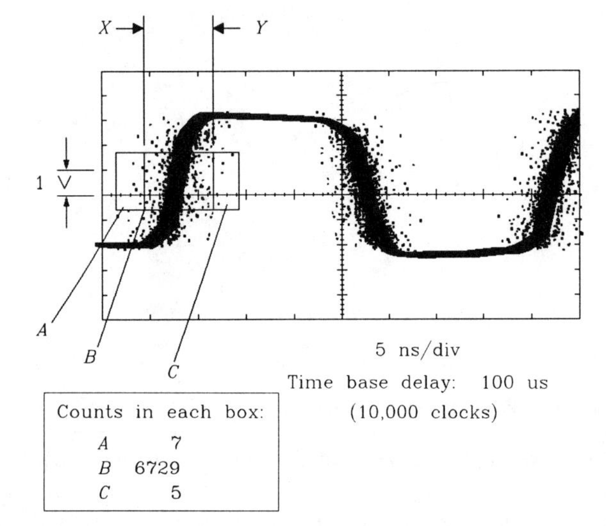

Figure 12.6 Point accumulation of a clock jitter waveform.

inside the oscillator. The poor immunity curve shown in Figure 12.7 displays symptoms of insufficient power filtering.

Another, more serious, effect is *squelching*. At some injected noise frequency the power supply filtering components internal to the oscillator may resonate. A low injected noise voltage at this frequency causes extreme amounts of jitter. A high injected noise voltage at this frequency may disrupt the action of the internal amplifier, stopping oscillation altogether. A stopped oscillator is said to be squelched.

12.2.4 Power Supply Filtering for Clock Sources

If your oscillator has poor power supply immunity or if it must work inside a noisy system, provide additional power supply filtering for it. The amount of filtering required depends on how much a reduction in jitter you must achieve. Determining a precise value for required jitter reduction is almost impossible because all the parameters vary:

- Jitter performance is not specified on many oscillators. When your purchasing department buys a different brand of oscillator, the jitter will change.
- Noise in a system changes when different brands of integrated circuits (perhaps faster-switching ones) are assembled.

Nevertheless, you need to do something, so try the circuit in Figure 12.8. It achieves 20 dB of power supply noise reduction in the frequency band above 14 MHz.

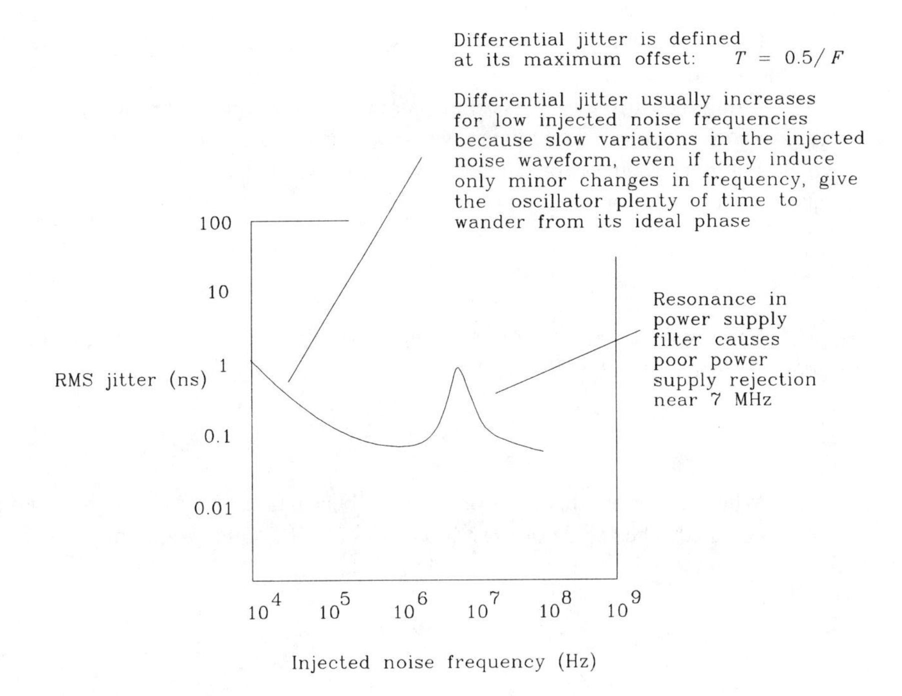

Figure 12.7 Power supply immunity of a clock oscillator.

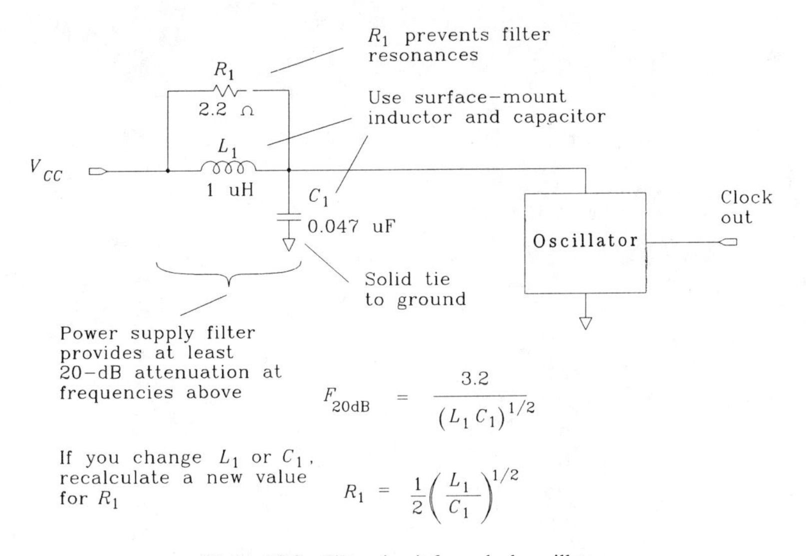

Figure 12.8 Filter circuit for a clock oscillator.

The attenuation slope above 14 MHz is 20 dB/decade. Cascading two sections yields double the attenuation.

Using the differential phase measurement technique of Section 12.2.2, check your jitter before and after adding the power supply filter. Improvements will show clearly.

A bigger inductor or capacitor will enhance the filter's attenuation at lower frequencies. For any combination of L and C, the 20-dB attenuation frequency is

$$F_{20\text{dB}} = \frac{3.2}{(LC)^{1/2}} \quad [12.2]$$

For any new combination of L and C, recalculate the value of resistor R needed to prevent resonance in this filter circuit:

$$R = \frac{1}{2}\left(\frac{L}{C}\right)^{1/2} \quad [12.3]$$

When laying out this filter on a circuit board, take care to keep the input and output well separated. The capacitor must directly connect a solid ground plane with at least one

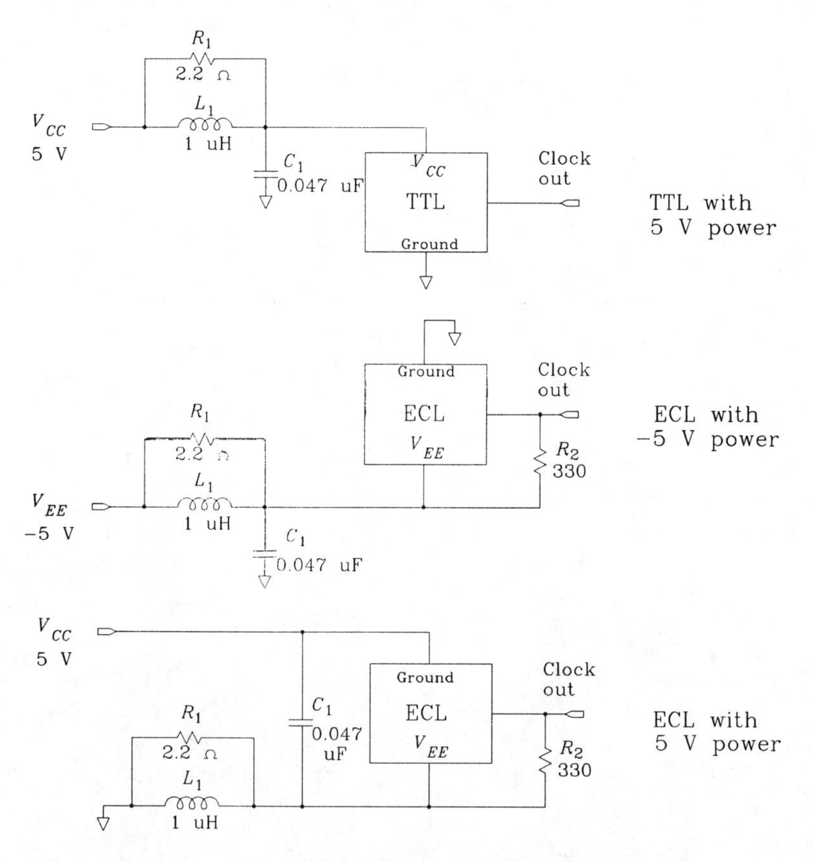

Figure 12.9 Applying filter circuit to clock oscillators.

large (0.035-in.-diameter) via. Keep all circuit traces as short as possible (less than 0.1 in.). Surface-mounted components work best.

Figure 12.9 shows three applications of this filter to canned oscillators. Note that when using positively biased ECL, it is the +5-V power supply rail that forms the common voltage reference between the clock oscillator and its associated buffers. Filter the ground input to a positively biased ECL oscillator, not the +5-V input.

POINTS TO REMEMBER:

Jitter is often not included on an oscillator's data sheet. For communications applications, it should be.

Differential phase measurement compares a jittery clock not to an ideal clock but to a delayed version of itself.

If your oscillator has poor power supply immunity, or if it must work in a noisy system, provide additional power supply filtering for it.

Collected References

Electromagnetic Compatibility

HENRY W. OTT, *Noise Reduction Techniques in Electronic Systems*, John Wiley, New York, 1988.

A first-class overview of noise problems, both radiation and susceptibility. Oriented toward analog electronics but now includes two sections on digital layout and radiation in the second edition.

RALPH MORRISON, *Grounding and Shielding Techniques in Instrumentation*, John Wiley, New York, 1986.

Material similar to Ott's book, but (like the title says) concentrated on analog instrumentation problems. If you are building distributed instrumentation, this is the book.

BERNHARD KEISER, *Principles of Electromagnetic Compatibility*, 3rd ed., Artech House, Norwood, Mass., 1987.

While Ott and Morrison treat the problems of both radiated and conducted interference (such as noise induced by ground loops), Keiser focuses mostly on radiated noise problems. That makes this a very intense reference work on electromagnetic compatibility (EMC) effects. It includes a good summary of EMC regulations around the world.

CLAYTON R. PAUL, *Introduction to Electromagnetic Compatibility*, John Wiley, New York, 1992.

A more theoretical, thorough, and newer review of EMC than Keiser's book. Begins with a motivational section on why we worry about EMC and then presents a quick review of electromagnetic fields, transmission lines, and antennas. The rest of the book covers practical aspects of EMC design. We think this text should be taught in colleges.

Beware of depending on this or Keiser's book for the latest information on government standards. Standards in the EMC world are changing very rapidly.

Transmission Lines and Digital Signals

WILLIAM R. BLOOD, JR., *MECL System Design Handbook*, 4th ed., Motorola Semiconductor Products, Inc., Phoenix, Ariz., 1988.

First published in 1980, this is the granddaddy of practical high-speed digital circuit books. Dealing exclusively with Motorola ECL circuits, this book discusses transmission lines, connectors, cables, power distribution, and thermal problems (a big issue with ECL). If you are using ECL, get this book. It has plenty of good ideas and is still very relevant.

B. L. HART, *Digital Signal Transmission Line Circuit Technology*, Van Nostrand Reinhold, New York, 1988.

This book has good hints about using long transmission lines. It is a very practical, readable text at the introductory level.

CHARLES S. WALKER, *Capacitance, Inductance and Crosstalk Analysis*, Artech House, Norwood, Mass., 1990.

A detailed and thorough guide to the calculation of mutual capacitance and inductance. If you need accuracy in your transmission line equations, get this book.

T. C. EDWARDS, *Foundations for Microstrip Circuit Design*, John Wiley, New York, 1981.

For those interested in the properties of ceramic substrates, Edwards goes into great detail. If you have a good background in analog circuit design and appreciate academic precision, this is a good book. Covers microstrips only (not striplines), including directional couplers and microwave filters.

HARLAN HOWE, *Stripline Circuit Design,* Artech House, Norwood, Mass., 1974

Complementing Edwards' book, this text focuses on stripline design. It includes coupler and microwave filter design information. A brief chapter at the end introduces the reader to microwave construction techniques.

S. R. SESHADRI, *Fundamentals of Transmission Lines and Electromagnetic Fields*, Addison-Wesley, Reading, Mass., 1971.

A highly technical work explaining the mathematical basis of transmission line theory. Includes both *RC* and low-loss transmission line examples.

Printed Circuit Board Manufacturing and Integrated-Circuit Packaging

CLYDE F. COOMBS, JR, ED., *Printed Circuits Handbook*, McGraw Hill, New York, 1988.

A giant compendium covering 35 topics in printed circuit board design, manufacture, and use. This book references and explains many military and IPC standards relating to printed circuit boards. An excellent overall reference work.

H. B. BAKOGLU, *Circuits, Interconnections, and Packaging for VLSI*, Addison Wesley, Reading, Mass., 1990.

Covers topics important to VLSI packaging including heat transfer and electrical and mechanical properties. This book describes chips used in the highest-performance computer systems. A real glimpse into the future of computer technology.

BERNARD S. MATISOFF, *Handbook of Electronics Packaging Design and Engineering*, 2nd ed., Van Nostrand Reinhold, New York, 1990.

A practical, broad description of standard packaging and assembly techniques used in the electronics industry. It includes data on the effectiveness of metallic connections between chassis components of vital interest to those with EMC problems. It also has a section on thermal transfer and heat problems.

RAYMOND H. CLARK, *Printed Circuit Engineering: Optimizing for Manufacturability*, Van Nostrand Reinhold, New York, 1989.

Clark presents this excellent set of guidelines and standards for printed circuit design and fabrication in a lucid, well-organized format. These standards are especially relevant if you manufacture in high volumes.

RAY P. PRASAD, *Surface Mount Technology*, Van Nostrand Reinhold, New York, 1989.

Ray Prasad works at Intel and really knows his subject. Written for manufacturing engineers, the practical guidelines and tips in this book will make you sound like an expert. Never underestimate the value of really understanding your manufacturing processes. If surface-mount technology is new to you, read this book.

Analog Issues

FREDERICK E. TERMAN, *Radio Engineers Handbook*, McGraw-Hill, New York, 1943.

That's right, published in 1943. All the really good work in engineering was done a long time ago, and this book proves the point as well as any. The section on transmission line effects is outstanding (especially the discussion of the proximity effect). It's fascinating to see how many problems we had back then we still haven't solved.

ROBERT A. PEASE, *Troubleshooting Analog Circuits*, Butterworth-Neinemann, Stoneham, Mass., 1991.

Those who read Bob Pease's column in Electronic Design magazine on circuit design are familiar with his easy, humorous approach to engineering. Even though this book deals with analog circuit problems, his technique and experience are so valuable that we recommend the book to digital people. It's fun, and you will probably learn something important.

IRVING M. GOTTLIEB, *Understanding Oscillators*, TAB Books, Blue Ridge Summit, Penn., 1987.

An easy-to-follow discussion of the oscillation process. Chapter 5 contains 17 pages treating various forms of oscillators built from digital integrated circuits or comparators.

DOUGLAS C. SMITH, *High Frequency Measurements and Noise in Electronic Circuits,* Van Nostrand Reinhold, New York, 1993.

A detailed study of accurate methods for making high-frequency measurements. This book delves into the operation of voltage probes, current probes, and probes designed to work in conjunction with electrostatic discharge testing. It also contains useful tips for debugging high-frequency problems.

A

Points to Remember

This appendix collects all the Points to Remember from throughout the book. Each item is listed alongside the section in which it is described. You may use this section as a checklist for system design or as an index to the text when facing a difficult problem.

CHAPTER 1: FUNDAMENTALS

1.1 The response of a circuit at high frequencies affects its processing of short-time events.

The response of a circuit at low frequencies affects its processing of long-term events.

Most energy in digital pulses concentrates below the knee frequency:

$$F_{\text{knee}} = \frac{0.5}{T_r}$$

The behavior of a circuit at the knee frequency determines its processing of a step edge.

The behavior of a circuit at frequencies above F_{knee} hardly affects digital performance.

1.2 Propagation delay is proportional to the square root of the dielectric constant.

The propagation delay of signals traveling in air is 85 ps/in.

Outer-layer PCB traces are always faster than inner traces.

1.3

$$\text{Length of rising edge } l = \frac{\text{rise time (ps)}}{\text{delay (ps/in.)}}$$

Circuits smaller than $l/6$ are lumped circuits.

1.6 A capacitance test jig is easy to build using a pulse source and an oscilloscope.

1.8 The area under an exponential L/R decay provides an accurate measure of the decay time constant.

A slow pulse generator rise time, or a slow scope, does not change the area measured with our inductance test jig.

1.10 Among high-speed digital circuits, mutual inductance is often a worse problem than mutual capacitance.

CHAPTER 2: HIGH-SPEED PROPERTIES OF LOGIC GATES

2.1 Today, just as in the days of relay logic, power and packaging have a big impact on system performance.

2.2 Always include active power dissipation and power dissipated driving heavy loads in device power calculations.

2.3 Given two logic families with identical maximum propagation delay statistics, the family with the slowest output switching time will be cheaper and easier to use.

We can figure the dI/dt in an output circuit given the voltage rise time and the load.

When we halve the rise time, we quadruple the amount of dI/dt flowing into capacitive loads.

A complete voltage margin budget in a system accounts for the effects of power supply variations, ground shifts, signal crosstalk, ringing, and thermal differences.

2.4 At high speeds, the inductance of logic device packaging is critical.

Output switching currents flowing through a ground pin cause ground bounce, which can cause double-clocking of flip-flops.

Thermal resistance is the ratio of temperature rise to power dissipation.

Heat flows from a silicon die to its package, and from the package to the ambient surroundings: $\Theta_{JA} = \Theta_{JC} + \Theta_{CA}$

400 ft/min is a lot of air.

CHAPTER 3: MEASUREMENT TECHNIQUES

3.1 When figuring a composite rise time, the squares of 10–90% rise times add.

3.2 A 3-in. ground wire used with a 10-pF probe induces a 2.8-ns 10–90% rise time. In addition, the response will ring when driven from a low-impedance source.

Fattening the ground wire hardly helps with ringing.

Radically shortening the ground loop improves ringing and reduces rise time.

3.3 Ground the probe near the signal of interest to reduce the ground wire pickup loop area.

Keep the probe ground wire as short as possible or use a knife blade to short the probe shield directly to the circuit board ground.

Make a magnetic field detector to test for noise induced by mutual inductive coupling.

3.4 A 10-pF probe looks like 100 Ω to a 3-ns rising edge.

Less probe capacitance means less circuit loading and better measurements.

3.5 A shop-built 21:1 probe has a terrific rise time.

3.6 A single-ended scope probe responds to shield voltage as if it were a real signal.

To see if you are getting noise induced by shield currents shield the tip of your probe with foil and then touch the probe and its ground to the circuit board ground.

Tie both differential scope probes temporarily to a common signal point while adjusting the gain balance between probes to best cancel their wave-forms.

3.7 View a serial data stream by triggering on the clock.

3.8 During testing, a sufficiently slow clock allows all signal transients to decay before starting the next clock cycle.

3.9 Amplify the visible effects of crosstalk by temporarily changing your system.

3.10 Measure how much stress a system absorbs before failing a go-nogo test. This procedure converts a simple go-nogo test into a quantitative measure of product quality.

3.11 All flip flops exhibit metastability.

The probability that a flip-flop output will delay more than T seconds goes down exponentially with increasing time T.

CHAPTER 4: TRANSMISSION LINES

4.1 Distributed circuits always ring if unterminated. Lumped circuits can also ring if their Q is too high.

Point-to-point wiring has a lot of inductance. This inductance, working into a heavy load capacitance, makes a high-Q circuit.

Large current loops carrying quickly changing currents generate big magnetic fields. Reducing current loop area reduces EMI.

Straight point-to-point wiring, pressed down as close to the ground plane as possible, is much better than gathered or bundled wiring.

Systems with thousands of connections warrant extra attention to crosstalk.

4.2 The input to an *infinite* transmission line looks resistive, not capacitive.

Handy relations for the inductance and capacitance of a transmission line:

$$L = Z_0 T_p$$

$$C = \frac{T_p}{Z_0}$$

Total wiring resistance is usually a small fraction of transmission line impedance for ordinary digital applications.

The skin effect seriously limits the frequency response of long transmission lines.

For short-haul digital applications, transmission line attenuation in decibels is proportional to the square root of frequency (skin effect).

The proximity effect has only a minor effect on transmission line attenuation.

For digital applications below 1 GHz, ignore dielectric losses.

4.3 Any combination of practical source and load impedances connected to a transmission line degrade its performance.

The frequency response of a transmission line system is

$$S_{\infty}(w) = \frac{A(w)H_X(w)T(w)}{1 - R_2(w)H_X{}^2(w)R_1(w)}$$

Only when the round-trip delay exceeds your signal rise time will overshoot and ringing arise.

Eliminate reflections by reducing R_2 (end termination), reducing R_1 (series termination), or making sure the line is short ($H_x = 1$).

4.4 Capacitive loads degrade the rise time of passing signals and reflect pulses back upstream.

Uniformly distributed capacitive loads reduce a transmission line's effective impedance and slow it down.

A printed circuit trace makes an effective small delay line.

4.5 For printed circuit board traces, the most critical dimension is the ratio of trace width to height above ground.

Double the reflection budget to find the allowed mismatch between the characteristic impedance and the terminating resistors.

Large variations in physical dimensions make a small impact on the resulting impedance.

The slope of any function plotted on log-log paper is equal to the sensitivity of the function to changes in its argument.

All formulas for transmission velocity are inversely proportional to the *effective* square root of electric permittivity.

CHAPTER 5: GROUND PLANES AND LAYER STACKING

5.1 High-speed current follows the path of least inductance.

Returning signal currents tend to stay near their signal conductors, falling off in intensity with the square of increasing distance.

5.2 Returning signal currents generate magnetic fields, which in turn induce voltages in other circuit traces.

The induced noise coupled into adjacent traces falls off with the square of increasing distance.

5.3 Slots in a ground plane create unwanted inductance.

Slot inductance slows down rising edges.

Slot inductance creates mutual inductive crosstalk.

5.4 If you must work with only two layers, use the power and ground grid system.

5.5 For high-speed logic, avoid the ground fingers layout.

5.6 A solid ground plane provides most of the benefit of grounded guard traces.

5.7 Regarding long transmission lines:

Over solid grounds, inductive and capacitive crosstalk are equal. Forward-crosstalk components cancel, while reverse crosstalk reinforces.

Over a slotted or imperfect ground plane, the inductive coupling exceeds capacitive coupling, making forward crosstalk large and negative.

Forward crosstalk is proportional to the derivative of the input signal and to line length.

Reverse coupling looks like a square pulse, with a constant height and duration equal to $2T_p$. For short lines, reverse coupling does not climb to its full value.

Reverse crosstalk, when it hits a low-impedance driver, reflects toward the far end.

5.8.3 As a rule, the greater your circuit wiring density, the greater your production costs per square inch.

Printed circuit board cost is proportional to the number of layers and to the board surface area.

Design the power and ground layers first.

For mechanical reasons, lean toward using a symmetric arrangement of ground and power planes in your layer stack.

Smaller, more closely spaced traces yield more crosstalk.

5.8.4 Don't count on filling more than half the spaces between through-hole pins.

When all else fails, use Rent's reasoning to figure the average trace length.

5.8.5 Core and prepreg layers alternate.

Outer layers, if plated, have greater trace width variation than inner traces.

Traces on routing layers tend to sink into the prepreg mixture. Their thickness doesn't add to the total board thickness.

The thickness of solid ground plane layers always adds to the total board thickness.

5.8.6 At the highest speeds, keep ground and power planes directly adjacent.

Use extra ground planes, not power planes, to isolate routing layers.

CHAPTER 6: TERMINATIONS

6.1 The rise time of an end-terminated circuit, when capacitively loaded, is half that of a series-terminated line driving the same load.

Most TTL or CMOS logic gates can't source enough current to drive end terminators.

You can daisy-chain receivers on an end-terminated line.

6.2 Source terminators have a slower rise time and usually smaller residual reflections than end terminators.

Do not daisy-chain receivers on lines having source terminators.

Subtract from the ideal source-termination value the output impedance of your driver.

At low-pulse repetition rates, source terminators dissipate little power.

The peak drive power for a source-terminated line and an end-terminated line (biased at the halfway point) are the same.

6.3 Middle terminations can improve system step response only at the expense of signal attenuation.

6.4 Combination RC circuits can terminate DC-balanced lines with no wasted quiescent power.

6.5 Specify both a resistance value tolerance and a power rating on terminating resistors.

Parasitic inductance in terminating resistors causes unwanted reflections.

6.6 The physical layout of terminating resistors affects crosstalk between signal paths.

CHAPTER 7: VIAS

7.1 The finished diameter of routing vias depends on drilling and plating technology. Smaller holes cost more.

Pad sizes are determined by drilling tolerances and the annular ring requirement. The annular ring controls breakout.

The minimum air gap is determined by line width tolerances and nominal pad positions. The air gap controls solder bridging.

Sacrificing pad size and air gap increases tracks but reduces yield.

7.2 Via capacitance is a measurable, but small, effect.

A scale model of a via or trace has X times the capacitance of a real via, where X is the model scale.

7.3 Via inductance degrades the shunting capability of bypass capacitors.

An array of bypass capacitors is more effective than a single bypass capacitor.

Power supply bypassing gets progressively more difficult as rise times shrink.

CHAPTER 8: POWER SYSTEMS

8.1 Three power system design rules:

(1) Use low-impedance ground connections between gates.
(2) The impedance between power pins on any two gates should be just as low as the impedance between ground pins.
(3) There must be a low-impedance path between power and ground.

8.2.1 Sense wires correct for resistance in power distribution wiring.

8.2.2 It is almost impossible to reduce wiring inductance by simply using a bigger wire.

Wide, flat parallel structures work much better as distribution wiring than round wires.

Differential transmission is practically immune to power supply fluctuations.

8.2.3 A power supply provides low impedance at low frequencies.

Local bypass capacitors provide low impedance at higher frequencies.

8.2.4 The best way to get very low inductance is to parallel a lot of small capacitors.

8.2.5 Power and ground planes separated by 0.01 in. of FR-4 have a capacitance of 100 pF/in.^{2}

8.2.6 A simple test jig measures power supply step response.

8.3 Combining TTL and ECL in one system without considering the system design consequences is not a good idea.

If the power distribution wiring has too much resistance, the local voltage at each card will differ.

When a card plugs into a working backplane, it draws a huge surge of current as its local bypass capacitor charges to full potential.

Changing currents in the power wiring easily radiate away from a digital product.

8.4.1 Lead inductance acts like an inductor in series with a capacitor.

ESR acts like a resistor in series with a capacitor.

Together they degrade a capacitor's effectiveness as a bypass element.

8.4.2 For large-valued capacitors, smaller packages have higher series inductance and ESR than larger packages.

Capacitor performance varies widely.

8.4.3 Ask whether your board will be assembled with wave or reflow soldering.

8.4.5 Higher-dielectric-constant materials pack more capacitance into a smaller space but have poor temperature coefficients and aging instability.

Aluminum electrolytics do not work well in cold applications.

8.4.6 Failure in capacitors is a statistical phenomenon, accelerating at high voltages.

CHAPTER 9: CONNECTORS

9.1 Mutual inductance, not mutual capacitance, is mostly responsible for crosstalk generated by connectors.

Spreading grounds across the connector reduces crosstalk.

9.2 EMI emanates from signal current flowing in large loops.

Provide a low-inductance return-current path for every connector.

Disrupt or eliminate remote return-current paths.

9.3 Multidrop bus applications place more of a burden on the connector system than point-to-point applications.

For multidrop applications we want a connector with very low parasitic capacitance, even at the expense of higher inductance.

9.4 A simple test setup characterizes the crosstalk in a connector.

9.5 If returning signal current from a connector must flow around a ground plane hole, it doesn't matter how many grounds we use; the performance gets no better than a connector with a single ground pin at either end.

9.6 Exposed wiring carrying high-speed digital signals between circuit boards always fails FCC and VDE radiated emission tests.

If you can tolerate the degradation in rise time, filter all outgoing digital signals before they exit the chassis.

A common mode choke reduces current flowing in remote return loops.

The drain wire loop marked (A) in Figure 9.14 is often large enough to radiate more than FCC/VDE limits.

9.7 Crosstalk and EMI control require extraordinary connectors at high speeds.

9.8 Differential reception is unaffected by ground voltage shifts between the transmitter and receiver.

Common mode current from a well-designed differential driver is a factor of 100 less than the primary current.

9.9 Variable-height pins help sequence soft-start power and reset operations as a card is inserted into a live backplane.

CHAPTER 10: RIBBON CABLES

10.1 The rise time of any ribbon cable varies with the square of its length.

All cables, whether they be coaxial, twisted, or ribbon, share the same basic frequency response. Their frequency response, in decibels, is proportional to the inverse square root of frequency.

A ribbon cable's dielectric configuration impacts both signal velocity and attenuation.

10.2 Any level of crosstalk attenuation is achievable given enough grounds.

Crosstalk falls off as $1/X^2$. It is also proportional to the separation Δ_1 between each transmitting wire and its nearest return wire, and the separation Δ_2 between any receiving wire and its nearest ground.

In the ground-signal-ground, or G-S-G, configuration the reverse crosstalk coefficient between nearest neighbors is about 2–5%.

In a twisted cable, if the rising edge spreads out over N precessions of the twist cycle, we can expect coupling to be about $1/N$ the amount in an ordinary parallel wire cable.

Far-end crosstalk accumulates as it propagates down a cable.

Near-end crosstalk stays the same amplitude regardless of cable length but elongates in time as the cable is lengthened.

10.3 Ribbon cables proliferate today because, in conjunction with mass termination, they are very cheap.

Whatever the connector scheme, it always introduces parasitic inductance and capacitance.

10.4 Shields provide a very-low-inductance return path for signal currents.

With a spiral wrap, be sure that each joint between overlapping spiral layers makes firm electric contact with the previous layer.

A drain wire is a weak spot in any shield.

CHAPTER 11: CLOCK DISTRIBUTION

11.1 *Timing margin* measures the slack, or excess time, remaining in each clock cycle.

Timing margin protects your circuit against signal crosstalk, miscalculation of logic delays, and later minor changes in the layout.

11.2 Clock skew has as much of an impact on overall operating speed as any other propagation delay.

11.3 Two or more driver outputs connected in parallel make a convenient and simple high-powered driver.

The total drive power required for TTL clock signals is 25 times that of ECL circuits.

11.4 A 20-Ω clock line is 2.5 times less sensitive to the capacitance of clock taps than a 50-Ω line.

11.5 A single driver can service two or more source-terminated lines under restricted circumstances.

11.6 The physical means of providing extra crosstalk protection are simple; the logistical means are complex.

11.7 Delay elements are built from three basic building blocks: transmission lines, logic gates, and passive lumped circuits.

A fixed delay cannot cancel variations in board fabrication or active component delay.

An adjustable delay compensates for actual delays, not just nominal delays, elsewhere in the circuit.

Whatever form of delay you choose, incorporate its uncertainty in delay into your timing margin calculations.

11.8 Crosstalk, as long as it affects the two parts of a differential clock line equally, induces no timing jitter.

11.9 A clock's duty cycle changes as it progresses along a chain of repeaters.

Clock signals propagate much farther, and with a better duty cycle, in an inverting repeater chain than in a noninverting one.

11.10 An inductor can partially cancel the parasitic capacitance of a clock receiver.

11.11 An attenuating network can increase the effective input impedance of a clock receiver.

CHAPTER 12: CLOCK OSCILLATORS

12.1 Any design involving the frequency difference between two clocks may require especially accurate or stable oscillators.

If you don't need special screening, don't specify a part that has it.

To combat temperature drift, piezoelectric digital oscillators come in (at least) three grades of increasing performance: noncompensated, temperature-compensated, and oven-controlled.

Since all manufacturers use the same material (quartz) and all know about the cutting angles needed to produce various temperature functions, we can expect similar specifications for temperature drift across all manufacturers.

12.2 Jitter is often not included on an oscillator's data sheet. For communications applications, it should be.

Differential phase measurement compares a jittery clock not to an ideal clock but to a delayed version of itself.

If your oscillator has poor power supply immunity, or if it must work in a noisy system, provide additional power supply filtering for it.

B

Calculation of Rise Time

$$T_{\text{composite}} = \left(T_{r1}^{\,2} + T_{r2}^{\,2} + \cdots + T_{rN}^{\,2}\right)^{1/2} \qquad \text{[B.1]}$$

Equation B.1 relates the rise time of a complete system $T_{\text{composite}}$, to the rise time of each of its parts. This equation applies to cascaded linear systems (like pulse generators, probes and oscilloscopes) that process digital steps in a linear fashion.

Pulse generators, probes, and oscilloscopes contain linear elements that deteriorate the sharpness of a rising edge as it passes through. The advantage of a linear processing element is that you may see at its output subtle details of the input waveform. Its disadvantage is that it always introduces a little distortion. Equation B.1 models that rise-time distortion.

Linear elements stand in contrast to nonlinear elements like logic gates. Logic gates contain saturating amplifiers that regenerate a fresh rising edge at their output. Equation B.1 does not apply to saturating amplifiers.

If we know the rise time of each part of a cascaded system, we can use Equation B.1 to predict the overall system rise time. A good application for this use of Equation B.1 is in fiber optic system design. There we must develop separate rise-time specifications for the optical transmitter, fiber characteristics, and receiver electronics such that the rise time of the overall system meets some predetermined goal. Equation B.1 shows how the component specifications combine to affect the whole system.

Alternately, if we know the overall rise time as well as the rise time for all but one part of the system, we can deduce the rise time for the remaining part. A good application for this clever trick occurs in component testing. Let's say that through prior measurement you know the rise time of your oscilloscope. You may then measure the output rise time of a chip and use Equation B.1 to subtract out the effect of your scope. The result will more accurately reflect the actual chip rise time. Using a scope only a little faster than the chip rise time, you can make quite accurate measurements.

The test conditions for measuring the rise time of one part of an overall system are easy to define but not so easy to actually implement. First, you must disconnect the input of the part under test. Then inject into it a perfectly square-edged input and measure the rise time at its output (the step response).

Note that we have discussed the test conditions for rise-time measurement but not how to define precisely the rise time. The wonder of Equation B.1 is that it hardly matters! For calculations in most digital systems you may choose the 10–90% rise time, the 20–80% rise time, the inverse of the slope at the center of a step, or the mathematically precise measure T_σ (defined below). Almost any measure works as long as you use the same measure for all terms in Equation B.1.

The robustness of Equation B.1 stems from a general property of the convolution operator:[1]

When impulse responses convolve, their variances add.

We may relate the convolution property to Equation B.1 by using this chain of logic:

(1) Variance is the square of standard deviation.
(2) The standard deviation of a pulse is proportional to its width.
(3) The width of an impulse response is proportional to the rise time of its corresponding step response.
(4) Therefore, variance is proportional to the square of rise time.

Substituting the words *squared rise time* for variances in the convolution property leads us to Equation B.1.

We can now discuss why it hardly matters what rise-time definition you use in Equation B.1. First, recognize that any measure of rise time that is proportional to the standard deviation of impulse response will exactly follow Equation B.1.

Rise-time measures based on parameters other than the standard deviation of impulse response will approximately follow Equation B.1, to the degree that they track the standard deviation of impulse response. Fortunately, all popular rise-time measures happen to closely mimic the standard deviation. This is a fortunate coincidence indeed because, as discussed in Section B.2.1 below, the standard deviation is very difficult to measure. Table B.1 shows that the correspondence between easy, practical rise-time measures and the standard deviation measure is so close that, unless you are planning a thesis in this area, it hardly matters which rise-time definition you use.

To illustrate this fact, in Table B.1 we present rise-time measurements for three different waveforms: single pole, double pole critically damped, and gaussian. All three wave shapes are normalized so that their standard deviation rise times are equal. Detailed definitions of the pulse shapes appear following the table.

Table B.1 lists five different rise-time measurements. As you can see, most of the measurements agree pretty well. A discussion of how the measurements are defined, and the strengths and weakness of each definition, follows.

[1]When cascading systems together, their impulse responses convolve. The properties of the convolution operator therefore determine how the complete system behaves.

TABLE B.1 RISE-TIME MEASUREMENTS FOR THREE EXAMPLE WAVEFORMS

Impulse type	T_σ	T_{10-90}	T_{20-80}	$T_{\text{center slope}}$	$T_{\text{max slope}}$	$F_{3\text{dB}}$	F_{RMS}
One-pole, $RC = 0.399$	1.00	0.877	0.553	0.798	0.399	0.399	0.626
Two-pole critically damped, $(LC)^{1/2} = 0.282$	1.00	0.947	0.612	0.900	0.767	0.363	0.443
Gaussian, $t_3 = 0.281$	1.00	1.02	0.672	1.00	1.00	0.332	0.354

Table B.1 also lists two different measures of system bandwidth. Because oscilloscope manufacturers insist on quoting performance in terms of system bandwidth, instead of rise time, we must occasionally convert a system bandwidth specification into a corresponding rise time. If you have some knowledge of the waveform type (oscilloscope step responses tend to look gaussian), Table B.1 can help you make the conversion.

In addition to the data in Table B.1, I have seen in the fiber optic industry a tendency to define bandwidth according to the 6-dB points on a gaussian frequency response. The 6-dB bandwidth for line 3 in Table B.1 is 0.47 Hz.

When rationalizing a set of rise-time measurements made using different techniques, it is convenient to convert everything to one standard rise-time definition. Using Table B.1 and a knowledge of the pulse type, you can convert from one type of measurement to another.

If you plan to apply Equation B.1, it makes the most sense to convert everything to the T_σ format. For example, let's say that you know your circuit responds with a simple *RC* time constant τ. As explained in Section B.2.5, the time constant τ equals the maximum slope rise time. Our handy chart shows the ratio of T_σ to $T_{\text{max slope}}$ for a single-pole circuit to be

$$\frac{T_\sigma}{T_{\text{max slope}}} = \frac{1.00}{0.399} = 2.506 \approx 2.5 \qquad \text{[B.2]}$$

Multiplying this ratio by τ yields the desired value of T_σ.

Had we wanted the 10–90% rise time instead of T_σ, we could have used the ratio

$$\frac{T_{10-90}}{T_{\text{max slope}}} = \frac{0.877}{0.399} = 2.197 \approx 2.2 \qquad \text{[B.3]}$$

The difference between the 10–90% rise time and T_σ is about 12% (1–2.2/2.5). For the greatest accuracy in applying Equation B.1, use T_σ. For casual measurements or for ease in interpreting your lab reports later, just use T_{10-90}.

Conversions between bandwidth and rise time may be made according to the principle that, for each signal type, the product of bandwidth and rise time is constant. For example, from the table we see that the product of 3-dB bandwidth and 10–90% rise time for a gaussian impulse response is

$$F_{3\text{dB}} T_{10-90} = (0.332)(1.02) = 0.339 \approx \tfrac{1}{3} \qquad \text{[B.4]}$$

Dividing the actual 3-dB frequency into this product yields the 10–90% rise time. Notice that the bandwidth–rise time products for other wave shapes are not exactly the same but are pretty close. This illustrates the point that the bandwidth–rise time product is fairly independent of wave shape.

From Table B.1 we may deduce that when you cannot identify your wave shape, the least satisfactory measurements are the maximum slope method and the RMS bandwidth method. Consistently better results, independent of wave shape, result from using the 10–90% slope-at-center, and 3-dB measurements because these measurements more accurately track the standard deviation method.

B.1 THREE PULSE SHAPES USED IN TABLE B.1

Relations between the various rise-time measurements vary according to the waveform shape. To illustrate the differences among these various measures we will be using three example wave shapes: single pole, double pole, and gaussian.

B.1.1 Single-Pole Pulse

The single-pole pulse shape is the exponential relaxation which results from passing a perfectly square-edged step through a single-pole RC low-pass filter. The equations below list the impulse response of this filter, the step response resulting from its excitation with a perfectly square-edged input pulse, and the Fourier transform of its impulse response.

$$h_1(t) = U(t)\frac{1}{t_1}e^{-(t/t_1)} \qquad \text{[B.5]}$$

$$\text{h step}_1(t) = U(t)\left(1 - e^{-(t/t_1)}\right) \qquad \text{[B.6]}$$

$$H_1(w) = \frac{1}{1 + jwt_1} \qquad \text{[B.7]}$$

where $h_1(t)$ = single-pole impulse response
$\text{h step}_1(t)$ = single-pole step response
$H_1(w)$ = Fourier transform of $h_1(t)$
$U(t)$ = unit step which is 0 for $t < 0$ and 1 otherwise
t_1 = exponential decay constant which equals RC for R and C low-pass filters and equals L/R for L and R low-pass filters; RC in Table B.1, = 0.399

Of the three waveforms, the single-pole wave shape has the fastest initial slope. The inverse of its initial slope equals the RC time constant t_1. A key characteristic of this wave shape is the abrupt corner at its rising edge, where the output suddenly changes from a rest condition to a quickly rising slope. You may recall from Fourier analysis that this abrupt corner causes the spectrum at high frequencies to fall off only as fast as $1/f$.

B.1.2 Two-Pole Critically Damped Pulse

The second example pulse shape is a two-pole decay which results from passing a perfectly square-edged step through a two-pole *critically damped RLC* low-pass filter.[2] The characteristics of this pulse shape lie in between the single-pole and the gaussian characteristics.

The equations below list the impulse response of this filter, the step response resulting from its excitation with a perfectly square-edged input pulse, and the Fourier transform of its impulse response:

$$h_2(t) = U(t)\frac{1}{(t_2)^2} t e^{-(t/t_2)} \quad \text{[B.8]}$$

$$\text{h step}_2(t) = U(t)\left[1 - \left(1 + \frac{t}{t_2}\right)e^{-(t/t_2)}\right] \quad \text{[B.9]}$$

$$H_2(w) = \left(\frac{1}{1 + jwt_2}\right)^2 \quad \text{[B.10]}$$

where $h_2(t)$ = two-pole critically damped impulse response
h $\text{step}_2(t)$ = two-pole critically damped step response
$H_2(w)$ = Fourier transform of $h_2(t)$
$U(t)$ = unit step which is 0 for $t < 0$ and 1 otherwise
t_2 = exponential decay constant which equals $(LC)^{1/2}$;
in Table B.1, $(LC)^{1/2} = 0.282$

A key characteristic of the two-pole critically damped wave shape is the softened corner at its rising edge. The output gracefully curves up from a rest condition to a rising slope. The maximum slope is less than for the single-pole wave shape, but more than for the gaussian. The trailing portion of the step extends somewhat longer than its rising portion.

You may recall from Fourier analysis that a two-pole low-pass filter causes the spectrum at high frequencies to fall off as $1/f^2$.

B.1.3 Gaussian Pulse

The final example pulse shape is a gaussian pulse. This pulse shape is the natural result of most complex systems. The central limit theorem tells us that whenever we convolve together a lot of similar impulse responses, the result looks gaussian. That's exactly what happens in the design of an oscilloscope amplifier chain. To avoid spending too much money on any single part, each section of the amplifier chain is designed for a bandwidth just high enough to pass the signals of interest. The resulting system is a concatenation of several filter stages having similar bandwidths, whose overall impulse response tends toward gaussian.

[2] Severely underdamped filters exhibit ringing, for which it becomes very difficult to measure rise time.

The equations below list the impulse response of the gaussian filter, the step response resulting from its excitation with a perfectly square-edged input pulse, and the Fourier transform of its impulse response:

$$g(t) = \frac{1}{2\pi^{1/2} t_3} e^{-(t/2t_3)^2} \tag{B.11}$$

$$\text{g step }(t) = \frac{1}{2}\left[1 + \operatorname{erf}\left(\frac{t}{2t_3}\right)\right] \tag{B.12}$$

$$G(w) = e^{-(t_3 w)^2} \tag{B.13}$$

where $g(t)$ = gaussian impulse response
g step (t) = gaussian step response
$G(w)$ = Fourier transform of $g(t)$
$U(t)$ = unit step which is 0 for $t < 0$ and 1 otherwise
t_3 = gaussian time decay constant; in Table B.1, $t_3 = 0.281$
erf( ) = error function; integral of gaussian function

A key characteristic of the gaussian step response is its symmetry. Both rising and trailing edges have graceful, sweeping curves. The maximum slope occurs right in the center. In some sense, this curve seems to achieve a good slope in the center while minimizing all its higher derivatives. The resulting spectrum falls off extremely rapidly with increasing frequency.

B.2 FIVE PULSE RISE-TIME MEASUREMENTS USED IN TABLE B.1

B.2.1 Standard Deviation Measure of Step Rise Time T_σ

Here we are measuring the rise time of a *step response*, and we do it by looking at the *impulse response* (derivative of the step response).

T_σ equals the standard deviation of an impulse response times the scaling factor $(2\pi)^{1/2}$. The standard deviation is a measure of the impulse response width, which in turn determines the rise time of the step response.

$$\sigma^2 = \int_{-\infty}^{+\infty} t^2 \frac{h(t)}{H(0)} dt - \left[\int_{-\infty}^{+\infty} t \frac{h(t)}{H(0)} dt\right]^2 \tag{B.14}$$

$$t_\sigma = \left(2\pi\sigma^2\right)^{1/2} \tag{B.15}$$

where $h(t)$ = impulse response; this is the system output resulting from injecting into the input a tall, skinny spike; the step response (integral of impulse response) is the system output resulting from injecting into the input a perfectly square-edged step.

σ^2 = variance of $h(t)$
σ = standard deviation of $h(t)$ (square root of variance)
t_σ = scaled version of standard deviation
$H(0)$ = value of Fourier transform of $h(t)$ at frequency zero (DC)

The scaling factor $(2\pi)^{1/2}$ ensures that, for gaussian pulses, the four rise-time measures T_σ, 10–90%, center slope, and max slope all yield the same result.

Let's try a quick sanity check. First note that $(2\pi)^{1/2}$ equals about 2.5. If you mark the points 1.25 standard deviations either way from the origin of a gaussian impulse, you have covered 79% of the area. That leaves 21% of the area outside the marks (10.5% each side). When we integrate the gaussian impulse response to find the step response, we will find that between the marks the step response traverses from 10.5 to 84.5% of its full value. Rounding off to 10 and 90%, we conclude that for gaussian wave shapes, T_σ and the 10–90% rise time are the same.

The primary advantage of using T_σ as a measure of rise time is that Equation B.1 always works exactly. The primary disadvantage of T_σ is that its definition involves an impulse response. Given a step response, we must first differentiate it into an impulse response before applying the definitions in Equations B.14 and B.15. Practically speaking, this differentiation, as well as the calculations of Equations B.14 and B.15, are quite difficult to compute.

Fortunately, we can substitute any of the other rise-time measurements for T_σ in Equation B.1 and get results accurate enough for practical lab use.

B.2.2 The 10–90% Rise Time

This rise time is defined as the difference in time between when a step response crosses the 10% threshold and when it reaches 90% of its final value. Many digital oscilloscopes can automatically make this measurement.

$$T_{10-90} = T(\text{step at } 90\%) - T(\text{step at } 10\%) \qquad [B.16]$$

The 10–90% rise time hinges on measurements taken at only two points on a step response. Its principal strength is its ease of application. Its principal weakness is its sensitivity to noise or ringing present at the sample points.

B.2.3 The 20–80% Rise Time

This rise time is defined as the difference in time between when a step response crosses the 20% threshold and when it reaches 80% of its final value. Manufacturers wishing to improve the appearance of their specification sheets sometimes quote a 20–80% rise time, which makes the rise-time figures look faster.

$$T_{20-80} = T(\text{step at } 80\%) - T(\text{step at } 20\%) \qquad [B.17]$$

The 20–80% rise time hinges on measurements taken at only two points on a step response. Its principal strength is its ease of application. Its principal weakness is its sensitivity to noise or ringing present at the sample points.

B.2.4 Center Slope Rise Time

Determined only by the slope of the step response as it crosses the *middle level*, this measurement is easy to make from photographs or oscilloscope plots. Just line up a ruler tangent to the pulse at its middle level and measure the difference in time between where the ruler intersects the initial and 100% final values.

$$T_{\text{center slope}} = \frac{\Delta v}{dV/dt\ (50\%)} \qquad [\text{B.18}]$$

where Δv = difference between 0 and 100% voltage levels
dV/dt (50%) = derivative of voltage, measured at 50% point (middle level)

This measure involves the region near the center of a step response. Its principal strength is that it removes the measurement from either asymptotic end point, rendering it less sensitive to ringing than the 10–90% measurement. Also, a degree of visual integration can compensate for funny bends or kinks in a pulse shape.

Its principal weakness is that few oscilloscopes support this measurement directly. Unlike the 10–90% or 20–80% rise-time measures, which can be directly read from a differential cursor spacing on a digital scope, the slope method requires some hand calculations or a paper oscilloscope plot.

B.2.5 Maximum Slope Rise Time

Determined only by the *maximum slope* of the step response, this measurement is easy to make from photographs or oscilloscope plots. Just line up a ruler tangent to the pulse at its *fastest moving point* (usually near the start) and measure the difference in time between where the ruler intersects the initial and 100% final values.

$$T_{\text{max slope}} = \frac{\Delta v}{dV/dt\ (\text{max})} \qquad [\text{B.19}]$$

where Δv = difference between 0 and 100% voltage levels
dV/dt (max) = derivative of voltage, measured at the fastest moving point

For simple *RC* or *LR* filters, the maximum slope rise time equals the decay constant τ. The maximum slope method excels as a method of determining an *RC* or *LR* decay constant.

Its principal weakness is that few oscilloscopes support this measurement directly. Unlike the 10–90% or 20–80% rise-time measures, which can be directly read from a differential cursor spacing on a digital scope, the slope method requires some hand calculations or a paper oscilloscope plot. Even worse, this measurement requires that you first determine the point of maximum slope.

B.3 TWO BANDWIDTH MEASUREMENTS USED IN TABLE B.1

B.3.1 Three-Decibel Bandwidth

This measurement is just what is sounds like. Look at the Fourier transform of the impulse response of a system and find that frequency at which the amplitude falls 3 dB below its DC value.

For oscilloscopes, a sine wave fed into the vertical amplifier at the 3-dB frequency will appear on the screen at only 70.7% of its actual amplitude.

B.3.2 RMS or Noise Bandwidth

Often used in amplifier noise analysis, this measurement takes into account the overall spectral shape. The RMS bandwidth of a low-pass filter is

$$F_{\text{RMS}} = \frac{1}{2\pi} \int_0^{+\infty} \frac{|H(w)|^2}{H(0)^2}\, dw \qquad \text{[B.20]}$$

where F_{RMS} = RMS bandwidth, Hz
$H(w)$ = frequency response of system component; w is in units of radians/s
$H(0)$ = frequency response at DC; if H is a bandpass filter, use the frequency response in the center of the band instead

The noise bandwidth of a frequency response *H(w)*, or RMS bandwidth, is the cut-off frequency at which a box-shaped frequency response would pass the same amount of white noise energy as *H(w)*. RMS bandwidth is not the same as the measure of "RMS duration" used in the proof that gaussian filters offer the best tradeoff of compact time duration versus compact frequency response.

This measure appears here because oscilloscope manufacturers often use it when specifying equipment performance.

C

MathCad Formulas

The following pages list standard formulas for computing the resistance, capacitance, and inductance of physical structures.

To ensure complete accuracy, the formulas have been implemented as MathCad spreadsheets. For those of you not familiar with mathematical spreadsheets, this electronic format allows you to enter equation definitions, evaluate the equations, and plot the results.

To ensure the greatest possible accuracy, the authors have used MathCad to evaluate each of the following formulas in well-known test cases and then checked the results manually. Once each equation was verified as correct, its final form was printed directly to hard copy without the benefit of manual typesetting. The hard copy pages were then reproduced lithographically in this appendix. This procedure doesn't guarantee perfection, but it's the best way we could think of to ensure accuracy.

The formulas may be used manually, or, if you have access to a computer, entered into a math spreadsheet program. We recommend the spreadsheet approach because it leaves a written record of your work that can be printed out and easily bound into a lab notebook.

For your convenience, these formulas are available from the authors in magnetic form. See the order form in the back of this book.

PHYSICAL CONSTANTS USED IN TRANSMISSION LINE WORK file: constants.mcd

Electric permittivity of free space (metric)	$E0\_meters := 8.854 \cdot 10^{-12}$ $C^2/N\text{-}m^2$
Recalculate in in.	$E0\_inches := E0\_meters \cdot .0254$
Display calculated value	$E0\_inches = 2.249 \cdot 10^{-13}$
Magnetic permeability of free space (metric)	$U0\_meters := 4 \cdot \pi \cdot 10^{-7}$ Wb/A-m
Recalculate in in.	$U0\_inches := U0\_meters \cdot .0254$
Display calculated value	$U0\_inches = 3.192 \cdot 10^{-8}$
We often need this number	$\frac{U0\_inches}{2 \cdot \pi} = 5.08 \cdot 10^{-9}$
Speed of light (metric)	$C\_meters := 2.998 \cdot 10^{8}$ m/s
Recalculated in in.	$C\_inches := \frac{C\_meters}{.0254}$
Display calculated value	$C\_inches = 1.18 \cdot 10^{10}$
Propagation delay at light speed (ps/in.)	$\frac{10^{12}}{C\_inches} = 84.723$

```
DC RESISTANCE OF COPPER WIRES AND TRACES          file: resist.mcd


Conversion formulas included in this spreadsheet:

  Diameter to AWG                           AWG()
  AWG to diameter                           DIAMETER()
  Thickness to copper plating weight        CPW()
  Copper plating weight to thickness        THICKNESS()


Resistance formulas included in this spreadsheet:

  DC resistance of round wires
    From diameter                        RROUND()
    From AWG wire size                   RROUND_AWG()
    At room temperature only             RROUND_RT()

  DC resistance of printed circuit board traces
    From trace thickness and width       RTRACE()
    Using copper plating weight          RTRACE_CPW()
    At room temperature only             RTRACE_RT()

  DC resistance of power or ground planes
    Using thickness and via diameter RPLANE()
    Using copper plating weight          RPLANE_CPW()


Variables used:
                                                            -7
  ρ  Bulk resistivity of copper                ρ := 6.787·10

     This coefficient is slightly different from the bulk
     resistivity of pure copper (6.58E+07) owing to the
     annealing process used in making wire, and chemical
     imperfections in the  copper used for making practical wires.

     In practice, the resistance of two wires making up a twisted
     pair may often be matched as well as 10%, but almost
     never as well as 1%.

  δρ Thermal coefficient of resistance        δρ := .0039

     If the resistance of a copper wire is R at room
     temperature, then at a temperature 1°C higher it will
     be R(1 + δρ).  This coefficient applies to standard annealed
     copper wires.  The coefficient for pure copper in its bulk state
     varies slightly.

     Over a temperature range 0-70°C the resistance of copper
     wires varies 28%.
```

x Length of wire (in.)
(or separation between contact points on ground plane)

d Diameter of wire (in.)
(or diameter of contact point on ground plane)

AWG American wire gauge (English units)

temp Temperature (°C)

w Width of printed circuit board trace (in.)

t Thickness of printed circuit board trace (in.)

cpw Thickness of printed circuit board traces, in units of copper plating weight (oz/ft^2)

Conversions between American
Wire Gauge (AWG) and diameter (in.):

$$AWG(d) := -10 - 20 \cdot \log(d) \qquad DIAMETER(awg) := 10^{-\left[\frac{awg+10}{20}\right]}$$

General formula for
resistance of a round wire (Ω):

$$RROUND(d,x,temp) := \frac{4 \cdot \rho \cdot x}{\pi \cdot d^2} \cdot (1 + (temp - 20) \cdot \delta\rho)$$

Resistance of a round wire specified
by AWG size instead of diameter (Ω):

RROUND_AWG(awg,x,temp) := RROUND(DIAMETER(awg),x,temp)

Resistance of a round wire
at room temperature (Ω):

RROUND_RT(d,x) := RROUND(d,x,20)

Conversion between thickness,
t (in.) and copper plating
weight, cpw (oz):

$$\mathrm{THICKNESS(cpw)} := .00137 \cdot \mathrm{cpw} \qquad \mathrm{CPW(t)} := \frac{t}{.00137}$$

Resistance of a
circuit trace (Ω):

$$\mathrm{RTRACE(w,t,x,temp)} := \frac{x \cdot \rho}{w \cdot t} \cdot (1 + (\mathrm{temp} - 20) \cdot \delta\rho)$$

Resistance of a trace specified
by plating weight instead of thickness (Ω):

$$\mathrm{RTRACE\_CPW(w,cpw,x,temp)} := \mathrm{RTRACE(w,THICKNESS(cpw),x,temp)}$$

Resistance of a circuit trace
at room temperature (Ω):

$$\mathrm{RTRACE\_RT(w,t,x)} := \mathrm{RTRACE(w,t,x,20)}$$

Resistance of a power or ground plane (Ω):

When using long, skinny traces or wires, the approximations above work extremely well. Each formula assumes a uniform distribution of current throughout the conducting body, for which resistance is directly proportional to length.

Currents circulating in a large ground or power plane are not uniform. Consequently, the resistance measured between two points on a ground or power plane is not directly proportional to the separation between measurement points.

The following equation models the resistance between two contact points on a ground plane. This model assumes each contact point touches the ground plane over some finite area. The approximate diameter of the contact point determines the overall resistance.

If the contact points lie near any edge of the plane, the resistance between them may go up by a factor of 2. The resistance near corners may rise even higher.

d1	Diameter of 1st contact point (in.)
d2	Diameter of 2nd contact point (in.)
t	Thickness of plane (in.)
cpw	Thickness of plane, copper plating weight (oz)
x	Separation between contact points (in.)
temp	Temperature (°C)

Resistance of a power or ground plane (Ω):

$$\mathrm{RPLANE}(d1,d2,t,x,temp) := \frac{\rho}{2\cdot\pi\cdot t}\cdot\left[\ln\left[\frac{2\cdot x}{d1}\right]+\ln\left[\frac{2\cdot x}{d2}\right]\right]\cdot(1+(temp-20)\cdot\delta\rho)$$

Resistance of a power or ground plane specified by plating weight instead of thickness (Ω):

RPLANE_CPW(d1,d2,cpw,x,temp) := RPLANE(d1,d2,THICKNESS(cpw),x,temp)

CAPACITANCE OF TWO PARALLEL PLATES file: capac.mcd

Formulas included in this spreadsheet:

Capacitance of two plates	CPLATE()
Impedance magnitude of capacitor at one frequency	XCF()
Impedance magnitude of capacitor as seen by rising edge	XCR()

Variables used:

w	Width of plate overlap (in.)
x	Length of plate overlap (in.)
h	Height of one plate above the other (in.)
er	Relative dielectric constant of material between plates

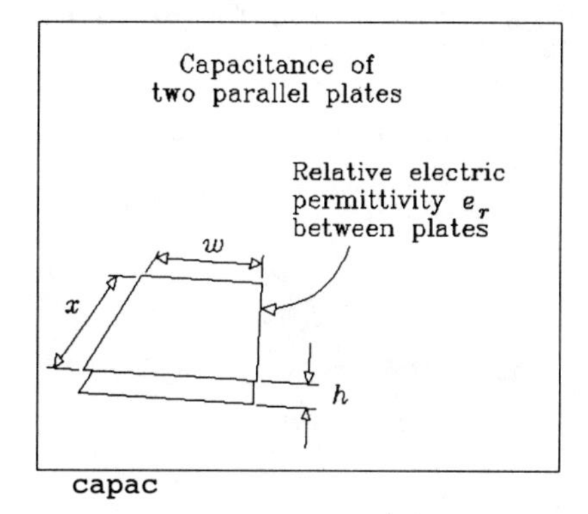

capac

Capacitance of two plates (F):

$$CPLATE(w,x,h,er) := 2.249 \cdot 10^{-13} \cdot \frac{er \cdot x \cdot w}{h}$$

A power and ground plane separated by 0.010 in. of FR-4 dielectric (er = 4.5) share a capacitance of 100 pF/in.^{2}

Halving the separation doubles the capacitance.

Impedance magnitude of capacitor at frequency f (Ω):

c	Capacitance (F)
f	Frequency (Hz)

$$XC(c,f) := \frac{1}{2 \cdot \pi \cdot f \cdot c}$$

The impedance, at 100 MHz, of a 100-pF capacitor is 16 Ω.

Impedance magnitude of capacitor as seen by rising edge (Ω):

c Capacitance (F)

tr 10-90% rise time (s)

$$XC(c,tr) := \frac{tr}{\pi \cdot c}$$

The impedance, as seen by a 5-ns rising edge of a 100-pF capacitor is 16 Ω.

INDUCTANCE OF CIRCULAR LOOPS file: circular.mcd

Formulas included in this spreadsheet:

Inductance of circular wire loop	LCIRC()
Impedance magnitude of inductor at one frequency	XLF()
Impedance magnitude of inductor as seen by rising edge	XLR()

Variables used:

d Diameter of wire (in.)

x Diameter of wire loop (in.)

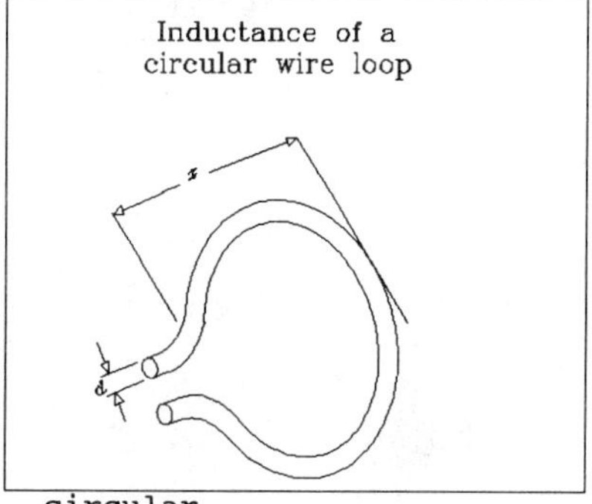

circular

Inductance of wire loop (H):

$$LCIRC(d,x) := 1.56\cdot 10^{-8}\cdot x\cdot\left[\ln\left[\frac{8\cdot x}{d}\right] - 2\right]$$

A loop of 24-gauge wire the size of the loop between your thumb and forefinger has about 100 nH of inductance.

Changing the wire diameter from AWG 30 to AWG 10 makes little difference. The log function is very insensitive to wire size.

Impedance magnitude of inductor at frequency f (Ω):

l Inductance (H)

f Frequency (Hz)

$$XLF(l,f) := 2\cdot\pi\cdot f\cdot l$$

The impedance, at 100 MHz, of a 100-nH inductor is 62 Ω.

Impedance magnitude of inductor as seen by rising edge (Ω):

l	Inductance (H)
tr	10-90% rise time (s)

$$XLT(l,tr) := \frac{\pi \cdot l}{tr}$$

The impedance, as seen by a 5-ns rising edge, of a 100-nH inductor is 62 Ω.

INDUCTANCE OF RECTANGULAR LOOPS file: rectangl.mcd

Formulas included in this spreadsheet:

Inductance of rectangular wire loop	LRECT()
Impedance magnitude of inductor at one frequency	XLF()
Impedance magnitude of inductor to rising edge	XLR()

Variables used:

d Diameter of wire (in.)

x Length of wire loop (in.)

y Breadth of wire loop (in.)

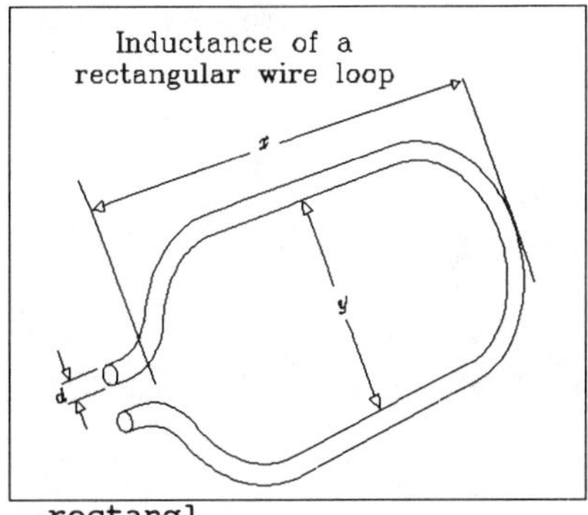

rectangl

Inductance of wire loop (H):

$$LRECT(d,x,y) := 10.16 \cdot 10^{-9} \cdot \left[x \cdot \ln\left[\frac{2 \cdot y}{d}\right] + y \cdot \ln\left[\frac{2 \cdot x}{d}\right]\right]$$

A loop of 24-gauge wire 1 in.² has about 100 nH of inductance.

Changing the wire diameter from AWG 30 to AWG 10 makes little difference. The log function is very insensitive to wire size.

If your loop consists of different-sized conductors, use the diameter of the smallest one.

Impedance magnitude of inductor at frequency f (Ω):

l Inductance (H)

f Frequency (Hz)

$$XLF(l,f) := 2 \cdot \pi \cdot f \cdot l$$

The impedance, at 100 MHz, of a 100-nH inductor is 62 Ω.

Impedance magnitude of inductor as seen by rising edge (Ω):

l Inductance (H)

tr 10-90% rise time (s)

$$XLT(l,tr) := \frac{\pi \cdot l}{tr}$$

The impedance, as seen by a 5-ns rising edge, of a 100-nH inductor is 62 Ω.

MUTUAL INDUCTANCE OF TWO LOOPS file: mloop.mcd

Formulas included in this spreadsheet:

Mutual inductance of two loops MLOOP()

Variables used:

r Separation between loop centers (in.)

A1 Surface area of loop 1 (in.²)

A2 Surface area of loop 2 (in.²)

(We assume the loops are flat, and that their faces are oriented parallel to each other for maximum coupling)

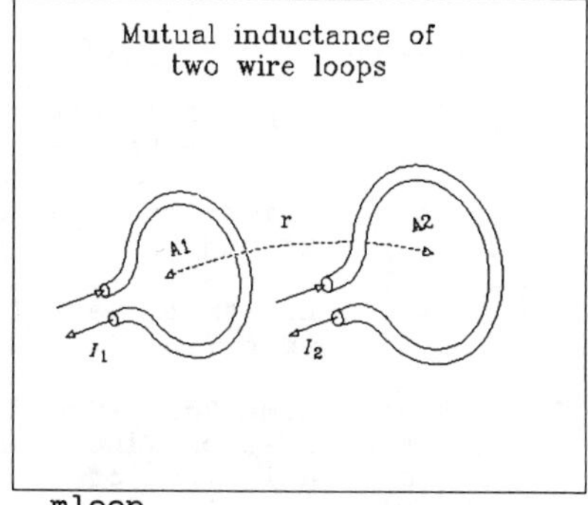

mloop

The loops must be well separated for the MLOOP() approximation to work:

$$r > \sqrt{A1} \qquad \text{and} \qquad r > \sqrt{A2}$$

Mutual inductance of two well-separated loops (nH):

$$MLOOP(r, A1, A2) := 5.08 \cdot \frac{A1 \cdot A2}{r^3}$$

MUTUAL INDUCTANCE OF PARALLEL TRANSMISSION LINES file: mline.mcd

Formulas included in this spreadsheet:

Mutual inductance of two lines MLINE()

Variables used:

s Separation between wire centers (in.)

h Height of wires above ground (in.)

x Length of parallel span (in.)

(We assume that two identical transmission lines share a parallel run of length x, with a horizontal separation s.)

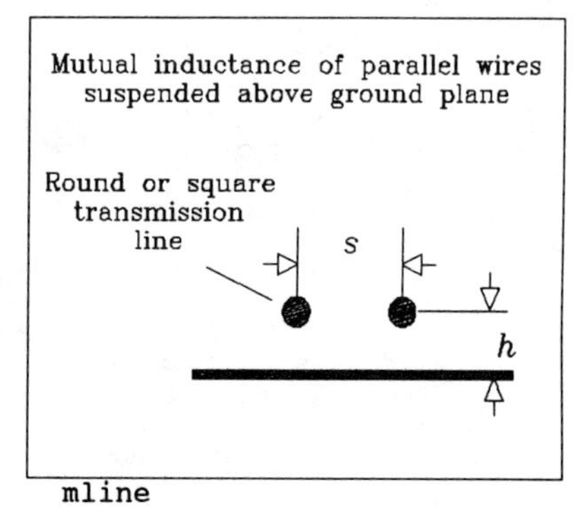

mline

Let L equal the inductance (H) of the first transmission line of length x (use formula for round, microstrip, or stripline geometry as appropriate):

$$MLINE(L,s,h) := L\cdot\left[\frac{1}{1+\left[\frac{s}{h}\right]^{2}}\right]$$

GENERAL RELATIONS AMONG TRANSMISSION LINE PARAMETERS file: general.mcd

Conversion formulas included in this spreadsheet:

Inductance and capacitance to impedance	Z0()
Inductance and capacitance to propagation delay	PDLY1()
Effective permittivity to propagation delay	PDLY2()
Impedance and propagation delay to capacitance	CPI()
Impedance and propagation delay to inductance	LPI()

Variables used:

lpi	Inductance per inch (H)
cpi	Capacitance per inch (F)
pdly	Propagation delay (s/in.)
z0	Line impedance (Ω)
eeff	Effective relative permittivity

Given inductance per inch and capacitance per inch, find the characteristic impedance in ohms:

$$Z0(lpi,cpi) := \sqrt{\frac{lpi}{cpi}}$$

Given inductance per inch and capacitance per inch, find the propagation delay per inch:

$$PDLY1(lpi,cpi) := \sqrt{lpi \cdot cpi}$$

Given the effective electric permittivity of the surrounding medium, find the propagation delay per inch:

$$PDLY2(eeff) := 84.72 \cdot 10^{-12} \cdot \sqrt{eeff}$$

Given impedance and propagation delay, find the capacitance per inch:

$$CPI(zo,pdly) := \frac{pdly}{zo}$$

Given impedance and propagation delay, find the inductance per inch:

$$LPI(zo,pdly) := zo \cdot pdly$$

COAXIAL TRANSMISSION LINE file: coax.mcd

Formulas included in this spreadsheet:

Coaxial cable characteristic impedance	ZCOAX()
Coaxial cable propagation delay	PCOAX()
Coaxial cable inductance	LCOAX()
Coaxial cable capacitance	CCOAX()

Variables used:

- d1 Diameter of inner wire (in.)
- d2 Diameter of outer shield (in.)
- x Length of cable (in.)
- er Relative dielectric constant of material surrounding the inner wire

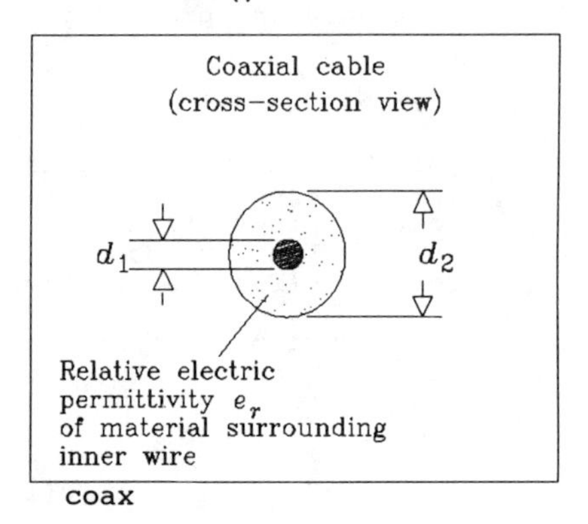

coax

Characteristic impedance of coaxial cable (Ω):

$$ZCOAX(d1,d2,er) := \frac{60}{\sqrt{er}} \cdot \ln\left[\frac{d2}{d1}\right]$$

Propagation delay per in. for coaxial cable (s/in.):

$$PCOAX(er) := 84.72 \cdot 10^{-12} \cdot \sqrt{er}$$

Inductance of coaxial cable (H):

$$LCOAX(d1,d2,x) := x \cdot 5.08 \cdot 10^{-9} \cdot \ln\left[\frac{d2}{d1}\right]$$

Capacitance of coaxial cable (F):

$$CCOAX(d1,d2,er,x) := \left[\frac{x \cdot 1.41 \cdot 10^{-12}}{\ln\left[\frac{d2}{d1}\right]}\right] \cdot er$$

Example coaxial cable calculations

Diameter of AWG 30 inner wire (in.)	D1 := .01
Inside diameter of shield (in.)	D2 := .1
Length of cable (in.)	X := 20.000
Relative dielectric constant	er := 2.2

Characteristic impedance (Ω):

$$ZCOAX(D1,D2,er) = 93.144$$

Total inductance (H):

$$LCOAX(D1,D2,X) = 2.339 \cdot 10^{-7}$$

Same result in nH:

$$LCOAX(D1,D2,X) \cdot 10^{9} = 233.943$$

Inductance per in. (H):

$$LCOAX(D1,D2,1) = 1.17 \cdot 10^{-8}$$

Total capacitance (F):

$$CCOAX(D1,D2,er,X) = 2.694 \cdot 10^{-11}$$

Same result in pF:

$$CCOAX(D1,D2,er,X) \cdot 10^{12} = 26.944$$

Capacitance per in. (F):

$$CCOAX(D1,D2,er,1) = 1.347 \cdot 10^{-12}$$

TRANSMISSION LINE MADE FROM ROUND WIRE (WIRE-WRAP) file: round.mcd

Formulas included in this spreadsheet:

Round wire characteristic impedance	ZROUND()
Round wire propagation delay	PROUND()
Round wire inductance	LROUND()
Round wire capacitance	CROUND()

Variables used:

d Diameter of wire (in.)

h Height of wire above ground (in.)

x Length of wire (in.)

(We assume the wire is suspended in air, for which the relative dielectric constant is 1.00.)

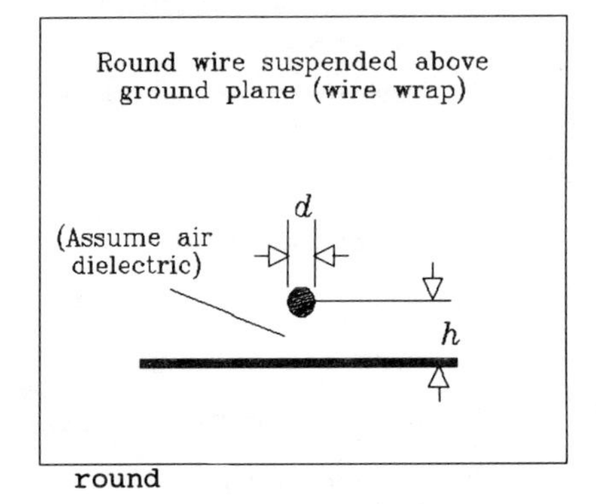

round

Characteristic impedance of round wire above ground plane (Ω):

$$ZROUND(d,h) := 60 \cdot \ln\left[\frac{4 \cdot h}{d}\right]$$

Propagation delay per in. of round wire above ground plane (s/in):

$$PROUND(d,h) := 84.72 \cdot 10^{-12}$$ (assume air dielectric)

Inductance of round wire above ground plane (H):

$$LROUND(d,h,x) := x \cdot 5.08 \cdot 10^{-9} \cdot \ln\left[\frac{4 \cdot h}{d}\right]$$

Capacitance of round wire above ground plane (F):

$$CROUND(d,h,x) := \left[\frac{x \cdot 1.413 \cdot 10^{-12}}{\ln\left[\frac{4 \cdot h}{d}\right]}\right]$$

Example round wire calculations

Diameter of AWG 30 wire (in.) D := .01

Length of wire (in.) X := 2.000

Height above ground (in.) H := .100

Characteristic impedance (Ω):

$ZROUND(D,H) = 221.333$

Total inductance (H):

$LROUND(D,H,X) = 3.748 \cdot 10^{-8}$

Same result in nH:

$LROUND(D,H,X) \cdot 10^{9} = 37.479$

Inductance per in. (H):

$LROUND(D,H,1) = 1.874 \cdot 10^{-8}$

Total capacitance (F):

$CROUND(D,H,X) = 7.661 \cdot 10^{-13}$

Same result in units pF:

$CROUND(D,H,X) \cdot 10^{12} = 0.766$

Capacitance per in. (F):

$CROUND(D,H,1) = 3.83 \cdot 10^{-13}$

TRANSMISSION LINE MADE FROM TWISTED PAIR WIRE file: twist.mcd

Formulas included in this spreadsheet:

Twisted-pair characteristic impedance	ZTWIST()
Twisted-pair propagation delay	PTWIST()
Twisted-pair inductance	LTWIST()
Twisted-pair capacitance	CTWIST()

Variables used:

- d Diameter of wire (in.)
- s Separation between wires (in.)
- x Length of wire (in.)
- er Effective relative dielectric constant of medium between wires

Twisted-pair transmission line

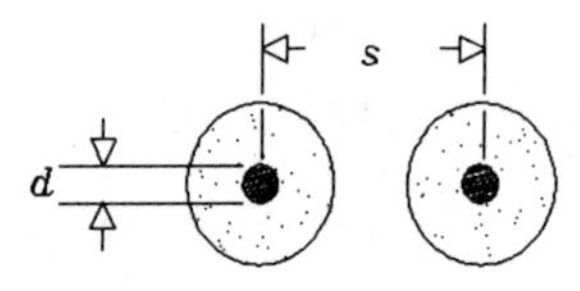

Effective relative electric permittivity e_r lies between the permittivity of the wire's insulator and the permittivity of air (1.00).

twist

Characteristic impedance of twisted pair (Ω):

$$\text{ZTWIST}(d,s,er) := \frac{120}{\sqrt{er}} \cdot \ln\left[\frac{2 \cdot s}{d}\right]$$

Propagation delay per in. twisted pair (s/in.):

$$\text{PTWIST}(er) := 84.72 \cdot 10^{-12} \cdot \sqrt{er}$$

Inductance of twisted pair (H):

$$\text{LTWIST}(d,s,x) := x \cdot 10.16 \cdot 10^{-9} \cdot \ln\left[\frac{2 \cdot s}{d}\right]$$

Capacitance of twisted pair (F):

$$\text{CTWIST}(d,s,er,x) := \left[\frac{x \cdot .7065 \cdot 10^{-12}}{\ln\left[\frac{2 \cdot s}{d}\right]}\right] \cdot er$$

Example twisted-pair calculations

Diameter of AWG 24 wire (in.)	D := .02
Length of wire (in.)	X := 2.000
Separation between wire centers (in.)	S := .038
Relative dielectric constant	er := 2.5

Characteristic impedance (Ω):

ZTWIST(D,S,er) = 101.319

Total inductance (H):

LTWIST(D,S,X) = $2.713 \cdot 10^{-8}$

Same result in nH:

LTWIST(D,S,X)$\cdot 10^{9}$ = 27.127

Inductance per in. (H):

LTWIST(D,S,1) = $1.356 \cdot 10^{-8}$

Total capacitance (F):

CTWIST(D,S,er,X) = $2.646 \cdot 10^{-12}$

Same result in pF:

CTWIST(D,S,er,X)$\cdot 10^{12}$ = 2.646

Capacitance per in. (F):

CTWIST(D,S,er,1) = $1.323 \cdot 10^{-12}$

MICROSTRIP TRANSMISSION LINES file: mstrip.mcd

Formulas included in this spreadsheet:

Effective relative permittivity	EEFF()	(used internally)
Effective electrical trace width	WE()	(used internally)
Microstrip characteristic impedance	ZMSTRIP()	
Microstrip propagation delay	PMSTRIP()	
Microstrip trace inductance	LMSTRIP()	
Microstrip trace capacitance	CMSTRIP()	

Formulas from: I. J. Bahl and Ramesh Garg, "Simple and accurate formulas for microstrip with finite strip thickness", Proc. IEEE, 65, 1977, pp. 1611-1612.

This material is nicely summarized in T. C. Edwards, "Foundations of Microstrip Circuit Design," John Wiley, New York, 1981, reprinted 1987.

(Watch out for Edward's error in Equation 3.52b, where he omits a ln() function.)

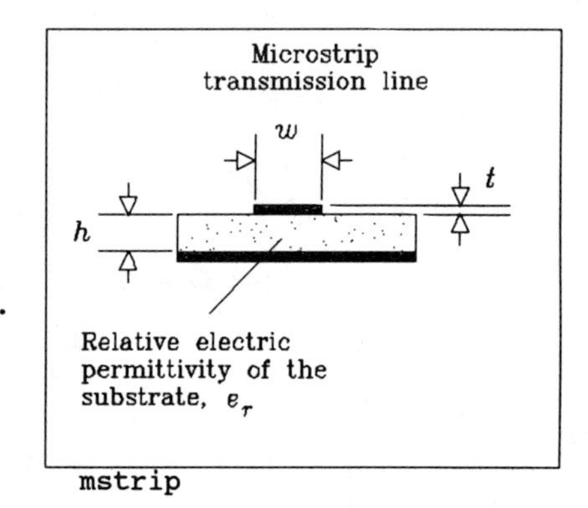

mstrip

Variables used:

h Trace height above ground (in.)

w Trace width (in.)

t Trace thickness (in.)

er Relative permittivity of material between trace and ground plane (dimensionless)

x Trace length (in.)

Effective relative permittivity
as a function of microstrip trace geometry:

For skinny traces (w < h)

$$E\_skny(h,w,er) := \frac{er + 1}{2} + \left[\frac{er - 1}{2}\right] \cdot \left[\left[1 + \frac{12 \cdot h}{w}\right]^{-.500} + .04 \cdot \left[1 - \frac{w}{h}\right]^{2}\right]$$

For wide traces (w > h)

$$E\_wide(h,w,er) := \frac{er + 1}{2} + \left[\frac{er - 1}{2}\right] \cdot \left[1 + \frac{12 \cdot h}{w}\right]^{-.500}$$

Composite formula picks skinny or wide model depending on w/h ratio:

E_temp(h,w,er) := if(w > h,E_wide(h,w,er),E_skny(h,w,er))

Special adjustment to account for trace thickness:

$$EEFF(h,w,t,er) := E\_temp(h,w,er) - \frac{(er - 1) \cdot \left[\frac{t}{h}\right]}{4.6 \cdot \sqrt{\frac{w}{h}}}$$

When w/h is skinny, you get the average of the
PCB permittivity, er, and the permittivity of air.
When w/h is wide, (the trace is very close to the
ground plane) you get er.

Effective trace width as a function
of other parameters (in.):

For skinny traces ($2\pi w < h$)

$$\text{WE\_skny}(h,w,t) := w + \frac{1.25 \cdot t}{\pi} \cdot \left[1 + \ln\left[\frac{4 \cdot \pi \cdot w}{t}\right]\right]$$

For wide traces ($2\pi w > h$)

$$\text{WE\_wide}(h,w,t) := w + \frac{1.25 \cdot t}{\pi} \cdot \left[1 + \ln\left[\frac{2 \cdot h}{t}\right]\right]$$

Composite formula picks skinny or wide model depending on w/h ratio:

$$\text{WE}(h,w,t) := \text{if}\left[w > \frac{h}{2 \cdot \pi}, \text{WE\_wide}(h,w,t), \text{WE\_skny}(h,w,t)\right]$$

Characteristic impedance as a function
of trace geometry (Ω):

Accuracy of better than 2 percent is
obtained under the following conditions:

$$0 < t/h < 0.2$$
$$0.1 < w/h < 20$$
$$0 < er < 16$$

For skinny traces (w < h)

$$\text{ZMS\_skny}(h,w,t) := 60 \cdot \ln\left[\frac{8 \cdot h}{\text{WE}(h,w,t)} + \frac{\text{WE}(h,w,t)}{4 \cdot h}\right]$$

For wide traces (w > h)

$$\text{ZMS\_wide}(h,w,t) := \frac{120 \cdot \pi}{\frac{\text{WE}(h,w,t)}{h} + 1.393 + .667 \cdot \ln\left[\frac{\text{WE}(h,w,t)}{h} + 1.444\right]}$$

Composite formula picks skinny or wide model depending on w/h ratio:

$$\text{ZMSTRIP}(h,w,t,er) := \frac{\text{if}(w > h, \text{ZMS\_wide}(h,w,t), \text{ZMS\_skny}(h,w,t))}{\sqrt{\text{EEFF}(h,w,t,er)}}$$

Microstrip propagation delay (s/in.):

$$PMSTRIP(h,w,t,er) := 84.72 \cdot 10^{-12} \cdot \sqrt{EEFF(h,w,t,er)}$$

Inductance of microstrip (H):

$$LMSTRIP(h,w,t,x) := PMSTRIP(h,w,t,1.) \cdot ZMSTRIP(h,w,t,1.) \cdot x$$

(Use a dummy er value of 1. It doesn't matter for inductance calculations.)

Capacitance of microstrip (F):

$$CMSTRIP(h,w,t,er,x) := \frac{PMSTRIP(h,w,t,er)}{ZMSTRIP(h,w,t,er)} \cdot x$$

Example microstrip wire calculations

Height above ground (in.)	H := .006	
Width of trace (in.)	W := .008	
Thickness of trace (in.)	T := .00137	(1-oz copper plating weight)
Length of wire (in.)	X := 11.000	
Relative electric permittivity (affects capacitance, but not inductance)	er := 4.5	

Impedance (Ω):

ZMSTRIP(H,W,T,er) = 56.4435

Total inductance (H):

$\text{LMSTRIP}(H,W,T,X) = 9.3401 \cdot 10^{-8}$

Same result in nH:

$\text{LMSTRIP}(H,W,T,X) \cdot 10^{9} = 93.4008$

Inductance per in. (H):

$\text{LMSTRIP}(H,W,T,1) = 8.491 \cdot 10^{-9}$

Total capacitance (F):

$\text{CMSTRIP}(H,W,T,er,X) = 2.9317 \cdot 10^{-11}$

Same result in pF:

$\text{CMSTRIP}(H,W,T,er,X) \cdot 10^{12} = 29.3172$

Capacitance per in. (F):

$\text{CMSTRIP}(H,W,T,er,1) = 2.6652 \cdot 10^{-12}$

STRIPLINE TRANSMISSION LINES file: sline.mcd

Formulas included in this spreadsheet:

Stripline characteristic impedance	ZSTRIP()
Offset stripline characteristic impedance	ZOFFSET()
Stripline propagation delay	PSTRIP()
Stripline trace inductance	LSTRIP()
Offset stripline inductance	LOSTRIP()
Stripline trace capacitance	CSTRIP()
Offset stripline capacitance	COSTRIP()

Formulas are from Seymour Cohn, "Problems in Strip Transmission Lines," MTT-3, No. 2, March 1955, pp. 199-126.

This material is summarized in Harlan Howe, Stripline Circuit Design, Artech House, Norwood, MA, 1974.

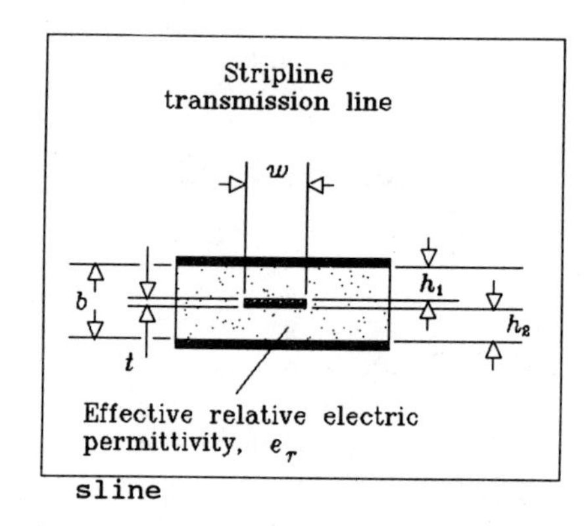

sline

Variables used:

h1	Trace height above lower ground plane (in.)
h2	Trace headroom below upper ground plane (in.)
b	Separation between ground planes, b = h1 + h2 + t (in.)
w	Trace width (in.)
t	Trace thickness (in.)
er	Relative dielectric constant
x	Trace length (in.)

Stripline characteristic impedance (Ω:)

Accuracy of better than 1.3% is
obtained under the following conditions:

t/b < 0.25
t/w < 0.11
er unrestricted

For skinny traces (w/b < 0.35)

$$ZSTR\_K1(w,t) := \left[\frac{w}{2}\right]\cdot\left[1+\frac{t}{\pi\cdot w}\cdot\left[1+\ln\left[\frac{4\cdot\pi\cdot w}{t}\right]\right]+0.255\cdot\left[\frac{t}{w}\right]^2\right]$$

$$ZSTR\_skny(b,w,t,er) := \frac{60}{\sqrt{er}}\cdot\ln\left[\frac{4\cdot b}{\pi\cdot ZSTR\_K1(w,t)}\right]$$

For wide traces (w/b > 0.35)

$$ZSTR\_K2(b,t) := \left[\frac{2}{1-\frac{t}{b}}\cdot\ln\left[\frac{1}{1-\frac{t}{b}}+1\right]-\left[\frac{1}{1-\frac{t}{b}}-1\right]\cdot\ln\left[\frac{1}{\left[1-\frac{t}{b}\right]^2}-1\right]\right]$$

$$ZSTR\_wide(b,w,t,er) := \frac{94.15}{\dfrac{\frac{w}{b}}{1-\frac{t}{b}}+\dfrac{ZSTR\_K2(b,t)}{\pi}}\cdot\frac{1}{\sqrt{er}}$$

Composite formula picks skinny or
wide model depending on w/b ratio:

ZSTRIP (b,w,t,er) := if (w > .35·b, ZSTR_wide (b,w,t,er), ZSTR_skny (b,w,t,er))

Rarely are the two parameters h1 and h2 equal in practice. The more common case is an assymetric stripline having the conducting trace offset to one side.

Offset, or asymmetric, stripline characteristic impedance (Ω) (no accuracy guaranteed):

$$\mathrm{ZOFFSET}(h1,h2,w,t,er) := \frac{2\cdot \mathrm{ZSTRIP}(2\cdot h1 + t,w,t,er)\cdot \mathrm{ZSTRIP}(2\cdot h2 + t,w,t,er)}{\mathrm{ZSTRIP}(2\cdot h1 + t,w,t,er) + \mathrm{ZSTRIP}(2\cdot h2 + t,w,t,er)}$$

Propagation delay of stripline (s/in.):

$$\mathrm{PSTRIP}(er) := 84.72\cdot 10^{-12}\cdot\sqrt{er}$$

(same formula for centered or offset stripline)

Inductance of stripline (H):

$$\mathrm{LSTRIP}(b,w,t,x) := \mathrm{PSTRIP}(1.)\cdot \mathrm{ZSTRIP}(b,w,t,1.)\cdot x$$

In the equation above, we can assume a relative permittivity of 1.; it doesn't affect the answer.

Inductance of offset stripline (H):

$$\mathrm{LOSTRIP}(h1,h2,w,t,x) := \mathrm{PSTRIP}(1.)\cdot \mathrm{ZOFFSET}(h1,h2,w,t,1.)\cdot x$$

Capacitance of stripline (F):

$$\mathrm{CSTRIP}(b,w,t,er,x) := \frac{\mathrm{PSTRIP}(er)}{\mathrm{ZSTRIP}(b,w,t,er)}\cdot x$$

In the equations above and below, we must use the relative permittivity.

Capacitance of offset stripline (F):

$$\mathrm{COSTRIP}(h1,h2,w,t,er,x) := \frac{\mathrm{PSTRIP}(er)}{\mathrm{ZOFFSET}(h1,h2,w,t,er)}\cdot x$$

Example stripline calculations

Ground plane separation (in.)	B := .020	
Width of trace (in.)	W := .006	
Thickness of trace (in.)	T := .00137	(1-oz copper plating weight)
Length of wire (in.)	X := 11.000	
Relative electric permeability (affects capacitance, but not inductance)	er := 4.5	

Impedance (Ω):

$$\mathrm{ZSTRIP}(B,W,T,er) = 51.5858$$

Total inductance (H):

$$\mathrm{LSTRIP}(B,W,T,X) = 1.0198 \cdot 10^{-7}$$

Same result in nH:

$$\mathrm{LSTRIP}(B,W,T,X) \cdot 10^{9} = 101.98$$

Inductance per in. (H):

$$\mathrm{LSTRIP}(B,W,T,1) = 9.2709 \cdot 10^{-9}$$

Total capacitance (F):

$$\mathrm{CSTRIP}(B,W,T,er,X) = 3.8323 \cdot 10^{-11}$$

Same result in pF:

$$\mathrm{CSTRIP}(B,W,T,er,X) \cdot 10^{12} = 38.3226$$

Capacitance per in. (F):

$$\mathrm{CSTRIP}(B,W,T,er,1) = 3.4839 \cdot 10^{-12}$$

Index

D

E

F

G

H

I

J

K

L

Q

R

S

T

V

W

X

Z